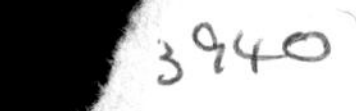

FOOD ANTIOXIDANTS

ELSEVIER APPLIED FOOD SCIENCE SERIES

Biotechnology Applications in Beverage Production
C. CANTARELLI and G. LANZARINI

Progress in Sweeteners
T. H. GRENBY (Editor)

Food Refrigeration Processes: Analysis, Design and Simulation
A. C. CLELAND

Development and Application of Immunoassay for Food Analysis
J. H. R. RITTENBURG (Editor)

Microstructural Principles of Food Processing and Engineering
J. M. AGUILERA and D. W. STANLEY

Forthcoming titles in this series:

Food Gels
P. HARRIS (Editor)

Food Irradiation
S. THORNE

FOOD ANTIOXIDANTS

Edited by

B. J. F. HUDSON

Honorary Research Fellow and Consultant,
Department of Food Science and Technology,
University of Reading, UK

ELSEVIER APPLIED SCIENCE
LONDON and NEW YORK

ELSEVIER SCIENCE PUBLISHERS LTD
Crown House, Linton Road, Barking, Essex IG11 8JU, England

Sole Distributor in the USA and Canada
ELSEVIER SCIENCE PUBLISHING CO., INC.
655 Avenue of the Americas, New York, NY 10010, USA

WITH 48 TABLES AND 44 ILLUSTRATIONS

British Library Cataloguing in Publication Data

Food antioxidants.
1. Processed food. Antioxidants
I. Hudson, B. J. F.
664'.06

ISBN 1-85166-440-8

Library of Congress Cataloging-in-Publication Data

Food antioxidants/edited by B. J. F. Hudson.
p. cm.
Includes bibliographical references.
ISBN 1-85166-440-8
1. Antioxidants. 2. Food additives. 3. Antioxidants-Physiological effect.
I. Hudson, B. J. F.
[DNLM: 1. Antioxidants—pharmacology. 2. Food Additives—pharmacology.
WA 712 F6858]
TX553.A73F66 1990
664'.028—dc20
DNLM/DLC 89-25875
for Library of Congress CIP

Typeset and printed in Northern Ireland by The Universities Press (Belfast) Ltd.

PREFACE

Antioxidants are present naturally in virtually all food commodities, providing them with a valuable degree of protection against oxidative attack. When food commodities are subjected to processing, such natural antioxidants are often depleted, whether physically, from the nature of the process itself, or by chemical degradation. In consequence, processed food products usually keep less well than do the commodities from which they originated. Ideally, food producers would like them to keep better. This objective can often be achieved by blending natural products rich in antioxidants with processed foods, or by using well recognised antioxidants as food additives.

In order to understand their action, and hence to apply antioxidants intelligently in food product formulation, some knowledge of the mechanisms by which they function is necessary. This is complex and may rely on one or more of several alternative forms of antioxidative intervention. Accordingly, the various mechanisms that may be relevant are discussed in Chapter 1, in each case including the 'intervention' mechanism.

When present in, or added to, foods antioxidants are functional in very small quantities, typically, perhaps, at levels of 0·01% or less. Indeed, at higher concentrations, since they themselves are susceptible to oxidation, they can behave as pro-oxidants. Identification and estimation of quantities of this order demand sophisticated, precise and unambiguous analytical techniques. It is important, both for quality assurance and to ensure that the regulations that apply in different countries and for different classes of food products are complied with. Chapter 2 provides comprehensive information on suitable analytical techniques. It also discusses a variety of methods available for evaluating the efficacy of antioxidants for foods.

A topic that seldom receives adequate attention in food antioxidant reviews is dealt with in Chapter 3. This serves to answer the question—'What happens chemically to antioxidants when they are in the process of exerting their protective function?' It is insufficiently recognised that, throughout the period during which they are extending the shelf-life of a food product, they are undergoing chemical changes. Sometimes the products of these changes are themselves still active as antioxidants, sometimes not. As the shelf-life comes to its close there may be little or none of the original antioxidant left. Such considerations are relevant both to the biological evaluation of antioxidants and to analytical methodology, especially if the food product under investigation, though still acceptable in terms of quality, is no longer fresh but is approaching the end of its shelf-life.

The benefits, from the quality standpoint, that can arise from the inclusion of antioxidants in foods, are dealt with in Chapters 4 and 5. Chapter 4 is concerned with the practical benefits arising from the uses of three very important groups of natural products, the tocopherols, carotenoids and ascorbic acid and its derivatives. Skilled technology has provided, in these cases by chemical synthesis, the more significant of these substances in forms identical with those found in nature. Though they function by different mechanisms, all three have proved highly effective in appropriate situations.

Not all natural antioxidants have proved to be commercially viable. However, because many of those so far unexploited are virtually ubiquitous, especially in plant products, including most, if not all, food plants, they must receive attention in this text. Chapter 5 discusses the properties and relevance of these natural antioxidants, with special attention to the flavonoids.

In considering the biological aspects of food antioxidants, one must be concerned with the part they play in the living organism. So far as food is concerned this means after ingestion of the food products in which they are present. Chapter 6, again mainly stressing naturally-occuring antioxidants, shows that they have positive roles in maintaining the health of the organism and guarding against the depletion of essential nutrients. We return here to mechanisms, not only those of the types reviewed in Chapter 1, but adding a biological dimension which now becomes highly relevant. Chapter 6 also seeks to emphasise the relationship between antioxidant and vitamin activity, especially in vitamins A, C and E but perhaps also extending to the flavonoids, which once included 'Vitamin P'.

Finally, Chapter 7 deals with toxicity. Few, if any, food components can be given unreservedly a 'clean bill of health' in this respect. If no adverse effects can be observed at low levels, one has only to increase intake sufficiently, if not always realistically, to produce undesirable effects. If acute toxicity trials show no adverse reactions, a component or additive may show chronic effects when ingested over a very long period. This applies both to 'natural' and 'artificial' antioxidants as much as to other additives. Toxicity data concerning the most important of the commercially available products in both of these classes of food additives are presented.

This text does not attempt to provide a complete or detailed coverage of the subject. The informed reader may well detect gaps in the presentaion. It does, however, offer an up-dated broad overview of antioxidants integral to, or used as additives for, food products, stressing natural as distinct from artificial classes. The continuing trend towards new forms of food processing, long-term food storage, added to informed consumer concern, ensures that these protective food ingredients will assume increasing importance to the food industry, in food distribution and at the point of sale.

My special thanks are due to the ten authors who have contributed specialised material for this volume and to the publishers for their patience and encouragement.

B. J. F. HUDSON

CONTENTS

Preface v

List of Contributors xi

1. The Mechanism of Antioxidant Action *in vitro* 1
   M. H. GORDON

2. Detection, Estimation and Evaluation of Antioxidants in Food Systems 19
   S. P. KOCHHAR and J. B. ROSSELL

3. Chemistry and Implications of Degradation of Phenolic Antioxidants 65
   K. KIKUGAWA, A. KUNUGI and T. KURECHI

4. Natural Antioxidants Exploited Commercially 99
   P. SCHULER

5. Natural Antioxidants not Exploited Commercially 171
   D. E. PRATT and B. J. F. HUDSON

6. Biological Effects of Food Antioxidants 193
   P. BERMOND

7. Toxicological Aspects of Antioxidants Used as Food Additives . . 253
   S. M. BARLOW

Index 309

LIST OF CONTRIBUTORS

SUSAN M. BARLOW
Toxicology and Environmental Health Division, Department of Health, Hannibal House, Elephant and Castle, London SE1 6TE, UK

P. BERMOND
Binningerstrasse 12, 4123-Allschwil, Switzerland

M. H. GORDON
Department of Food Science and Technology, University of Reading, Whiteknights, Reading RG6 2AP, UK

BERTRAM J. F. HUDSON
Department of Food Science and Technology, University of Reading, Whiteknights, Reading RG6 2AP, UK

KIYOMI KIKUGAWA
Tokyo College of Pharmacy, 1432-1 Horinouchi, Hachioji, Tokyo 192-03, Japan

S. P. KOCHHAR
Oils and Fats Section, Leatherhead Food Research Association, Leatherhead, Surrey KT22 7RY, UK. Present address: *SPK Consultancy Services, 48 Chiltern Crescent, Early, Reading, Berkshire RG6 1AN, UK*

AKIRA KUNUGI
Tokyo College of Pharmacy, 1432–1 Horinouchi, Hachioji, Tokyo 192-03, Japan

TSUTAO KURECHI
Tokyo College of Pharmacy, 1432-1 Horinouchi, Hachioji, Tokyo 192-03, Japan

DAN E. PRATT
Department of Foods and Nutrition, Purdue University, West Lafayette, Indiana 47907, USA

J. B. ROSSELL
Oils and Fats Section, Leatherhead Food Research Association, Leatherhead, Surrey KT22 7RY, UK

PETER SCHULER
Department VM/H, F. Hoffmann-La Roche & Co., 4002-Basle, Switzerland

Chapter 1

THE MECHANISM OF ANTIOXIDANT ACTION *IN VITRO*

M. H. GORDON

Department of Food Science and Technology, University of Reading, Whiteknights, Reading RG6 2AP, UK

INTRODUCTION

The spontaneous reaction of atmospheric oxygen with organic compounds leads to a number of degradative changes that reduce the lifetime of many products of interest to the chemical industry, especially polymers, as well as causing the deterioration of lipids in foods. The importance of oxygen in the deterioration of rubber was demonstrated over a century ago,[1] and this finding led chemists to investigate the chemistry of oxidative deterioration and its inhibition.

Several observations including the detection of an induction period, the pro- or antioxidant effects of small concentrations of impurities, and the effect of UV light clearly led to the conclusion that the reaction between lipids and oxygen, which is termed autoxidation, is a free radical chain reaction. Like all chain reactions, the mechanism can be discussed in terms of initiation reactions during which free radicals are formed, propagation reactions during which free radicals are converted into other radicals, and termination reactions which involve the combination of two radicals with the formation of stable products (Fig. 1).

Initiation of Autoxidation

The direct reaction of a lipid molecule, RH, with a molecule of oxygen (1) would contravene the principle of conservation of spin angular momentum, since the lipid molecule is in a singlet electronic state and the oxygen molecule has a triplet ground state. As a consequence of this principle, the initiation of autoxidation by the reaction of a lipid

Initiation	$ROOH^* \rightleftharpoons ROO^\cdot + H^\cdot$
	$ROOH \rightleftharpoons RO^\cdot + {}^\cdot OH$
	$2\,ROOH \rightleftharpoons RO^\cdot + H_2O + ROO^\cdot$
Propagation	$R^\cdot + O_2 \rightleftharpoons ROO^\cdot$
	$ROO^\cdot + R^1H \rightleftharpoons ROOH + R^{1\cdot}$
Termination	$ROO^\cdot + R^1OO^\cdot \rightleftharpoons ROOR^1 + O_2$
	$RO^\cdot + R^{1\cdot} \rightleftharpoons ROR'$

* Formed by various pathways including reaction of $^1O_2$ with unsaturated lipids or lipoxygenase-catalysed oxidation of polyunsaturated fatty acids.

FIG. 1. Mechanism of autoxidation.

molecule with oxygen in its ground state (1) is highly improbable.

$$RH + {}^3O_2 \not\rightarrow ROOH \tag{1}$$

The energy of activation for this reaction is very high, values in the range of 146–272 kJ/mol being quoted.[2]

There are two more probable processes to be considered in the formation of free radicals. Chain initiation may occur by a direct reaction between a metal catalyst and a lipid molecule (2). This process has been shown to be exothermic for methyl linoleate ($\Delta H = -63$ kJ/mol).[3]

$$M^{(n+1)+} + RH \rightarrow M^{n+} + H^+ + R^\cdot \tag{2}$$

However, the major initiation process is likely to involve the decomposition of hydroperoxides. Hydroperoxides are formed during the propagation reactions (12), but they may also be formed by the reaction of a lipid molecule with an oxygen molecule in its singlet excited state or by an enzyme-catalysed reaction. The conversion of triplet oxygen to singlet oxygen may occur when a photosensitiser, such as chlorophyll, haematoporphyrins or flavins, including riboflavin, is present.[4] Photosensitisers absorb light in the visible or near UV region, becoming electronically excited. They may then transfer their excess energy to an oxygen molecule in an allowed reaction (3).

$$^1\text{Sens} \rightarrow {}^1\text{Sens (excited)} \xrightarrow[\text{crossing}]{\text{Intersystem}} {}^3\text{Sens (excited)} \xrightarrow{^3O_2} {}^1\text{Sens} + {}^1O_2 \tag{3}$$

The singlet oxygen thus formed may then react with a lipid molecule to yield a hydroperoxide (4).

$$RH + {}^1O_2 \rightarrow ROOH \tag{4}$$

An alternative pathway to hydroperoxides involves the reaction of a polyunsaturated fatty acid such as linoleic acid with oxygen in the presence of the enzyme lipoxygenase (5).[5]

$$RH + {}^3O_2 \xrightarrow{\text{Lipoxygenase}} ROOH \quad (5)$$

The O—O bond in a hydroperoxide is relatively weak with an activation energy for homolytic cleavage (6) of 184 kJ/mol,[6,7] but the bimolecular process (7) is the more important initiation process.[8,9]

$$ROOH \rightarrow RO^{\cdot} + {}^{\cdot}OH \quad (6)$$

$$2ROOH \rightarrow RO^{\cdot} + H_2O + RO_2^{\cdot} \quad (7)$$

In the presence of metal ions, the metal catalysed decomposition of hydroperoxides is likely to be the main source of radicals. Metal ions such as iron or copper catalyse the decomposition of hydroperoxides both in their lower (8) and in their higher oxidation state (9).[10–12]

$$ROOH + M^{n+} \rightarrow RO^{\cdot} + OH^- + M^{(n+1)+} \quad (8)$$

$$ROOH + M^{(n+1)+} \rightarrow ROO^{\cdot} + H^+ + M^{n+} \quad (9)$$

These two reactions may operate as a cycle, so that even trace amounts of metal ions may be effective in generating radicals. Waters[13] concluded that reaction (9) was unlikely to be significant in autoxidation in aqueous solution, although the metal ion in its higher valence state may oxidise the lipid RH or a reaction product, e.g. ROH or $R^{\cdot}$.

Propagation and Termination of Autoxidation

Lipid radicals are highly reactive species and they can readily undergo propagation reactions either by abstraction of a hydrogen atom (10, 12) or by reaction with an oxygen molecule in its ground state (11).

$$RH + R^{1\cdot} \rightleftharpoons R^{\cdot} + R^1H \quad (10)$$

$$R^{\cdot} + {}^3O_2 \rightleftharpoons ROO^{\cdot} \quad (11)$$

$$ROO^{\cdot} + RH \rightleftharpoons ROOH + R^{\cdot} \quad (12)$$

The oxygenation reaction (11) is very fast, with almost zero activation energy, and therefore the concentration of $ROO^{\cdot}$ is much higher than that of $R^{\cdot}$ in most food systems where oxygen is present. The hydroperoxide formed by reaction (12) is able to take part in the initiation of radicals, as indicated earlier.

The combination of two radicals is a process with a very low enthalpy of activation, but the occurrence of termination reactions (e.g. (13), (14)) is limited both by the low concentration of radicals, which makes an encounter between two radicals uncommon, and also by steric factors, when radicals are required to collide with the correct orientation.

$$ROO^{\cdot} + ROO^{\cdot} \rightleftharpoons ROOR + O_2 \tag{13}$$

$$ROO^{\cdot} + R^{\cdot} \rightleftharpoons ROOR \tag{14}$$

Termination reactions may become important in edible oils heated at elevated temperatures, as indicated by the formation of polymers in frying oils. Hydroperoxides decompose spontaneously at 160°C[14] and the radical concentration can become relatively high under such conditions.

PRIMARY (CHAIN-BREAKING) ANTIOXIDANTS

Ingold[15] classified all antioxidants into two groups, namely primary or chain-breaking antioxidants, which can react with lipid radicals to convert them to more stable products, and secondary or preventive antioxidants which reduce the rate of chain initiation by a variety of mechanisms. As stated earlier, the major lipid radical at normal oxygen pressures is the alkylperoxy radical $ROO^{\cdot}$ which is an oxidising agent and is readily reduced to the related anion and thence converted to a hydroperoxide by an electron donor (15), or which may be directly converted to a hydroperoxide by a hydrogen donor, AH (16).

$$ROO^{\cdot} \xrightarrow{+e} ROO^{-} \xrightarrow{H^{+}} ROOH \tag{15}$$

$$ROO^{\cdot} + AH \longrightarrow ROOH + A^{\cdot} \tag{16}$$

Alkyl radicals, on the other hand, are in general reducing agents and are scavenged by electron acceptors (17).[16]

$$R^{\cdot} \xrightarrow{-e} R^{+} \longrightarrow \text{Alkene} + H^{+} \tag{17}$$

Inhibition by electron acceptors is not significant in most food systems, but it can become important in biological tissues since the oxygen pressure is much lower in healthy tissues (*ca.* $2{\cdot}67 \times 10^2$ Pa) than it is in the atmosphere. The antioxidant properties of Vitamin K and tocopheroquinone in biological tissues have been ascribed to their electron-acceptor properties.[17–19]

The most common food antioxidants interfere with lipid autoxidation by rapid donation of hydrogen atoms to lipid radicals according to reactions (16) or (18).

$$RO^{\cdot} + AH \rightleftharpoons ROH + A^{\cdot} \tag{18}$$

Alternative mechanisms only become important for primary antioxidants at very low oxygen pressures, very high concentrations of antioxidant or very low rates of chain initiation.

A molecule will be able to act as a primary antioxidant if it is able to donate a hydrogen atom rapidly to a lipid radical and if the radical derived from the antioxidant is more stable than the lipid radical, or is converted to other stable products. Phenol itself is inactive as an antioxidant but substitution of alkyl groups into the 2, 4 or 6 positions increases the electron density on the hydroxyl group by an inductive effect and thus increases the reactivity with lipid radicals. Substitution at the 4-position with an ethyl or *n*-butyl group rather than a methyl group improves the activity of a phenolic antioxidant but longer chain or branched alkyl groups in this position decrease the activity.[20] The electron donating, or inductive effects of substituents can be quantified according to the Hammett equation by their effects on the dissociation of benzoic acid. The substituent constants, σ for H, CH_3, OCH_3 and OH are 0, $-0{\cdot}17$, $-0{\cdot}268$ and $-0{\cdot}37$ respectively where σ is a measure of the inductive effect and is quantified by:

$$\sigma = \log(K/K_0)$$

where K is the dissociation constant of a para substituted benzoic acid and K_0 is the dissociation constant of benzoic acid (19).

$$C_6H_5COOH \underset{}{\overset{K_0}{\rightleftharpoons}} C_6H_5COO^- + H^+ \tag{19}$$

Hence, the strong electron donating effect of the methoxy substitutent is an important contributor to the effectiveness of 2-*t*-butyl-4-methoxyphenol (BHA) as an antioxidant.

The radical formed by reaction of a phenol with a lipid radical is stabilised by delocalisation of the unpaired electron around the aromatic ring as indicated by the valence bond isomers (20).

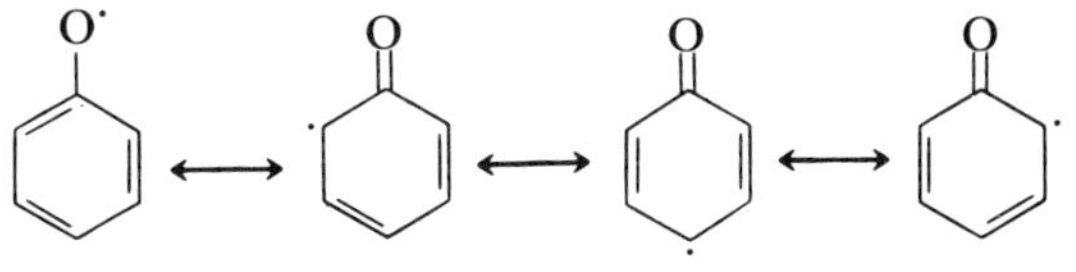

O˙ ⟷ O ⟷ O ⟷ O (20)

The stability of the phenoxyl radical ($A^{\cdot}$) reduces the rate of propagation of the autoxidation chain reaction since propagation reactions such as (21)–(23) are very slow compared with (10), (11) and (12).

The stability of the phenoxyl radical is further increased by bulky groups in the 2 and 6 positions as in 2,6-di-*t*-butyl-4-methylphenol (BHT),[21] since these substituents increase the steric hindrance in the region of the radical and thereby further reduce the rate of propagation reactions involving the antioxidant radical ((21)–(23))

$$A^{\cdot} + O_2 \rightleftharpoons AOO^{\cdot} \tag{21}$$

$$AOO^{\cdot} + RH \rightleftharpoons AOOH + R^{\cdot} \tag{22}$$

$$A^{\cdot} + RH \rightleftharpoons AH + R^{\cdot} \tag{23}$$

However, the presence of bulky substituents in the 2,6 positions also reduces the rate of reaction of the phenol with lipid radicals (reactions (16) or (18)). This steric effect opposes the increased stabilisation of the radical and both effects must be considered in assessing the overall activity of an antioxidant.

The introduction of a second hydroxyl group into the 2 or the 4 position of a phenol increases the antioxidant activity. The effectiveness of a 1,2-dihydroxybenzene derivative is increased by the stabilisation of the phenoxyl radical through an intramolecular hydrogen bond (24):[22]

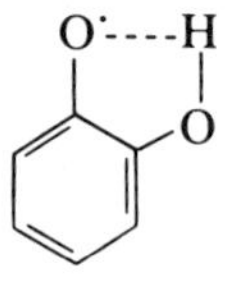

O˙----H
O
(24)

The antioxidant activity of dihydroxybenzene derivatives is partly due to the fact that the semiquinonoid radical produced initially can be further oxidised to a quinone by reaction with another lipid radical, or it may disproportionate into a quinone and a hydroquinone molecule (25).[23]

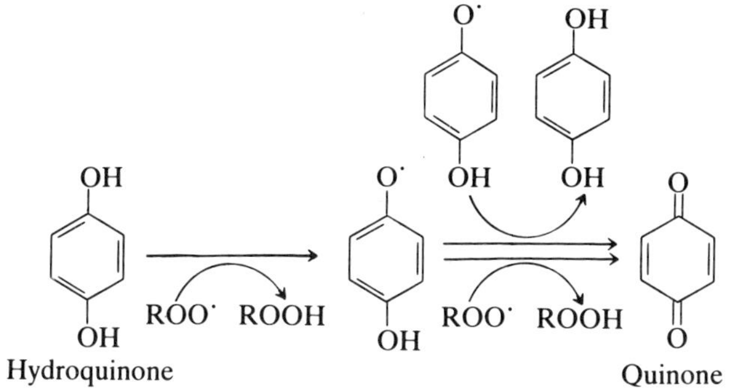

O· OH

OH O· OH OH O

ROO· ROOH ROO· ROOH (25)

OH OH O

Hydroquinone Quinone

The activity of 2-methoxyphenols is much lower than that of the corresponding pyrocatechol possessing two free hydroxyl groups[24] because 2-methoxyphenols are unable to stabilise the phenoxyl radical by hydrogen bonding as in (24).

Phenolic antioxidants are effective in extending the induction period when added to an oil which has not deteriorated to a great extent, but they are virtually ineffective in retarding deterioration of lipids when added to severely deteriorated systems.[25,26] The radical concentration swamps any inhibitory effect of antioxidants under these conditions.

The effect of antioxidant concentration on autoxidation rate depends on several factors including antioxidant structure, oxidation conditions and sample being oxidised. Often the antioxidant activity of phenolic antioxidants is lost at high concentrations and they may become pro-oxidant.[27] This is due to their involvement in initiation reactions such as (26) or (27) at high concentrations.[28]

$$AH + O_2 \rightarrow A^{\cdot} + HOO^{\cdot} \tag{26}$$

$$AH + ROOH \rightarrow RO^{\cdot} + H_2O + A^{\cdot} \tag{27}$$

Antioxidant action by donation of a hydrogen atom is unlikely to be limited to phenols. Endo *et al.*[29] have suggested that the antioxidant effects of chlorophyll in the dark occur by this mechanism.

Radical Trapping Antioxidants

Although antioxidants such as nitroxides that trap lipid radicals are useful in chemical studies of free radical reactions, these antioxidants are not used as food additives. However, recently attention has focused on the effects of *β*-carotene at low oxygen pressures.[18] It is recognised that *β*-carotene is an effective quencher of singlet oxygen,[30,31] and hence may act as an antioxidant by preventing the formation of hydroperoxides in the presence of singlet oxygen (see p. 12). However, it has also been proposed that *β*-carotene, as well as other carotenoids and retinoids, may be an effective antioxidant at low oxygen pressures under conditions where singlet oxygen is not formed because it reacts rapidly with the chain-carrying peroxyl radicals to produce a resonance-stabilised carbon-centred radical[18] as shown in reaction (28).

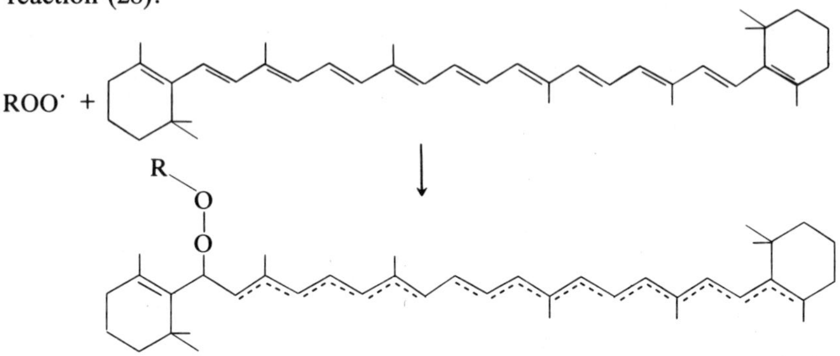

ROO˙ +

(28)

The antioxidant behaviour of *β*-carotene is limited to oxygen partial pressures significantly less than 150 Torr, which is the partial pressure of oxygen in the atmosphere. At higher oxygen pressures, *β*-carotene loses its antioxidant activity and shows an autocatalytic pro-oxidant effect especially at concentrations higher than 5×10^{-4} M.

SECONDARY ANTIOXIDANTS

Compounds which retard the rate of autoxidation of lipids by processes other than that of interrupting the autoxidation chain by converting free radicals to more stable species are termed secondary antioxidants. These may operate by a variety of mechanisms including compounds that bind metal ions, scavenge oxygen, decompose hydroperoxides to non-radical species, absorb UV radiation or deactivate singlet oxygen. Secondary antioxidants usually only show antioxidant activity if a second minor component is present in the sample. This can be seen in the case of sequestering agents which are effective in the presence of metal ions, and reducing agents such as ascorbic acid which are effective in the presence of tocopherols or other phenolic antioxidants.

Sequestering Agents

Food lipids generally contain trace amounts of metal ions. These may arise from the presence of metal activated enzymes or their decomposition products. Further metal ions may be picked up from refining equipment, metal containers or processes such as hydrogenation. Heavy metals, particularly those possessing two or more valency states with a suitable oxidation–reduction potential between them (e.g. Co, Cu, Fe, Mn, etc.) reduce the length of the induction period and increase the maximum rate of oxidation of lipids. Pro-oxidant effects of porphyrin complexes of metals and metal salts have been shown to occur at levels of 1 part in 10^8.[32,33] Hence it is likely that no food system can be considered to be free of metal ions which may be of significance in promoting autoxidation. Metals act as pro-oxidants by electron transfer, liberating radicals from fatty acids or hydroperoxides by reactions such as (2), (8) and (9). The superoxide radical, $O_2^{\cdot -}$, may also be formed by a metal-catalysed reaction[3] and this may cause lipid oxidation by several pathways.[34,35]

The effectiveness of metal ions in catalysing autoxidation varies, copper being particularly active at low concentrations (Table 1). The presence of only 0·02 mg/kg copper has been shown to catalyse off-flavour production in butter.[36] Pokorny[37] showed that copper was

TABLE 1
CONCENTRATION OF METAL REQUIRED TO REDUCE THE KEEPING TIME OF LARD BY 50% AT 98°C[73,74]

Metal	*Concentration (ppm)*
Copper	0·05
Manganese	0·6
Iron	0·6
Chromium	1·2
Nickel	2·2
Vanadium	3·0
Zinc	19.6
Aluminium	50·0

very effective as a catalyst of hydroperoxide decomposition. Cu^{I} has been found to reduce hydroperoxides by reaction (8)[38] but Cu^{II} is inactive in linoleic acid.[33,39]

Chelation of metal ions by food components reduces the pro-oxidative effect of these ions and raises the energy of activation of the initiation reactions considerably. Chelating agents which form σ-bonds with a metal are effective as secondary antioxidants because they reduce the redox potential thereby stabilising the oxidised form of the metal ion whereas chelating agents such as heterocyclic bases which form π-complexes raise the redox potential and may therefore accelerate reaction (9)[13] and hence have a pro-oxidant effect.

Citric acid, ethylenediaminetetraacetic acid (EDTA) and phosphoric acid derivatives may extend the shelf-life of lipid-containing foods to a great extent because of their chelating properties. EDTA forms thermodynamically stable complexes with all the transition metal ions[40] and its effectiveness in opposing the pro-oxidant effect of copper has been shown in studies on the development of off-flavours in margarines during storage.[41,42] The spatial structure of the anion of EDTA, which has six donor atoms, allows it to satisfy the coordination number of six frequently encountered among metal ions. Strainless five-membered rings are formed on chelation as in Fig. 2. This structure may be formed when the complex ion exhibits the maximum chelating power as a hexa-dentate ligand, but it may not be correct for all metal–EDTA complexes.

Citric acid, which is widely used in foods, is a weaker chelating agent than EDTA. However, it is very effective in retarding the oxidative deterioration of lipids in foods and is commonly added to vegetable oils after deodorisation. The efficiency of citric acid in reducing the pro-oxidant effect of a range of metals is shown in Table 2. The lack of an antioxidant effect in the sample containing copper was ascribed to the powerful pro-oxidant effect of the metal at 3 ppm, and an antioxidant effect is expected at lower metal concentrations. The antioxidant effect of citric acid is decreased on esterification of the carboxyl groups, while esterification of the hydroxyl group has no effect.[43] Hence, it is clear that the carboxyl groups are involved in chelation of metal ions.

Polyphosphates which are derivatives of phosphoric acid also sequester metal ions. Phosphates are nearly inactive but oligophosphates show good metal-chelating properties, the activity increasing with the number of phosphate residues in the range 2–6.[44]

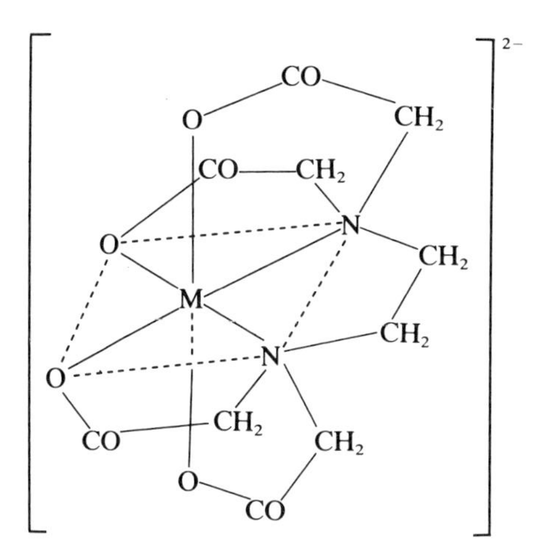

FIG. 2. Structure of chelate formed between EDTA and metal ion.

Oxygen Scavengers and Reducing Agents

Ascorbic acid, ascorbyl palmitate, erythorbic acid (isoascorbic acid) or sodium erythorbate are commonly used in stabilising fatty foods. According to Cort[45] ascorbyl palmitate at 0·01% is more effective than BHA or BHT in retarding development of rancidity in vegetable oils. Ascorbic acid is oxidised to dehydroascorbic acid when it functions as an oxygen scavenger and therefore the 2- and 3-positions must be unsubstituted. The oxygen scavenging activity of ascorbic acid makes it very useful in canned or bottled products with a headspace of air but

TABLE 2

EFFECT OF 100 ppm CITRIC ACID ON OIL CONTAINING 3 ppm OF A PRO-OXIDANT METAL[75]

Metal as chloride (3 ppm)	*Peroxide value (after 8 h at 100°C) (meq . kg^{-1})*	
	Control	*+Citric acid*
Control	46·6	10·7
Cu (II)	294·0	291·0
Co (II)	239·0	9·1
Mn (II)	85·4	13·1
Fe (III)	293·0	125·0
Cr (III)	153·0	17·8

in fatty foods ascorbyl palmitate is often more effective as an antioxidant because of its increased solubility in the fat phase. Although ascorbic acid may function as an antioxidant by oxygen scavenging in some foods, it is particularly effective in combination with primary antioxidants, such as tocopherols. One explanation is that ascorbic acid acts by decomposing lipid hydroperoxides into non-radical products,[46] but it may also be effective in regenerating the tocopherol by reducing the tocopheroxyl radical or the tocoquinone.[47]

Enzymes as Antioxidants

Various radicals can contribute to lipid oxidation in biological systems. Superoxide radicals $O_2^{\cdot -}$ which may be produced by the enzyme xanthine oxidase and hydrogen peroxide[48] may be removed by the enzyme superoxide dismutase according to (29).

$$2\,O_2^{\cdot -} + 2\,H^+ \xrightarrow{\text{Superoxide dismutase}} H_2O_2 + {}^3O_2 \qquad (29)$$

The enzyme catalase may alo play an important role in converting the hydrogen peroxide to water and oxygen (30).

$$2\,H_2O_2 \xrightarrow{\text{Catalase}} 2\,H_2O + {}^3O_2 \qquad (30)$$

Since milk contains xanthine oxidase and superoxide dismutase, Korycka–Dahl & Richardson[49] concluded that these reactions may contribute to the stability of milk.

Singlet Oxygen Quenchers

As mentioned earlier, β-carotene may act as an antioxidant by quenching of singlet oxygen according to reaction (31).

$${}^1O_2 + {}^1\beta\text{-carotene} \rightleftharpoons {}^3O_2 + {}^3\beta\text{-carotene} \qquad (31)$$

Clements *et al.*[50] have shown that β-carotene at a level of 0·46 ppm reduces the rise in peroxide value of soybean oil after 6 h at 20°C under photo-oxidising conditions from 36 meq/kg for the control to 30 meq/kg for the sample containing β-carotene. Kellogg & Fridovich[48] also showed that β-carotene inhibits lipid oxidation that has been initiated by xanthine oxidase, and this probably occurs because of its effect on singlet oxygen.

Methyl Silicone and Sterol Antioxidants

Methyl silicone (polydimethyl siloxane) has been shown to retard the oxidative deterioration of oils when heated on a hot plate but not in an oven.[51] Methyl silicone was more effective than TBHQ in reducing the room odour of heated soybean oil in combination with citric acid.[52] A concentration of 0·03 ppm methyl silicone was found to be sufficient to inhibit oxidative changes in frying oils,[53] and it was found that the critical concentration for antioxidant activity of methyl silicone corresponded to a monolayer on the surface.[54] It was therefore concluded that the protective effects could be due to one of four reasons: (1) the methyl silicone may represent a physical barrier preventing the penetration of oxygen into the oil from the atmosphere, although this is unlikely; (2) it may present an inert surface to the atmosphere inhibiting oxidation at the surface; (3) it may act as a chemical antioxidant being oxidised and inhibiting the propagation of free radical chains with its effectiveness arising from its concentration in the surface where oxidation occurs; or (4) it may inhibit convection currents in the surface layer.[54] The importance of convection currents in the oxidation of oils was demonstrated by Rock and Roth,[55] who showed that the rate of oil oxidation was strongly dependent on the extent of convection currents.

Δ^5-Avenasterol, fucosterol and citrostadienol have been shown to exhibit antioxidant properties in oils heated at 180°C (Sims *et al.*;[56] Boskou & Morton[57]) whereas most sterols are ineffective as antioxidants. It has been suggested[58] that the donation of a hydrogen atom from the allylic methyl group in the side chain (Fig. 3), followed by isomerisation to a relatively stable tertiary allylic free radical represents the mode of action of the sterol antioxidants. Recently it has been found that Δ^5-avenasterol is similar to methyl silicone in being effective at 100°C or 180°C when the oil is heated on a hotplate but not in an oven (Gordon & Williamson, unpublished data). In addition avenasterol appears to be increased in concentration in a layer at the

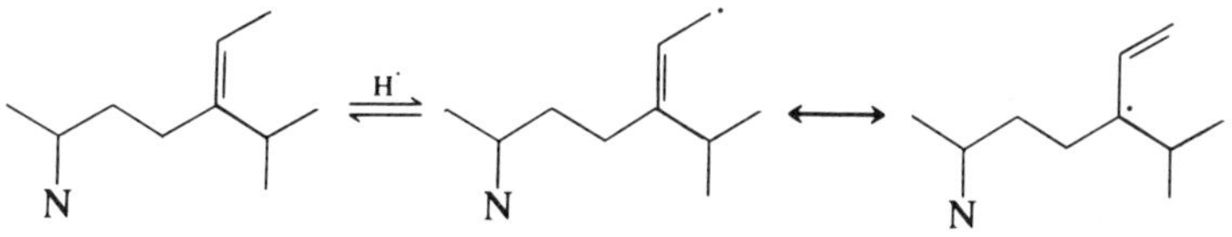

$$\text{N} \xrightleftharpoons{\text{H}^{\cdot}} \text{N} \longleftrightarrow \text{N}$$

FIG. 3. Mechanism for the antioxidant activity of sterols (N = sterol ring) (Gordon & Magos[58]).

surface, and it is ineffective at room temperature. These findings suggest that avenasterol acts as a chemical antioxidant as indicated in Fig. 3, its effectiveness arising from its concentration in the surface where oxidation occurs. This is similar to the third possible mechanism suggested above for methyl silicones.

Antioxidants with Multiple Functions

There has been much discussion in the literature about the mechanism of antioxidant action of phospholipids and Maillard reaction products. It appears likely that these antioxidants may retard autoxidation by more than one mechanism and therefore these components will be discussed individually.

Phospholipids

Secondary antioxidant effects of phospholipids have been observed on frequent occasions. This has been ascribed to their metal chelating character.[59,60] Inactive complexes with metals have been found with phosphatidyl inositol (PI) and other acidic phospholipids but not with phosphatidyl choline (PC) or phosphatidyl ethanolamine (PE).[37] Hudson & Ghavami[61] observed that the antioxidant properties of dipalmitoyl phosphatidyl ethanolamine in combination with α-tocopherol increased with phospholipid concentration up to 0·6% which was well above the level required for complexing with metal ions, and it was therefore concluded that metal complexing was not an important mechanism for this effect. PE but not PC was effective as a synergist in combination with polyhydroxyflavones,[62] and PE, PI, phosphatidyl serine and phosphatidic acid were effective in protecting α-tocopherol during the autoxidation of methyl linoleate, whereas PC was not effective.[63] Phospholipids may act by releasing protons and bringing about the rapid decomposition of hydroperoxides without the formation of free radicals.[64,65] An alternative mechanism may involve the regeneration of primary antioxidants.[59] Studies of synergism between propyl gallate and PE led Dziedzic *et al.*[66] to support the latter view.

Maillard Reaction Products

The Maillard Reaction is a complex reaction between reducing sugars and amino acids, peptides or proteins. Several Maillard products have antioxidant activity.[67,68] Eichner[67] showed that intermediate reductone compounds possessed antioxidant activity while Yamaguchi *et al.*[69]

found that high molecular weight melanoidins also possessed strong antioxidant properties.

The antioxidant effect of reductones has been attributed to the breaking of the radical chain by donation of a hydrogen atom.[70] Maillard reaction products have been found to have metal chelating properties[71,72] and they are also effective at reducing hydroperoxides to non-radical products.[67] Thus it appears that the antioxidant properties of Maillard reaction products vary with product structure, which is dependent on the components involved in the browning reaction, the processing conditions and the degree of browning. Antioxidant effects may occur by several mechanisms.

REFERENCES

1. Hoffman, A. W., Changes of gutta-percha under tropical influences. *J. Chem. Soc.*, **13** (1861) 87.
2. Privett, O. S. & Blank, M. L., The initial stages of autoxidation. *J. Am. Oil. Chem. Soc.*, **39** (1962) 465–9.
3. Heaton, F. W. & Uri, N., The aerobic oxidation of unsaturated fatty acids and their esters. *J. Lipid Res.*, **2** (1961) 152–60.
4. Korycka-Dahl, M. B. & Richardson, T., Activated oxygen species and oxidation of food constituents. *CRC Crit. Rev. Food Sci. Nutr.*, **10** (1978) 209–41.
5. Eskin, N. A. M., Grossman, S. & Pinsky, A., Biochemistry of lipoxygenase in relation to food quality. *Crit. Rev. Food Sci. Nutr.*, **9** (1977) 1–40.
6. Benson, S. W., Kinetics of pyrolysis of alkyl hydroperoxides and their O—O bond dissociation energies. *J. Chem. Phys.*, **40** (1964) 1007–13.
7. Hiatt, R. & Irwin, K. C., Homolytic decompositions of hydroperoxides. V. Thermal decompositions. *J. Org. Chem.*, **33** (1968) 1436–41.
8. Bateman, L. & Hughes, H., The thermal decomposition of cyclohexenyl hydroperoxide in hydrocarbon solvents. *J. Chem. Soc.* (1952) 4594–601.
9. Walling, C. & Heaton, L., Hydrogen bonding and complex formation in solutions of tert-butyl hydroperoxide. *J. Am. Chem. Soc.*, **87** (1965) 48–51.
10. Hiatt, R., Irwin, K. C., and Gould, C. W., Homolytic decompositions of hydroperoxides IV. Metal-catalysed decompositions. *J. Org. Chem.*, **33** (1968) 1430–5.
11. Hiatt, R., Mill, T. and Mayo, F. R., Homolytic decompositions of hydroperoxides I. Summary and implications for autoxidation. *J. Org. Chem.*, **33** (1968) 1416.
12. Bawn, C. E. H., Free-radical reactions in solution initiated by heavy metal ions. *Disc. Farad. Soc.*, **14** (1953) 181–99.
13. Waters, W. A., The kinetics and mechanism of metal-catalysed autoxidation. *J. Am. Oil Chem. Soc.*, **48** (9) (1971) 427–33.

14. Chan, H. W-S, Prescott, F. A. A. & Swoboda, P. A. T., Thermal decomposition of individual positional isomers of methyl linoleate hydroperoxide. *J. Am. Oil Chem. Soc.*, **53** (1976) 572–6.
15. Ingold, K. U., Inhibition of autoxidation. *Adv. Chem. Ser.*, **75** (1968) 296–305.
16. Scott, G., *Developments in Polymer Stabilisation*, Vol. 7, Chapter 2. Applied Science, London, 1984.
17. Harman, D., In *Free Radicals in Biology*, Vol. 5. ed. W. A. Pryor. Academic Press, New York, 1982, pp. 255–75.
18. Burton, G. W. & Ingold, K. U., β-Carotene: an unusual type of lipid antioxidant. *Science*, **224** (1984) 569–73.
19. Stryer, L., *Biochemistry*. Freeman, San Francisco, 1975, p. 72.
20. Ingold, K. U., Inhibition of oil oxidation by 2,6-di-t-butyl-4-substituted phenols. *J. Phys. Chem.*, **64** (1960) 1636–42.
21. Miller, G. J. & Quackenbush, F. W., A comparison of alkylated phenols as antioxidants for lard. *J. Am. Oil Chem. Soc.*, **34** (1957) 249–50, 404–7.
22. Baum, B. O. & Perun, A. L., Antioxidant efficiency versus structure. *Soc. Plast. Engrs Trans.*, **2** (1962) 250–7.
23. Täufel, K., Kretzschmann, F. & Franzke, C., *Fette seifen Anstrichm.*, **62** (1960) 1061–7.
24. Rosenwald, R. H. & Chenicek, J. A., Alkylhydroxyanisoles as antioxidants. *J. Am. Oil Chem. Soc.*, **28** (1951) 185–8.
25. Scott, G., *Atmospheric Oxidation and Antioxidants.* Elsevier, New York, 1965.
26. Mabrouk, A. F. & Dugan, L. R., Kinetic investigation into glucose-, fructose- and sucrose-activated autoxidation of methyl linoleate emulsion. *J. Am. Oil Chem. Soc.*, **38** (1961) 692–5.
27. Cillard, J., Cillard, P. & Cormier, M., Effect of experimental factors on the prooxidant behaviour of tocopherol. *J. Am. Oil Chem. Soc.*, **57** (1980) 255–61.
28. Lundberg, W. O., Dockstader, W. B. and Halvorson, H. O., The kinetics of the oxidation of several antioxidants in oxidising fats. *J. Am. Oil Chem. Soc.*, **24** (1947) 89–92.
29. Endo, Y., Usuki, R. & Kareda, T., Antioxidant effects on chlorophyll and pheophytin on the autoxidation of oils in the dark II. *J. Am. Oil Chem. Soc.*, **62** (9) (1985) 1387–90.
30. Foote, C. S. & Denny, R. W., Chemistry of singlet oxygen. VII. Quenching by β-carotene. *J. Am. Chem. Soc.*, **90** (1968) 6233–5.
31. Foote, C. S., Photosensitised oxidation and singlet oxygen: consequences in biological systems. In *Free Radicals in Biology*, Vol. II, ed. W. A. Pryor. Academic Press, New York, 1976, pp. 85–133.
32. Lea, C. H., *Rancidity in Edible Fats*. Chemical Publishing Co. Inc., New York, 1939.
33. Uri, N., Metal-ion catalysis and polarity of environment in the aerobic oxidation of unsaturated fatty acids. *Nature*, **177** (1956) 1177–8.
34. Singer, T. P. & Edmondson, D. E., Biological reduction of molecular oxygen to hydrogen peroxide. *Mol. Oxygen Biol.* (1974) 315–37.
35. Samuel, D. & Steckel, F., Physicochemical properties of molecular

oxygen. *Mol. Oxygen Biol.* (1974) 1–32.
36. Rogers, W. P. & Pont, E. G., Copper contamination in milk production and butter manufacture. *Aust. J. Dairy Technol.*, **20** (1965) 200–5.
37. Pokorny, J., Major factors affecting the autoxidation of lipids. In *Autoxidation of Unsaturated Lipids,* ed. H. W-S Chan. Academic Press, London, 1987, pp. 141–206.
38. Kochi, J. K., Oxidation-reduction reactions of free radicals and metal complexes. *Free Radicals,* **1** (1973) 591–683.
39. Uri, N., *Essential Fatty Acids,* 4th Int. Cong. Biochem. Prob. Lipids, ed. H. M. Sinclair. Academic Press, New York, 1958, p. 30.
40. Pribil, R., *Analytical Applications of EDTA and Related Compounds.* Pergamon Press, Oxford, 1972.
41. Mertens, W. G., Swindells, C. E. & Teasdale, B. F., Trace metals and the flavour stability of margarine. *J. Am. Oil. Chem. Soc.,* **48** (10) (1971) 544–6.
42. Melniek, D., US Patent 2,983,615, 1961.
43. Täufel, K. and Linow, F., *Fette Seifen Anstrichm.*, **65** (1963) 795–9.
44. Watts, B. M., Polyphosphates as synergistic antioxidants. *J. Am. Oil. Chem. Soc.*, **27** (1950) 48–51.
45. Cort, W. M., Antioxidant activity of tocopherols, ascorbyl palmitate, and ascorbic acid, and their mode of action. *J. Am. Oil Chem. Soc.*, **51** (7) (1974) 321–5.
46. Reinton, R. & Rogstad, A., Antioxidant activity of tocopherols and ascorbic acid. *J. Food Sci.*, **46** (1981) 970–1, 973.
47. Packer, J. E., Slater, T. F. & Willson, R. L., Direct observation of a free radical interaction between vitamin E and vitamin C. *Nature,* **278** (1979) 737.
48. Kellogg, E. W. III & Fridovich, I., Superoxide, hydrogen peroxide, and singlet oxygen in lipid peroxidation by a xanthine oxidase system. *J. Biol. Chem.*, **250** (1975) 8812–5.
49. Korycka-Dahl, M. B. & Richardson, T., Initiation of oxidative changes in foods. *J. Dairy Sci.*, **63** (7) (1980) 1181–98.
50. Clements, A. H., Van den Engh, R. H., Frost, D. T. & Hoogenhout, K., Participation of singlet oxygen in photosensitised oxidation of 1,4-dienoic systems and photooxidation of soybean oil. *J. Am. Oil Chem. Soc.*, **50** (8) (1973) 325–30.
51. Rock, S. P., Fisher, L. & Roth, H., Methyl silicone in frying fats—antioxidant or prooxidant? *J. Am. Oil. Chem. Soc.*, **44** (1967) 102A.
52. Warner, K., Mounts, T. L. & Kwolek, W. F., Effects of antioxidants, methyl silicone and hydrogenation on room odor of cooking oils. *J. Am. Oil Chem. Soc.*, **62** (10) (1985) 1483.
53. Martin, J. B., Stabilisation of fats and oils. US Patent 2,634,213, 1953.
54. Freeman, I. P., Padley, F. B., and Sheppard, W. L., Use of silicones in frying oils. *J. Am. Oil Chem. Soc.*, **50** (1973) 101–3.
55. Rock, S. P. & Roth, H., Factors affecting the rate of deterioration in the frying quality of fats II. *J. Am. Oil Chem. Soc.*, **41** (1964) 531–3.
56. Sims, R. J., Fioriti, J. A. & Kanuk, M. J., Sterol additives as

polymerisation inhibitors for frying oils. *J. Am. Oil Chem. Soc.*, **49** (1972) 298–301.

57. Boskou, D. & Morton, I. D., Effect of plant sterols on the rate of deterioration of heated oils. *J. Sci. Food Agric.*, **27** (1976) 928–32.
58. Gordon, M. H. & Magos, P., The effect of sterols on the oxidation of edible oils. *Food Chem.*, **10** (1983) 141–7.
59. Brandt, P., Hollstein, E. & Franzke, C., Pro- and antioxidative effects of phosphatides. *Lebensmittel Ind.*, **20** (1973) 31–3.
60. Linow, F. & Mieth, G., Zur fettstabilisierenden Wirkung von Phosphatiden. 3. Mitt. Synergistische Wirkung ausgewahlter Phosphatide. *Die Nahrung,* **20** (1) (1976) 19–24.
61. Hudson, B. J. F. & Ghavami, M., Phospholipids as antioxidant synergists for tocopherols in the autoxidation of edible oils. *Lebensm. Wiss. u-Technol.*, **17** (1984) 191–4.
62. Hudson, B. J. F. & Lewis, J. I., Polyhydroxy flavonoid antioxidants for edible oils. Phospholipids as synergists. *Food Chem.*, **10** (1983) 111–20.
63. Ishikawa, Y., Sugiyama, K. & Nakabayashi, K., Stabilisation of tocopherol by three component synergism involving tocopherol, phospholipid and amino compound. *J. Am. Oil Chem. Soc.*, **61** (5) (1984) 950–4.
64. Tai, P. T., Pokorny, J. & Janicek, G., Non-enzymic browning X. Kinetics of the oxidative browning of phosphatidylethanolamine. *Z. Lebensm. Unters. Forsch.*, **156** (5) (1974) 257–62.
65. Pokorny, J., Poskocilova, H. & Davidek, J., Effect of phospholipids on the decomposition of hydroperoxides. *Nahrung,* **25** (1981) K29–31.
66. Dziedzic, S. Z., Robinson, J. L. & Hudson, B. J. F., Fate of propyl gallate and diphosphatidylethanolamine in lard during autoxidation at 120°C. *J. Agric. Food Chem.*, **34** (1986) 1027–9.
67. Eichner, K., Antioxidative effects of Maillard reaction intermediates. *Prog. Food Nutr. Sci.*, **5** (1981) 441–51.
68. Lingnert, H. & Eriksson, C. E., Antioxidative effect of Maillard reaction products. *Prog. Food Nutr. Sci.*, **5** (1981) 453–66.
69. Yamaguchi, N., Koyama, K. & Fujimaki, M., Fractionation of antioxidative activity of browning reaction products between D-xylose and glycine. *Proc. Food Nutr. Sci.*, **5** (1981) 429–39.
70. Lundberg, W. O., *Autoxidation and Antioxidants,* Vol. 1, Interscience, New York, 1961.
71. Kajimoto, G. & Yoshida, H., Studies on the metal-protein complex VII. *Yukagaku,* **24** (1975) 297–300.
72. Gomyo, T. & Horikoshi, M., On the interaction of melanoidin with metal ions. *Agric. Biol. Chem.*, **40** (1976) 33–40.

Chapter 2

DETECTION, ESTIMATION AND EVALUATION OF ANTIOXIDANTS IN FOOD SYSTEMS

S. P. KOCHHAR* & J. B. ROSSELL

Oils and Fats Section, Leatherhead Food Research Association, Leatherhead, Surrey KT22 7RY, UK

GENERAL INTRODUCTION

It is well known that in foods, antioxidants retard oxidative rancidity caused by atmospheric oxidation and thus protect oils, fats and fat soluble components such as vitamins, carotenoids and other nutritive ingredients. In addition, they delay undesirable change brought about by oxidation in foods, for example discoloration[1,2] in meat and meat products, browning or 'scald' on fruit and vegetables,[3,4] etc. Care in food processing and treatment of the food packaging materials with antioxidants minimise such deteriorations. Antioxidants do not render a rancid fat or spoilt food palatable, nor do they suppress hydrolytic rancidity, which is an enzymically catalysed hydrolysis of fats. Furthermore, they cannot replace good raw material quality nor the careful handling and suitable storage conditions necessary for the manufacture of good wholesome foodstuffs.

Antioxidants may be added directly to the food system or as a solution in the food's oil phase, in a food grade solvent or in an emulsified form which may be sprayed onto the food product. The type of food[2,5–8] to which antioxidants may be added is variable, ranging from dry (e.g. cereal-based products), convenience and snack foods (such as instant potato granules, and crisps), biscuits, nuts, mayonnaise, fruit drinks, chewing gum and meat products, to oils and fats. It should be pointed out that antioxidants must not be added above a certain level not only due to legal restraints, but also because a pro-oxidant effect would occur. To be most effective,[9] antioxidants must be added, as soon as possible, to a fresh product because they cannot reverse any oxidation that has already occurred.

* Present address: SPK Consultancy Services, 48 Chiltern Crescent, Early, Reading, Berkshire RG6 1AN, UK.

Before discussing the detection and determination of naturally occurring antioxidants such as tocopherols (mainly present in vegetable oils) and related synthetic antioxidants, it would be useful to provide a short review of types of food antioxidants in general.

The term 'food antioxidants' is generally applied to those compounds that interrupt the free-radical chain reactions involved in lipid oxidation. However, the term should not be used in such a narrow sense because of the complexity of food systems. Broadly speaking, antioxidants are classified into five types:

(1) Primary antioxidants: those compounds, mainly phenolic substances, that terminate the free radical chains[9,10] in lipid oxidation. Natural and synthetic tocopherols, alkyl gallates, butylated hydroxyanisole (BHA), butylated hydroxytoluene (BHT), tertiary butyl hydroquinone (TBHQ), etc., belong to this group, and function as electron donors.
(2) Oxygen scavengers,[11,12] for example, ascorbic acid (vitamin C), ascorbyl palmitate, erythorbic acid (D-isomer of ascorbic acid) and its sodium salt, etc., which react with oxygen, and can thus remove it in a closed system. The regeneration of phenolic antioxidants,[12,13] an entirely different mechanism, by ascorbic acid (present in many fruits and vegetables) has also been proposed to explain the synergistic action of mixed antioxidants.
(3) Secondary antioxidants,[9,14] such as dilauryl thiopropionate and thiodipropionic acid, which function by decomposing the lipid hydroperoxides into stable end products.[7] These compounds, although approved by the American Food and Drug Administration (FDA), are not yet readily accepted for use in food within the EEC, and will in consequence not be discussed in depth in this chapter.
(4) Enzymic antioxidants,[15] for example, glucose oxidase,[7] superoxide dismutase, catalase, glutathione peroxidase, etc. These antioxidants function either by removing dissolved/headspace oxygen, e.g. with glucose oxidase, or by removing highly oxidative species (from food systems), e.g. with superoxide dismutase.
(5) Chelating agents or sequestrants[8,9,14] for example, citric acid, amino acids, ethylenediaminetetra-acetic acid (EDTA), etc., which chelate metallic ions such as copper and iron that

promote lipid oxidation through a catalytic action. The chelates are sometimes referred to as synergists since they greatly enhance the action of phenolic antioxidants. Most of these synergists exhibit little or no antioxidant activity when used alone, except amino acids[16] which can show antioxidant[17–19] or pro-oxidant activity.[20] Phospholipids such as cephalin act as antioxidant synergists[21] in some systems, perhaps also due to their chelating effect.

At least 30 food spices and herbs[22,23] have been shown to possess antioxidant properties. A variety of antioxidant compounds, mainly phenolics, from the extracts of various spices have been identified. For instance, carnosol, carnosic acid, rosmanol, rosmarinic acid, rosmaridiphenol and rosmariquinone have been isolated from rosemary leaves.[24] Two major antioxidant components, gallic acid and eugenol, in cloves have been identified by Kramer.[25] Also certain oilseeds (viz. sesame, chia, cocoa and cotton seeds) contain characteristic antioxidant components in addition to commonly occurring tocopherols. For example, the strong antioxidant activity of crude sesame seed oil is attributed to sesamol, sesamin, sesamolin[26] and γ-tocopherol. During the acid bleaching of the oil, part of sesamolin is converted to the more potent antioxidant, sesaminol and its epimers which remain present during deodorisation. Other natural antioxidants are β-carotene (i.e. carotenoids present in carrots and in all foods that contain chlorophyll), caffeic, quinic and ferulic acids, flavonoids,[27] (e.g. myricetin, quercitrin and rutin), coniferyl alcohol, guaiaconic and guaiaretic acids (from gum guaiac), etc. Some naturally occurring antioxidants, for example, gossypol (present in crude cotton seed oil) and nordihydroguaiaretic acid (NDGA) (from the bush *Larrea divaricata*) have toxic properties. The FAO/WHO Joint Expert Committee on Food Additives[14] (JECFA) has given a list of 29 antioxidant compounds, and surprisingly lecithin is listed as an emulsifier only. Within the limited pages of this chapter, it is not possible to describe the analytical determinations of this vast range of antioxidant compounds added or identified in various foods. We therefore limit the discussion to the procedures of detection and determination to the most common food antioxidants listed in Table 1. The majority of these antioxidant additives (as indicated by the ‘E’ codes) are permitted in the EEC countries.[28–30]

TABLE 1
ANTIOXIDANTS MOST COMMONLY USED IN FOODS

Antioxidant (common abbreviation)	*'E' number*[a]	*Typical applications*
L-Ascorbic acid	E300	Fruit juices, drinks, mayonnaise, cured meat, fish products, butter, etc.
Sodium L-ascorbate	E301	Meat products
Calcium L-ascorbate	E302	
Palmitoyl L-ascorbic acid (ascorbyl palmitate)	E304	Scotch eggs, sausages, milk fat
Mixed natural tocopherols concentrate	E306	Vegetable oils, milk fat, mayonnaise
Synthetic alpha-tocopherol (α-T)	E307	Infant foods, milk fat, mayonnaise
Synthetic gamma-tocopherol (γ-T)	E308	
Synthetic delta-tocopherol (δ-T)	E309	
Propyl gallate (PG)	E310	Chewing gum, vegetable oils
Octyl gallate (OG)	E311	
Dodecyl gallate (DG)	E312	
Butylated hydroxyanisole (BHA)	E320	Animal fat, cheese spread, biscuits, potato flakes, beef stock cubes
Butylated hydroxytoluene (BHT)	E321	Walnuts, chewing gum
Lecithins	E322	Low fat spread, milk fat, margarine
Citric acid	E330	Vegetable oils, mayonnaise
Others such as:		
Tertiary butylhydroquinone (TBHQ)[b]		Palm oil, frying oils
Ethoxyquin, diphenylamine		Antiscald agents for pears and apples, animal feeds
2,4,5-Trihydroxybutyrophenone (THBP)		
2,6-Di-*tert*-butyl-4-hydroxymethylphenol (Ionox-100)		
Nordihydroguaiaretic acid (NDGA)[c]		
3,3'-Thiodipropionic acid (TDPA)		
Citrate mixture		

[a] 'E' numbers are the European Economic Community (EEC) codes (see Ref. 28).
[b] Not permitted in the EEC countries at the time of writing.
[c] Found to have toxic properties and subsequently removed from the GRAS list by the FDA.

DETECTION AND DETERMINATION OF ANTIOXIDANTS

A number of review articles for the detection and determination of natural tocopherols[31–36] (vitamin E), ascorbic acid[37–39] (vitamin C) and common synethetic antioxidants[40–44] (such as BHT, BHA, TBHQ, etc.) in oils, fats and fatty foods have been published. Numerous methods based on colorimetry,[45] spectrophotometry, fluorimetry,[46] voltammetry,[47] polarography, thin-layer chromatography,[48,49] paper chromatography,[50] gel permeation chromatography[51] (GPC), gas–liquid chromatography,[52,53] (GLC), and high performance liquid chromatography[54,55] (HPLC) have been described for the qualitative and quantitative determination of antioxidants in various foods. With few exceptions (namely HPLC and polarographic/voltammetric methods) almost every procedure requires considerable sample preparation such as saponification and extraction prior to separation and estimation of individual antioxidants. Because of possible toxicity[56–58] and legal restrictions on the use of many antioxidants,[59,60] a considerable number of procedures for their determination have been developed and tested in collaborative studies. Table 2 lists the official, standard or tested methods for the identification and quantitative determination of several common antioxidants.

The EEC approach on safety and necessity of food antioxidants has been described by Haigh.[61] Recently, the WHO/FAO Codex Committee[60] on Fats and Oils revised the maximum levels of commonly used antioxidants and recommended that they be limited as follows; propyl gallate, BHT, BHA and TBHQ 100, 75, 175 and 120 mg/kg respectively, with any combination of these additives, limited to 200 mg/kg, with individual limits not exceeded; added natural or synthetic tocopherols, 500 mg/kg; ascorbyl palmitate and ascorbyl stearate, 500 mg/kg, individually or in combination; and dilauryl thiodipropionate, 200 mg/kg.

By virtue of their physical and chemical properties, losses of antioxidants may occur during processing and storage of foods. Likewise, losses may take place during analytical processes since antioxidants will also combine with peroxides in solvents such as diethyl ether, and suffer oxidation by light and heat. Hence, due care must be given to the purification of solvents used and to the conditions of clean-up, extraction and quantitation techniques employed. It should also be pointed out that although extract purification, separation and derivatisation remove interferences, each step also enhances

TABLE 2

OFFICIAL OR STANDARD METHODS FOR DETECTION AND DETERMINATION OF ANTIOXIDANTS

Component	*Product*	*Method*	*Procedure/technique*	*Reference*
α-T, β + γ-T, δ-T and their dimers	Oils and fats	IUPAC	Saponification/extraction/TLC/ colorimetry/spectrophotometry	62, Method 2.411 (A)
α-T, β + γ-T, δ-T	Oils and fats	IUPAC	Silylation/packed column GLC	62, Method 2.411 (B)[a]
α-T, β-T, γ-T, δ-T, α-T-3, β-T-3, γ-T-3, δ-T-3	Oils and fats	IUPAC	Silylation/capillary column GLC	62, Method 2.411 (C)[a]
α-T, β-T, γ-T, δ-T, α-T-3, β-T-3, γ-T-3, δ-T-3	Oils and fats	DGF; IUPAC[b] ISO[b]	Concentration/HPLC, UV or fluorescence detection	63, 64
α-T, β + γ-T, δ-T	Acid oils	AOCS	Saponification/extraction/GLC	65, Method Ce 3-74
RRR-α-T[c] (vitamin E), all-rac-α-T, and α-tocopheryl acetate	Several foods and feeds	AOAC	Extraction/saponification/TLC/ colorimetry/polarimetry	66, 43.129–43.151
PG, OG, DG, BHT, BHA and others	Oils and fats	IUPAC;	Extraction/detection/identification/ TLC	62, Method 2.621
		BS 684[d] ISO 5558[d]	Extraction/detection/identification/ TLC	67, Section 2.33 68

BHA and BHT	Oils and fats	BS 684[d]	Direct injection GLC, internal standard calibration	67, Section 2.36
		ISO 6463[d]		68
BHA and BHT	Oils and fats	IUPAC	Steam distillation/colorimetry/ spectrophotometry	62, Method 2.622
BHA and BHT	Breakfast cereals	AOAC	Extraction/GLC	66, 20.014–20.020
PG, OG and DG	Oils and fats	BS 684[d]	Selective extraction/ spectrophotometry	67, Section 2.44
		ISO 6464[d]		68
PG, NDGA, BHT and BHA	Oils and fats	AOAC	Extraction/clean up/detection/ colorimetry	66, 20.006–20.008 20.0018–20.0020
PG, THBP, TBHQ, NDGA, BHA, Ionox 100 and BHT	Oils, fats and shortenings	AOAC, AOCS	Extraction/HPLC/ UV detection at 280 nm	66, 20.009–20.013 65, Method Ce 6-86
Sesamol, sesamoline and sesamine	Sesame seed oil in oils and fats	AOCS, AOAC	Qualitative detection, colour reaction	65, Method Cb 2-40 66, 28.129

See Table 1 for abbreviations.
[a]Not applicable to virgin oils.
[b]These HPLC methods are being developed for the determination of free tocopherols and total tocopherols (free and esterified) in oils and fatty foods.
[c]RRR-α-T = 2R,4′R, 8′R-α-tocopherol, formerly d-α-tocopherol or natural form.
[d]BS 684 and ISO methods are technically identical.

the chances for losses of the antioxidant. Therefore, for control arbitration and dispute purposes, the tested and approved methods (given in Table 2) should be used for estimation of antioxidants, since any interferences, artefacts, and other problems in a particular foodstuff are better understood and corrected for where necessary.

Analytical Methods for Tocopherols (Vitamin E)

The natural occurrence of tocopherols, tocotrienols and related compounds (collectively termed 'tocols') is well-established.[25,31,69] The tocols are classified as a group of structurally related compounds that are individually named according to whether the phytyl side chain to the chroman ring has no double bonds (tocopherols: T), one double bond (tocomonoenols: T-1), or three double bonds (tocotrienols: T-3). Further classification is according to the substitution of methyl groups on the phenolic ring; thus 5,7,8-trimethyltocol (α-T); 5,8-dimethyltocol (β-T); 7,8-dimethyltocol (γ-T), and 8-methyltocol (δ-T). These various tocol compounds are sometimes called tocopherol 'isomers', but in the strict chemical sense they are not isomeric with one another as they have different molecular formulae. In this chapter, therefore, they are called tocols (or tocopherols, as the case may be).

These various tocopherols (vitamin E) may exist in free or esterified forms.[69] In seed oils, they are present mainly in the free state. Food materials such as nuts, seeds, some grains and vegetable oils are good sources of natural tocopherol antioxidants. The tocopherols vary in their activities. It is known that vitamin E activity[70,71] decreases, whereas the antioxidant activity[35] increases in the order α-, β-, γ- and δ-tocopherols. The corresponding tocotrienols possess slightly higher antioxidant activity but lower biological activity.[71]

Several methods for the analysis of tocopherols (vitamin E) in foodstuffs have been published.[35,72] Figure 1 gives a general analytical scheme for the determination of tocopherols in oils and foods. Broadly speaking, the techniques for tocopherol estimation can be classified as colorimetric/spectrophotometric, electrochemical (polarographic/voltammetric), or chromatographic. α-Tocopherol tends to become a pro-oxidant if present at concentrations much above 1000 mg/kg, while detection limits of most methods are about 2 mg/kg. Most samples of interest have tocol concentrations between these two values.

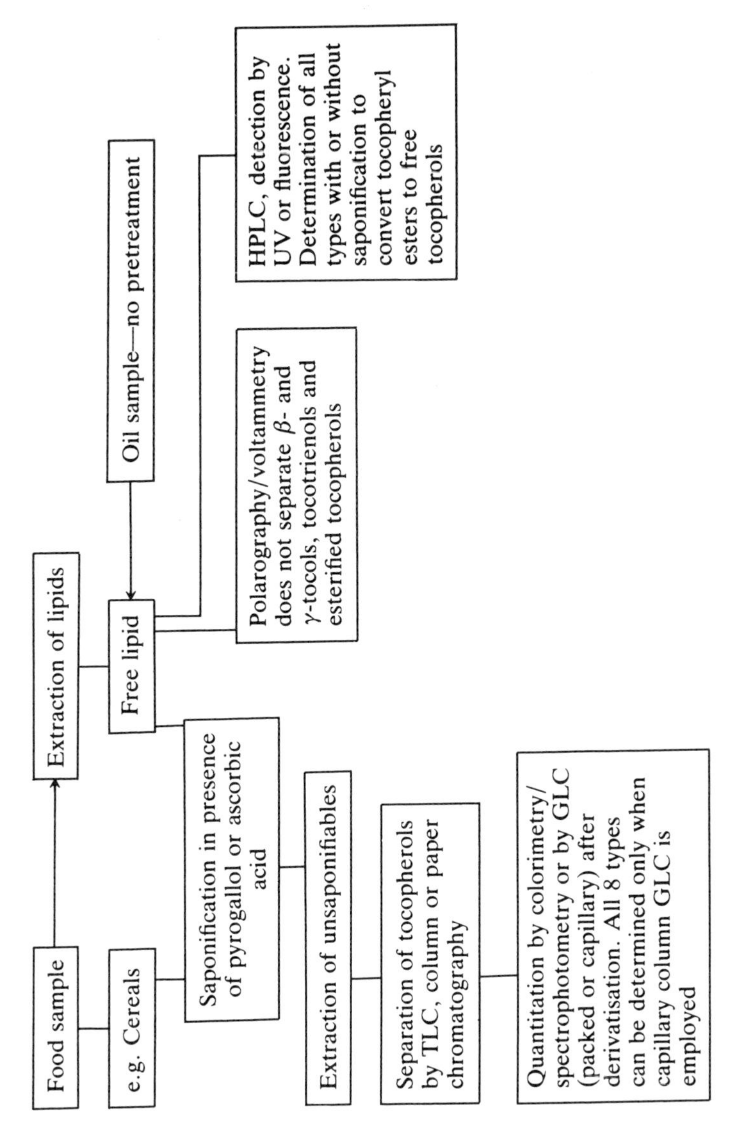

FIG. 1. General analytical scheme for determination of tocopherols.

(a) Colorimetric Methods

Among these methods which determine only total tocopherols, the classical procedure of Emmerie and Engel[73] was reported nearly 50 years ago. This method was based on the colour reaction between the reagent 2,2′-bipyridyl and Fe^{2+}, which results from the oxidation of total tocopherols with $FeCl_3$. Some years later, a modification[74] of this procedure was described for the estimation of total tocopherols in mixtures containing δ-tocopherol. A sensitive and highly reproducible method for tocopherol analysis in biological samples that involves saponification in the presence of ascorbic acid to remove interfering substances and protect tocopherols, extraction of the unsaponifiables with hexane, and fluorometric measurement of α-tocopherol, has been described by Taylor *et al.*[46] Apart from difficulties encountered due to the various rates of reactivities of the different tocopherols, the disadvantage of classical procedures is that they measure only total tocopherol content and yield no information on the levels of individual tocols. However, it is important to meausre the concentration of various forms of tocopherols in order to calculate the biological potency of vitamin E in foods, and to understand their antioxidant effect on the stability of oils. Also in many cases the information about the type of tocopherol distribution and their rates may be useful for establishing purity or adulteration of oils. Bearing these in mind, a number of separation procedures for tocopherols have been developed by many workers. For example, very recently a derivative spectrophotometric method for determining individual saturated tocopherols has been published by Bukovits & Lezerovich.[75]. The procedure measures the second derivative spectrum of the sample at four specified wavelengths and then calculates each concentration by means of a method of multicomponent analysis which uses standard mixtures as calibrants.

(b) Electrochemical Methods

Voltammetric/polarographic techniques have been successfully applied in measuring common individual tocopherols.[76–78] The earlier procedures[76,79,80] employed electrochemical reduction of tocopherylquinones on the dropping mercury electrode for the determination of the various tocopherols. After saponification of the sample, and then extraction of the unsaponifiable material with ether, the tocopherols were oxidised with Ce^{IV} to the tocopherylquinones required for the polarographic reduction. Later on, polarographic methods[77,78,81,82] involving direct oxidation of tocopherols at different carbon electrodes were described. Although the voltammetric method described by McBride & Evans[47] is a rapid one (requiring about 10 min), being uncomplicated and suitable for routine analysis, the technique suffers from the lack of separation of β- from γ-tocopherol, and the interferences of tocotrienols with the corresponding tocopherol peaks. Moreover, one has to take into account the possible interferences of other additives such as propyl gallate and ascorbyl palmitate which do not separate from α-tocopherol. Nevertheless, the recent, rapid and direct electrochemical methods are well-suited for the determination of the principal tocopherols (α, γ, and δ) for quality control purposes, e.g. monitoring tocopherol loss during fat processing or quality control in foods fortified with vitamin E. The detection imits of α, $\beta + \gamma$, and δ are estimated to be 10, 10 and 20 mg/kg oil respectively. The detection limit is defined as tocopherol concentration whose detectable signal is at least twice as high as the baseline noise. An improved electrochemical method for determining tocopherols in oils, fats and fat-containing foods has been described by Löliger & Saucy.[83] The application of an electrochemical micro flow-through detector has reduced the detection limit to 0·01 mg (for all three tocopherols) per kg fat. The reproducibility of the method is ±5% at the tocopherol level of 0·1–600 mg/kg.

(c) Chromatographic Methods

Chromatographic techniques[34,35] such as thin-layer chromatography (TLC), gas–liquid chromatography (GLC) and column (including high-performance) liquid chromatography have been used extensively for quantitation of individual tocopherols in oils and a variety of foods.

In the past, the TLC methodology was employed quite routinely as it was within the reach of many laboratories, and can be used for separation of tocopherols and tocotrienols. For example, a TLC procedure has been adopted by the Association of Official Analytical Chemists (AOAC) as the official method (Table II, Section 43.129–43.151) for the determination of α-tocopherol and α-tocopheryl acetate in foods and feed.[45] The Official International Union of Pure and Applied Chemistry (IUPAC) method (Table II, Method 2.411 A) is based on the saponification of the oil with alkali, extraction of unsaponifiables in the presence of pyrogallol to prevent oxidation of tocopherols, separation by TLC and then estimation spectrophotometrically. The method can also be used for the determination of tocopherol dimers, but it does not discriminate between tocopherols and tocotrienols (β- and γ-tocols are not separated either). As the saponification stage converts tocopheryl acetate to free tocopherols the total of free and combined tocopherol is determined in this procedure. Chow *et al.*[84] have described a TLC method for the estimation of free and esterified tocopherols. The bulk of the lipids extracted with acetone from several plant oils was removed by low temperature crystallisation instead of saponification. Solvent was evaporated from the filtrate and the eight vitamin E compounds obtained were then separated by two successive TLC systems. The individual tocopherols were then quantified spectrophotometrically by the modified Emmerie–Engel procedure. For calibration, standards of saturated tocopherols were obtained commercially, while tocotrienol (T-3) standards were prepared from natural sources: α-T-3 from barley, β-T-3 from wheat bran, and γ- and δ-T-3 from latex. The yields of individual tocopherols obtained from plant oils by this method were found to be higher than those by the saponification procedure. For the determination of tocopherol esters, which are solid at the fractionation temperature, the crystallised fraction was processed to liberate tocopherols which were then separated and determined quantitatively. A shorter procedure, which deserves further exploration, is based on extracting unsaponifiable material after direct saponification of cereal samples, instead of first extracting the lipids, and then separation of individual tocopherols by TLC on silica gel G, followed by the determination with a new reaction, involving cupric ion and a complexing agent (either cuporine or bathocuprine), and has been developed[85] in recent years.

Gas–liquid chromatography (GLC) has also been employed regularly for the quantification of tocopherols[86,87] after preliminary separation and clean-up of the samples by TLC. The official AOAC and IUPAC methods shown in Table 2 are similar in this analytical scheme. Hartman[88] has described a GLC procedure for quantitation of the vitamin E content of vegetable oils and oils extraced from snack foods. Vitamin E is determined by saponification of the oil, ether extraction of the saponified mixture, drying and evaporation of the extract, followed by esterification and determination of the butyrate esters using a gas chromatograph equipped with an FID detector and fitted with a glass column (1·83 m long, 5 mm i.d.) packed with 3% SE-30 on Gas Chrom Q, 100–120 mesh. The lower limit of detection of α-T was found to be about 5 mg/kg oil, but β- and γ-tocopherol compounds were not resolved. Palm oil and palm oil fractions contain a considerable amount of tocotrienols. A reliable TLC–GLC method which distinguishes between tocopherols and tocotrienols has been reported by Meijboom & Jongenotter.[89] The procedure involves separation of tocols from the oil by saponification and TLC, followed by direct packed column GLC of the recovered material with added internal standard (hexadecyl stearate). However, β- and γ-tocopherols were not separated. Slover *et al.*[90] have extensively employed capillary column gas–liquid chromatography (CCGLC) for quantitative determination of individual tocopherols in oils and fatty foods. The quantitation of all eight tocopherols can be achieved successively by temperature-programmed CCGLC analysis of derivatised tocopherols (trimethylsilyl ethers), which is also the basis of the IUPAC method 2.411 (C) (see Table 2).

Most of the procedure employed in GLC, TLC and colorimetric methods require long analysis times with extensive sample preparation and/or derivatisation of tocopherols prior to their estimation. Since tocopherols are known to be light sensitive and prone to oxidation, some losses due to oxidative degradation may occur during the lengthy treatment of the sample. Furthermore, the methods require saponification which precludes the possibility of determining the relative proportion of tocopherols present in the form of esters. These limitations and difficulties of the above analytical procedures prompted the development of a liquid chromatographic technique, and nowadays the HPLC technique described below is most often used.

A high-performance liquid chromatographic (HPLC) method for the determination of four free tocopherols in vegetable oils was first described by Van Niekerk.[91] Since then several papers have appeared on this versatile technique for the estimation of tocopherols and tocotrienols in vegetable oils and in foods. Many HPLC systems described for the determination of the eight tocopherols in vegetable oils employ a silica column. A solution of the oil in *n*-hexane or in the mobile phase is applied directly to the column. The chromatographic separation is then performed usually with a mixture of two or three solvents as mobile phase, hexane or heptane being major components, and at a flow rate of 1–2 ml/min, using either UV set at 280 nm (or at the maximum absorption at a wavelength between 280 and 297 nm) or fluorescence detection (excitation at between 290 and 296 nm; emission at between 325 and 340 nm). In the past decade, considerable developments in the stationary phases, particularly the introduction of microparticulate (commonly 5 μm) column packing, have taken place. This has led to great improvements in separating power and cut down the analysis time to about 10 min. Table 3 summarises typical conditions of a variety of HPLC procedures employed for the determination of tocopherols in foods.

In the authors' laboratory, an HPLC technique for the determination of tocopherols in vegetable oils is used routinely, and tocopherol composition data for ten commercially important oils, from different geographical origins, have been obtained as a means of establishing new oil purity criteria. A typical chromatographic elution profile of the tocopherol compounds in a crude soya bean oil, using fluorescence detection is shown in Fig. 2; the conditions used are given by Rossell *et al.*[92]

Most common vegetable oils contain tocopherols in the unesterified form and may be injected on to the column directly, without any pre-treatment, apart from dissolving the sample in the mobile phase or in hexane. Lipids containing tocopherols need to be extracted from most food samples prior to chromatography. Thompson and Hatina[93] have published an extraction procedure for a variety of food samples such as spinach, milk, beef and cereal products. The sample (10 g) was homogenised in isopropanol and acetone, which was then extracted twice with hexane. The pooled hexane extracts were washed two times with water and then evaporated under reduced pressure. The efficiency of the extraction procedure for vitamin E was found to be better than 97%.

As tocopherol esters are not fluorescent, fortified foods containing tocopheryl acetate or palmitate are analysed after saponification. Morever, a saponification step should also be employed to free vitamin E for extraction from the sample matrix when it has been added in an encapsulated form or when the sample contains a homogenised spray dried milk powder. The tocopherols, especially tocotrienols are very sensitive to oxygen in the presence of alkali and care should be taken to avoid destruction during saponification. Antioxidants such as ascorbic acid or pyrogallol are added to the saponification mixture, which is refluxed gently under nitrogen, in subdued light. The unsaponifiables may be extraced with diethyl ether, petroleum ether or hexane, but the hydrocarbon solvents have the advantage of having a low solubility for water. Furthermore, they are less likely to contain peroxides, which would react with the tocopherols, as might diethyl ether. When evaporating the solvent it should never be dried

TABLE 3
HPLC SYSTEMS USED FOR DETERMINATION OF TOCOPHEROLS IN FOOD SAMPLES

Tocol	*Sample: type; preparation*	*Stationary phase*	*Mobile phase (v/v)*	*Detection*	*Reference*
α-, β-, γ-, δ-T	Veg. oils; D, hexane	Corasil II	Hexane : diisopropyl ether (95 : 5)	Fluorescence, ex. 295 nm, em. 340 nm	91
α-, β-, γ-, δ-T	Veg. oils; D, hexane	Jascopack-WC-03-500	Hexane : diisopropyl ether (98 : 2)	Fluorescence, ex. 298 nm, em. 325 nm; UV 295 nm	102
α-, β-, γ-, δ-T, α-, β-, γ-, δ-T_3; α-Tocopheryl acetate	Cereals, veg. oils, milk; E, propan-2-ol and acetone, hexane and water, D, hexane; S for α-T acetate, D, hexane	5 μm LiChrosorb Si 60 (Merck)	Hexane : diethyl ether (95 : 5) or hexane : propan-2-ol (99·8 : 0·2)	Fluorescence, ex. 290 nm, em. 330 nm; UV 290	93
α-, β-, γ-, δ-T, α-, β-, γ-T-3	Veg. oils, D, direct injection	10 μm Merckosorb Si 60 (Merck)	Hexane : dioxane (96 : 4)	Fluorescence, ex. 295 nm, em. 330 nm	103
Tocopherols	Foods; S	LiChrosorb RP-8	Methanol : water (95 : 5)	Fluorescence, ex. 295 nm, em. 330 nm; UV 308	104
α-Tocopheryl acetate	Cereals; E, chloroform, ethanol and water	μ Porasil (Waters)	Hexane : chloroform (85 : 15)	UV 280 nm	105

(continued)

TABLE 3—*contd.*

Tocol	*Sample: type; preparation*	*Stationary phase*	*Mobile phase (v/v)*	*Detection*	*Reference*
α-T, α-Tocopheryl acetate	Milk and soya based products; enzymic hydrolysis, E, pentane	Zorbax ODS (Dupont)	Methanol, ethyl acetate, acetonitrile-gradient	UV 265 nm	106
α-, β-, γ-, δ-T, α-, β-, γ-, δ-T-3	Veg. oils; D, mobile phase	10 μm R sil (RSL)	Hexane: ethyl acetate (97:3)	Fluorescence, ex. 303 nm, em. 328 nm	107
α-, β-, γ-, δ-T	Veg. oils; D, mobile phase	μ Porasil	Hexane: isopropanol (98·5:1·5)	UV 295 nm	108
α-Tocopheryl acetate	Fortified oils; D, hexane	μ Porasil	Hexane: diethyl ether (97:3)	UV 280 nm	110
α-, β + γ-, δ-T	Veg. oils; S, chloroform	Micro Pak Si 10	Hexane: diethyl ether (87:13)	UV 294 nm	109
α-, β-, γ-, δ-T	Veg. oils; D, hexane	LiChrosorb Si 60	Hexane: diethyl ether (95:5)	Fluorescence, ex. 290 nm, em. 330 nm; UV 292 nm	111
α-, β-, γ-, δ-T	Veg. oils; D, hexane	Partisil 5	Hexane: dibutyl ether (90:10)	Fluorescence, ex. 290 nm, em. 330 nm; UV 292 nm	98
α-T	Veg. oils; S, petroleum ether	LiChrosorb Si 60	Hexane: isopropanol (99·9:0·1)	Fluorescence, ex. 296 nm, em. 330 nm	112

α-, β-, γ-, δ-T; α-, β + γ-, δ-T	Oils and fatty foods; S, diethyl ether	LiChrosorb Si 60, 5 μm (250 × 4 mm) or two RP connected in series, ODS-HC Sil X-1 (250 × 2·6 mm) and Supelcosil LC 18 (150 × 4·6 mm)	Hexane : isopropanol (99·6 : 0·4) or acetonitrile, methanol and water gradient	UV 295 nm	113
α-, β-, γ-, δ-T, α-, β-, γ-, δ-T-3	Veg. oils; D,	Partisil 5	Dry heptane : damp heptane : isopropanol (49·55 : 49·55 : 0·9)	Fluorescence, ex. 290 nm, em. 330 nm	92
α-T	Foods; S, diethyl ether	LiChrosorb Si 60, 5 μm	Hexane : dioxane (97 : 3)	Fluorescence, ex. 293 nm, em. 326 nm	114
α-, β-, γ-, δ-T, α-, β-, γ-, δ-T-3	Oils and fats; concentration in acetone, D, hexane	LiChrosorb Si 60, 5 μm	Heptane : dioxane (96 : 4)	Fluorescence, ex. 296 nm, em. 326 nm	64
α-T and other tocopherols	Milk based foods; S, petroleum ether : diethyl ether (1 : 1)	Partisil Si 60	Hexane : isopropanol (98·5 : 1·5)	UV 292 nm	71
α-, β-, γ-, δ-T α-T-3	Veg. oils; D, hexane; S, diisopropyl ether for free and esterified tocols	Polygosil 60-5, 5 μm (250 × 4.5 mm i.d.)	Hexane : diisopropyl ether (90 : 10)	Fluorescence, ex. 296 nm, em. 320 nm	94
All types of T, T-1, and T-3	Veg. oils; D, hexane	Partisil PAC 5 μm (polar material, amino-cyano)	Hexane : tetrahydrofuran (94 : 6)	Fluorescence, ex. 210 nm, em. 325 nm	101
α-T, γ-T	Butter; D, hexane or S, hexane	LiChrosorb Si 60, 5 μm	Hexane : 1,4-dioxane (96 : 4)	Fluorescence, ex. 295 nm, em. 330 nm	115

Abbreviations used:
D = no saponification, injection of sample diluted with the solvent listed.
E = extraction of tocopherols from food sample with the solvent.
S = saponification of sample, followed by extraction and dilution with the solvent.

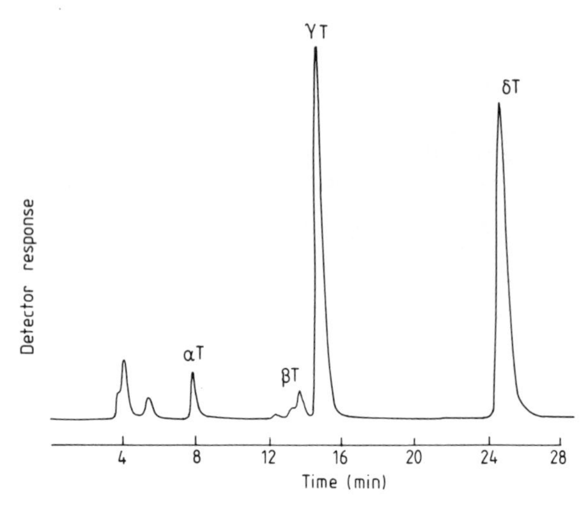

FIG. 2. Typical tocopherol profile of a crude soya-bean oil by HPLC using fluorescence detection, for conditions see Rossell *et al.*[92]

completely and antioxidants such as BHT may be added to the mobile phase to protect the tocopherols from oxidative degradation.[94] For the analysis of vitamin E in a variety of milk-based food products such as liquid and dried milks, ice-cream, yoghurt, soft and hard cheese, Reynolds[71] has published a general scheme of sampling, releasing the vitamin from the sample matrix by saponification and extraction and then quantitation using HPLC.

A variety of reversed phase HPLC systems have also been used for vitamin E analysis in foods, especially when other vitamins are determined simultaneously.[71,95–97] However, it should be pointed out that the use of C18 reverse phase column and mobile phase methanol:water (96:4, V/V) do not separate β- and γ-tocopherols using either UV at 292 nm or fluorescence detection (excitation at 290 nm, emission 330 nm). Therefore, a straight (normal) phase chromatographic system is employed when the separation of all eight common natural forms of tocopherol is desirable.

As mentioned earlier, free tocopherols exhibit strong fluorescence. Because of the inherent sensitivity and selectivity of a fluorescence detector, this is an obvious choice for their determination. Tocopherol esters do not fluoresce effectively and a UV detector should be used if the esters are not saponified before chromatography. A consequence of this, however, is that free and esterified tocopherols could be determined separately by simultaneous use of UV and fluorescence detectors, and comparison of the results. However, no reports of such separations have as yet reached the authors. In contrast, the separation of tocopheryl acetate from the eight free tocopherols has been reported by Taylor & Barnes,[98] using a partisil 5 silica column, with a mobile phase of dry hexane, water-saturated hexane:dibutyl ether:propan-2-ol (45:45:10:0·5, V/V) and fluorescence detection (excitation at 290 nm; emission at 330 nm). It has been observed[99,100] that the tocopherols fluoresce most strongly in solvents such as dioxane and diethyl ether and the inclusion of these solvents in the mobile phase increases the sensitivity of the detector. Using a mobile phase of 5% diethyl ether in moist hexane, a detection limit for tocopherols has been reported to be 2 mg/kg lipid.[93] Recently, an HPLC method for separating α-, β-, γ- and δ-tocopherols, the corresponding tocomonoenols and tocotrienols, and plastochromanol-8 (a tocol structurally related to γ-T-3) on a 5 μm amino-cyano column using as mobile phase hexane:tetrahydrofuran (94:6, V/V) has been demonstrated.[101] The detection limit (signal-to-noise ratio of 3) for α-T using fluorescence excitation at 210 nm is observed to be 0·5 mg/kg oil. Further, the presence of α-tocomonoenol (not identified previously) in palm oil at a level of 25 mg/kg has been discovered. These observations clearly show that the HPLC methods for determination of tocopherols (vitamin E) either in vegetable oils or in foods are preferred in comparison to the older methods involving colorimetric, TLC or GLC, because of their speed, ease of operation, precision and reliability. Moreover, as a minimum pre-treatment of the sample is performed, tocopherol losses are very low.

Analytical Methods for Synthetic Antioxidants

There are a considerable number of publications about the determination of one or more of the synthetic antioxidants added to oils, fats or food products. Very recently, an excellent review on the quantitative determination of five common synthetic antioxidants, namely BHA, BHT, NDGA, PG and TBHQ (see Table 1 for abbreviations) has been published by Robards & Dilli.[42] Various analytical methods used for their estimation include colorimetry/spectrophotometry, voltammetry/polarography, chromatography (paper, TLC, GPC, GLC or HPLC), etc. In principle, these techniques are similar to those used in the case of natural tocopherols, synthetic all-*rac*-α-tocopherol, tocopheryl acetate or palmitate. Generally, each of these techniques requires isolation of the antioxidants from the food matrix, prior to qualitative and quantitative analysis of the isolated material.

(a) Extraction of Antioxidants

Extraction or isolation procedures include steam distillation,[116–118] solvent distillation,[119] solvent extraction[120–122] and vacuum sublimation.[123] Solvent extraction has been widely used for the isolation of antioxidants, with or without subsequent clean-up of the extracted material. Liquid–liquid partition[124,125] and column clean-up[120] of the extracts are sometimes used to remove the co-extracted substances with the antioxidants. The extracting solvents used are numerous, for example, acetonitrile, petroleum ether, dimethyl sulphoxide, hexane, various alcohols and aqueous solutions of alcohols.[42] The extraction procedure employed also depends on the nature of the food product and the antioxidant present. Antioxidants such as BHT and BHA are steam volatile, insoluble in water but soluble in polar solvents, while PG is slightly soluble in water (0·3%). Anglin *et al.*[116] found steam distillation suitable for the isolation of BHA and BHT, and 72% aqueous ethanol solvent extraction for the recovery of PG and NDGA from edible oils. Stuckey & Osborne[43] have given two schemes for the extraction of antioxidants from foods. For fatty foods such as nuts, margarine, butter, prepared meat foods and some baked goods, the fat and antioxidants are removed by Soxhlet/direct extraction with petroleum ether or similar solvents. The extract is then either analysed directly or the solvent is first evaporated from the extract and antioxidants such as BHT and BHA are steam distilled at about 160°C. The initial extraction is not required for low fat foods such as potato flakes, rice and some cereals. For the analysis of BHT and BHA, the food is steam distilled by any of the procedures described.[33,117,118] A following technique is then used to determine individual antioxidants in the extract and/or steam distillate.

(b) Colorimetric and UV Visible Spectrophotometric Methods

In the past, several colorimetric and UV visible spectrophotometric[33,43,125] procedures have been applied to estimate antioxidants in the steam distillate/extract. For example, BHA was determined by the specific reaction with Gibbs' reagent, 2,6-dichloro-*p*-benzo-quinone-4-chlorimine. The development of blue colour was then measured by an absorption at 620 nm. The reaction is specific for phenolic compounds in the *ortho*- and *para*-positions and consequently the presence of BHT does not interfere. However, it has been reported[43] that the presence of small amounts of naturally occurring phenolic substances gives false positive results. The other commonly used reagent for BHA estimation is diazotised sulphanilic acid[126,127] which gives a red-purple colour reaction. BHT also reacts with the reagent very slowly. However, if the measurements are carried out within 5–10 min then little interference is encountered. The presence of PG inhibits the colour development, therefore it must be removed[43] before the analysis (e.g. by extraction with ammonium acetate solution). A colorimetric method for the specific estimation of BHT, based on its reaction with *o*-dianisidine (3,3′-dimethoxybenzidine) and nitrous acid, has been described by Szalkowski & Garber.[128] Absorbance of the coloured product extract in chloroform was measured at 520 nm. Recently, suitable reaction conditions for the analysis of BHT with dianisidine and BHA with the Gibbs' reagent are reported by Kobayashi *et al.* [129] Detection limits of these two antioxidants in food products are reported to be 2–4 mg/kg.

Gallates do not steam distil and therefore their determination is carried out on the extracts. They may be estimated by reaction with ferrous sulphate or ferrous tartrate to give characteristic purple-blue complexes.[117,130,131] For example, the level of PG has been determined in the colour reaction mixture[131] by measuring the absorption at 530 nm, but NDGA (if present) will interfere with this analysis. Other reagents used for colorimetric/visible spectrophotometric analysis of antioxidants include ferric chloride-2,2′-bipyridyl,[132] *p*-N-methylaminophenol (Metol) with periodate or permanganate[133,134] for BHA and PG in oils and fats. A new, rapid, spectrophotometric procedure for BHA determination in fats and fatty foods has been described by Komaitis & Kapel.[135] The purple-violet colour developed from the reaction of BHA with NN-dimethyl-*p*-phenylenediamine and potassium ferricyanide is extracted with CCl_4,and the absorption is measured at 550 nm. BHA, as low as 0·4 mg/kg food was detected. Very recently, a rapid, sensitive and accurate spectrophotometric method, based on the reagent 3-methyl-2-benzothiazolinone hydrazone hydrochloride (MBTH) and ceric ammonium sulphate, for the determination of TBHQ, BHA and gallic acid in oils has been described by Viplava Prasad *et al.*[136] The absorbances of coloured species were measured at 500, 480 and 440 nm respectively.

Numerous UV spectrophotometric procedures for the estimation of antioxidants have also been described by many earlier workers.[137–141]

For example, a UV method[140] for the determination of BHA and PG involved the absorption measurements of the sample solution, in propan-2-ol, at 232, 241 and 252 nm. The amount of BHA in lard was estimated[137] in the extract by UV absorption measurements at 270, 290 and 310 nm. However, possible interferences from other antioxidants and/or naturally occurring phenolic compounds in such UV procedures have been reported.[42] In general, because of the various limitations of spectrophotometric and colorimetric methods, chromatographic procedures such as GLC and HPLC are now being used for the synthetic antioxidants.

(c) Chromatographic Methods

A number of procedures based on thin-layer, paper and gel permeation[51,142] chromatography have been developed for the detection and estimation of most common antioxidants in oils and food products. In the past, thin-layer chromatography[33,124,143–151] (TLC) and paper chromatography[152–155] (PC) have been used extensively for the separation and determination of the antioxidants. The main reasons for their use are low cost and simplicity. It is reported[156] that the separation is more precise by PC and rapid by TLC. Seher[157,158] has described the relative merits of these two techniques for antioxidant analysis. A number of TLC absorbent phases, for example silica gel, Kieselgel, alumina, acetylated cellulose, polyamide powder, etc., for the separation of antioxidants have been used successfully. A variety of solvent systems and spray reagents employed for the detection of PG, BHT, BHA, TBHQ and NDGA are tabulated by Robards & Dilli.[42] The detection reagents include phosphomolybdic acid, dianisidine, Gibbs' reagent and diazotised sulphanilic acid. Endean[33] has listed about 20 spray reagents together with the colour responses for many common antioxidants. Quantitative estimation of the antioxidant has been obtained by comparison of the spot intensities run against 5 and 10 μl of 0·1% solution of the appropriate antioxidant in methanol. Recently, a TLC procedure for the identification of 11 antioxidants in oils and fats, involving two absorbents, three solvent systems and UV viewing at 254 and 366 nm, has been described.[159] Despite the limitation of TLC compared with GLC or HPLC, particularly regarding detection limits, the application of TLC for the identification of antioxidants in foods seems to continue because of its low cost. Further, high performance TLC appears to have potential application in the field of antioxidants.

Several gas–liquid chromatographic (GLC) methods have been developed to determine quantitatively those antioxidants present in various food products. For example, GLC procedures for the determination of BHT, BHA, TBHQ, PG or NDGA in oils and fats,[122,123,160–169] potatio granules,[121,170] chewing gum,[171] cereals,[172–176] seafoods,[177,178] dairy[179,180] and meat[181] products have been published. Concentrations of BHA and BHT added to oils, fats, butter, mayonnaise, cheese, curry coux and fried food products have been determined by GLC at 280°C, using SE-30 as a stationary phase. Distillation of a 5 g food sample mixed with heptane containing hexadecane as an internal standard was carried out prior to quantitation of the antioxidants in the heptane layer. The recoveries of added BHA and BHT were 80·1–84·0% and 92·6–98·6% respectively.[119] A rapid GLC method for the determination of BHT and BHA in fortified oils has been described by Hartman & Rose.[160] The procedure involves the addition of an internal standard, methyl undecanoate, to an oil solution in carbon disulphide. The antioxidants were isolated from non-volatile oil components by using a short pre-column located in the injector port. The recovery range of BHA and BHT was found to be 97–104%. A simple extraction method using acetonitrile solvent for the recovery of TBHQ from food oils has been reported.[182] After silylation of the extract, the antioxidant has been determined quantitatively by GLC fitted with a glass column packed with 10% versilube F-50 on 100–200 mesh Gas Chrom Q. Substantial losses in the quantitative determination of TBHQ by GLC have been reported by Stijve & Diserens.[179] The cause of low recoveries of TBHQ in the GLC analysis has been attributed to its conversion (due to oxidative breakdown) to tert-butylquinone. Possible remedies to overcome these problems either by exclusion of air from the preparation of samples and standards or by determining both TBHQ and its breakdown product have been described in the literature.[42] Table 4 gives some selected column packings and conditions used for GLC quantitation of antioxidants isolated from various food products. Kline *et al.*[183] have developed a GLC method, using two separate procedures with or without derivatisation, for the multidetermination of eight antioxidants (BHA, BHT, Ionox-100, TBHQ, TDPA, PG, THBP and NDGA) in oils, fats and other food products such as cereals, party snacks and flour product mixes. Simultaneous determination of antioxidants such as TBHQ, BHT, BHA and Ionox-100 in vegetable oils by means of high resolution gas chromatography fitted with a fixed silica capillary

TABLE 4
COLUMNS AND CONDITIONS EMPLOYED FOR QUANTITATIVE ANALYSIS OF SYNTHETIC ANTIOXIDANTS BY GLC

Antioxidant	*Product*	*Stationary phase (column packing)*	*Column temperature (°C)*	*Reference*
BHT, BHA, TBHQ, PG	Oils	10% GE-Versilube F-50 on Gas Chrom Q	150–210	168
BHT, BHA, NDGA	Lard	3% GE-XE-60 on Gas Chrom Q	150–250	123
BHT, BHA, TBHQ	Soya bean oil	10% polymetaphenoxylene on Tenax-GC	140–250	163
BHT, BHA	Oils and fats	3% SP2250 on Supelcoport	160	184
BHT	Oils	3% JXR on Gas Chrom Q	105–250	122
BHT, BHA, TBHQ[a]	Milk fat	7% SE-30/3% XE-60	135	179
BHT, BHA	Potato granules	20% Apiezon L on firebrick	220	121
BHT, BHA	Potato mashed	3% OV-17 on Gas Chrom Q	165	170
BHT, BHA	Chewing gum	DB-1 fused silic WCOT	65–200	171
BHT, BHA	Cereals	3% SE-30 on Chromosorb W	125–190	172
BHT, BHA	Cereals	10% silicone grease on chromosorb P	160	174
BHT, BHA	Cereals	10% QF-1 or 5% Apiezon L on Gas chrom Q	160	174, 175

BHT	Powdered milk	5% SE-30 on Chromosorb W	154	186
BHT, BHA	Meat products	5% GE-XE-60 on Gas Chrom Q	150	181
BHT, BHA, TBHQ, Ionox-100	Oils, fats, dried foods	3% OV-17 on Gas Chrom Q	85–175	183
BHT, BHA	Oils	10% DC 200 or 10% Carbowax 20M on Gas Chrom Q	160, 190	160
TMS[b] derivative of TBHQ	Oils	10% versilube F-50 on Gas Chrom Q	190	182
TMS[b] derivatives of BHT, BHA, PG, NDGA, TBHQ	Oils	3% JXR on Gas Chrom Q	105–250	122
TMS[b] derivatives of PG, NDGA, TDPA, THBP, TBHQ	Oils, fats, dried foods	3% OV-225 on Gas Chrom Q	100–250	183
HFB[c] esters of BHA, PG, TBHQ	Oils	3% OV-3 on Chromosorb WP	120	185

See Table 1 for abbreviations.
[a] Determined after its conversion into silylated derivative
[b] TMS = trimethylsilylation.
[c] HFB = heptafluorobutyrate.

TABLE 5

HPLC SYSTEMS USED FOR DETERMINATION OF SYNTHETIC PHENOLIC ANTIOXIDANTS

Antioxidant	*Product*	*Stationary phase*	*Mobile phase*	*Detection*	*Reference*
BHT, BHA, PG, TBHQ, Ionox-100, THBP, NDGA, OG, DG	Oil, lard, shortenings	10 μm LiChrosorb RP-18	Gradient elution: (A) 5% acetic acid in water and (B) 5% acetic acid in acetonitrile	UV 280 nm	200, 201
BHT, BHA, PG, TBHQ	Oils, fats	μ Bondapak C_{18}	Gradient mixture: 1% acetic acid in water and 1% acetic acid in methanol	UV 280 nm	187
BHT, BHA, PG, DG, OG	Oils, fats	10 μm Bondapak C_{18}	Gradient elution: aqueous acetic acid plus methanol	UV 280 nm	199
BHA, PG, TBHQ	Oils, foods	10 μm Bondapak C_{18}	Methanol: 0·1 M ammonium acetate (1:1, V/V)	Amperometric	202
TBHQ	Oils, butter	LiChrosorb NH_2	Hexane: ethanol (85:15)	UV 293 nm	191
TBHQ	Oils	5 μm LiChrosorb Si 60	Dioxane: hexane (24:76)	Fluorescence, 309 nm and 340 nm	198
BHT, BHA, TBHQ	Oils	Porasil or Rad-Pak cyano	Hexane: dichloromethane: acetonitrile (85:9·5:5·5)	UV 280 nm	195
BHA, PG, OG, DG, TBHQ, NDGA	Oils, fats	5 μm LiChrosorb DIOL	Hexane: 1,4-dioxane: acetonitrile (62:28:10)	UV 280 nm	203

See Table 1 for abbreviations.

column coated with cross-linked 5% phenylmethyl silicone has also been reported.[169] The lower limits of detection are estimated to be 0·5 mg/kg for TBHQ, and Ionox-100 and 0·3 mg/kg for BHT and BHA.

A variety of reversed phase chromatographic procedures[187–193] have been developed for the analysis of synthetic antioxidants, mostly using UV detection. Normal phase HPLC separations[194–197] of the antioxidants have been utilised less commonly and seem to be restricted to the analysis of BHT, BHA and TBHQ. Selected HPLC procedure conditions used for the determination of common antioxidants are listed in Table 5. A rapid (analysis time 5 min) and specific HPLC method for the quantitative determination of TBHQ in oils has been described by Van Niekerk & du Plessis.[198] Hammond[199] has determined BHT, BHA, PG, OG and DG in oils and fats on a C_{18} μ-Bondapak column using gradient elution with aqueous acetic acid and methanol solvents. Simultaneous determination of nine synthetic antioxidants, namely PG, THBP, TBHQ, NDGA, BHA, Ionox-100, OG, DG and BHT in oils, lard and shortenings has been reported by Page.[200] The antioxidants were partitioned from hexane oil into acetonitrile, concentrated under vacuum, and separated on a 10 μm LiChrosorb RP-18 column. Gradient elution from water:acetic acid (95:5, V/V) to acetonitrile:acetic acid (95:5, V/V) was employed at a flow rate of 1·0 ml/min. Using a similar procedure, the results of a collaborative study on the analysis of seven antioxidants in oil and lard samples have been published.[201] Losses in TBHQ were minimised by the use of proper evaporation and chromatographic conditions. King *et al.*[202] have described the extraction of PG, BHA and TBHQ from oils and foods prior to analysis on a 10 μm Bondapak C_{18} column, using methanol: 0·1 M ammonium acetate buffer (1:1 V/V) as the mobile phase and amperometric detection. Very recently, a simple HPLC procedure which requires neither extraction nor derivatisation for the direct determination of six antioxidants namely BHA, DG, TBHQ, OG, PG, and NDGA in oils and fats has been described by Anderson & Van Niekerk.[203] The antioxidant separations were achieved using a 5 μm LiChrosorb DIOL column (25 cm × 0·46 cm), mobile phase hexane:1,4-dioxane:acetonitrile (62:28:10) and UV detection at 280 nm. Under the HPLC conditions used, BHT is not resolved from the oil samples which are injected directly onto the column.

Analytical Methods for Ascorbic Acid and Oxygen Scavengers

As mentioned earlier, ascorbic acid (vitamin C), its sodium or calcium salt and oil soluble ascorbyl palmitate, which function by oxygen scavenging,[11,12] are used as antioxidants in food systems. Direct free-radical interaction between vitamin C (water soluble) and vitamin E (oil soluble) has also been reported.[12,13] The biological activity of vitamin C in foods is derived from both naturally occurring L-ascorbic acid and its oxidation product L-dehydroascorbic acid. The latter is not stable and readily converts into diketogulonic acid which does not possess biological activity. Iso-ascorbic acid (erythorbic acid or D-ascorbic acid) does not occur naturally and has biological activity about 20 times less than L-ascorbic acid. Numerous methods have been used for the determination of total ascorbic acid in citrus juices and other food products.[204–208] Among them are titrimetric,[209] spectrophotometric[210–11] and fluorometric[212–214] methods. The AOAC standard method[215] for determining ascorbic acid in foods is by titration with 2,6-dichlorophenol-indophenol. FAO/WHO[14] have also described titrimetric methods for the assay of ascorbic acid, ascorbyl palmitate, ascorbyl stearate, sodium ascorbate and sodium erythorbate. The procedures involving direct UV spectrophotometry[216–218] and differential pulse polarography[219–221] have been used to determine ascorbic acid at ppm levels. A Sephadex method, involving colorimetric quantification for the determination of vitamin C in a variety of foods has been described by Brubacher *et al.*[222]

High performance liquid chromatographic procedures using strong anion exchange, or reversed phase columns and UV detection, have also been developed for the analysis of ascorbic acid in food products.[223–228] For example, using reversed phased systems, food samples such as fruit juices,[229,230] infant foods and milk,[231] tomatoes,[232] potato[233] and meat[234] products have been analysed for their ascorbic acid contents. The total vitamin C content of a sample is determined after reduction of dehydroascorbic acid to ascorbic acid by homocysteine before chromatography. The dehydroascorbic acid content of the sample is then estimated by the difference in the concentration of ascorbic acid before and after reduction. The separation of L-ascorbic and D-ascorbic acid has been achieved[235] on a 10 μm LiChrosorb NH_2 column with a mobile phase consisting of 75% acetonitrile in 0·005 M phosphate buffer at a pH between 4·4 and 4·7. Lloyd *et al.*[229] have developed a rapid HPLC method for the separation of L-ascorbic acid from erythorbic acid using a 5 μm polymeric PLRP-S 100A column with a 0·2 M sodium dihydrogen phosphate mobile phase (pH adjusted to 2·14) and UV detection at 220 nm.

Recently, Wilson & Shaw[236] have measured ascorbic acid values in orange juice using UV and electrochemical (EC) detectors. The EC detector, which reduces interference from co-eluting compounds because of its specificity, has been reported to be 100 times more sensitive to ascorbic acid than the UV detector. In contrast, a rather limited number of procedures for the determination of ascorbate esters (palmitate and stearate) in fats and butter has been reported.[38] These include titrimetric,[237] colorimetric,[238,239] complexometric[240] and chromatographic methods. A direct potentiometric titration method, which does not require extraction, for the determination of ascorbyl palmitate in oils and fats has been described by Pongracz.[241] A few chromatographic procedures such as gel permetation,[51] thin-layer[41,159,242,243] and high performance liquid[244–246] chromatography have also been developed for the analysis of ascorbyl palmitate (AP). In one HPLC procedure,[245] the food sample was extracted with methanol and analysis of AP in the extract was carried out on a 5 μm chromegabond diamine column, using a mobile phase of 70:30 (V/V) methanol: 0·02 M monobasic phosphate buffer, pH 3·5, and UV detection at 255 nm. In addition to the analysis of AP in vegetable oils, the procedure has been used for accurate determination of ascorbyl palmitate in flour and yoghurt.

Analytical Methods for Synergists, Chelators and other Antioxidants

In this section various methods of analysis of antioxidant synergists such as phosphatides, chelating agents, e.g. citric acid and related compounds, and ethoxyquin are reported.

Phosphatides are widely distributed in foods.[247] The term 'lecithins' is a commercial name for several naturally occurring phospholipids, viz. phosphatidylcholine (lecithin), phosphatidylethanolamine (cephalin), phosphatidylinositol and phosphatidylserine. The pro- and antioxidant effects of various phospholipids have been reported in the literature.[248–252] For example, in part, the increased stability of crude soyabean oil is due to the presence of phosphatides.[253,254] It has also been suggested that phosphatides known in the trade as 'gums', can be very effective antioxidants, particularly in intermediate moisture foods.[255] It should, however, be emphasised that phospholipids, which are good emulsifiers, act more as synergists[21,256–260] and/or chelating agents than actual primary antioxidants. This effect of phosphatides also explains why crude vegetable oils, which contain lecithins, often travel and store better than refined oils, which, containing no lecithins, oxidise more quickly.

Various methods have been described[261–263] for the qualitative and quantitative determination of phosphorus (as lecithin or the equivalent phosphatide content). A conversion factor of 30 is conventionally applied to convert an elemental phosphorus content to the lecithin content, although it is known that this factor can in fact vary slightly, depending on the nature of the phosphatides.[264,265] Standard methods for determining phosphorus in vegetable oils[266] and cocoa fat[267] are also described. The official AOCS method,[266] involving ashing the oil sample in the presence of zinc oxide followed by colorimetric measurement of phosphorus as molybdenum blue, is relatively cumbersome and lengthy. Recently the AOCS has recommended a nephelometric (turbidimeter) method[268] for determining phosphatides (phosphorus levels) in oils. This rapid (10 min) procedure is 30 times faster than colorimetric methods.[269] An enzymic method for determination of lecithin in a variety of foods (e.g. mayonnaise, egg products) has been described by Beutler & Henniger.[270] The product is alkali hydrolysed and the liberated choline is phosphorylated by ATP in the presence of choline kinase, prior to measurement of the absorbance at 365 nm. Other methods for determining phospholipids include ashing of oil, using oxygen bomb combustion[271] followed by colorimetry, perchloric acid digestion followed by TLC or a saponification procedure[272–275] followed by colorimetric determination, flameless atomic absorption spectroscopy[276,277] and molecular emission cavity analysis[278] of phosphorus in oils. On comparing three chemical digestion-spectrophotometric procedures with the official AOCS method,[266] Daun *et al.*[262] reported that the oxygen bomb combustion method provided the best combination of accuracy, precision and speed.

A rapid and reliable method for quantitative determination of phospholipids after TLC separation followed by absorbance measurement at 830 nm has been described by Gentner & Haasemann.[279] Totani *et al.*[280] isolated phospholipids from vegetable oils with glacial acetic acid and determined the phospholipid content colorimetrically. Using plasma emission spectroscopy, very low levels of phosphorus (detection limit 0·5 mg/kg) in oils can be determined.[281] The use of high performance liquid chromatography in the analysis of phospholipids has also been described by several workers.[282–287] Rhee &

Shin[288] have developed a rapid HPLC method for the determination of phosphatidylcholine content in soya lecithins. The procedure uses a porasil column, a mobile phase of chloroform:methanol: acetate:water (14:14:1:1, V/V) and refractive index detection. The use of normal phase HPLC in the separation and quantification of lecithin in chocolate products has been described by Hurst & Martin.[289] The chocolate samples were extracted with a chloroform:methanol mixture (2:1, V/V) and the extract was purified using silica gel Seppak^{Tm} before analysis by HPLC with UV detection at 210 nm. Kaitaranta & Bessman[290] have combined the HPLC system with an automatic phosphorus analyser for the direct quantitative detection of phospholipids in a lipid extract.

The amount of chelating agent, citric acid, in oils and fats has been determined by some workers using a colorimetric method.[291–293] The procedure involves the Furth and Hermann reaction, in which colour is formed in the presence of pyridine and acetic anhydride. A direct titrimetric method for the quantification of small amounts of citric acid in refined oils and fats has been developed by Ohlson *et al.*[294] A method using isotachophosphoresis for the determination of citric acid or sodium citrate in meat products has been reported by Klein.[295] A GLC procedure for the quantitative analysis of citric acid and its decomposition products in edible oils has been described.[296] Citric acid and its derivatives were extracted with water from the oil sample, the extract was esterified with *n*-butanol and the butyl esters were analysed by GLC on a temperature programmed 10% DEGS column equipped with a flame ionisation detector. Mono-, di- and triisopropyl citrates, more soluble in oil than citric acid, are sometimes used as sequestrants. A simple and rapid GLC procedure for determining isopropyl citrates in many foods such as butter, oils and milk powder has been developed by Tsuji *et al.*[297] The detection limits of isopropyl citrates are reported to be 1–2 mg/kg of sample.[297–299]

Ethoxyquin is often used in many countries to treat apples and pears during storage to protect against scald (i.e. the formation of brown spots). Several methods have been described for the determination of ethoxyquin in food products[300–304] and animal feed.[305,306] Among them are included highly sensitive fluorometric methods,[307–312] GLC[3,305,313,314] and HPLC procedures.[303,315] As ethoxyquin is not commonly permitted in foods but is used in animal feeds, tolerances of ethoxyquin residues have been specified in various foods such as edible animal tissue, eggs and milk.[315] An official AOAC method[316] for determining ethoxyquin in animal tissues and eggs has been published. In the GLC procedure[3] described, ethoxyquin is extracted from a homogenised sample with hexane and the heptafluorobutyl derivative is then analysed by GLC using an electron capture detector, with tetrahydroquinoline as an internal standard. The limit of detection (defined as peak height equal to three times the base line noise) is reported to be 0·05 mg/kg. The application of an HPLC technique using UV,[303] fluorescence[315] or electrochemical[4] detection to the determination of ethoxyquin residues in foods such as spices, milk and apples has also been reported. The HPLC procedure combined with an electrochemical or fluorescence detector has the advantages of high sensitivity, and requires no derivatisation or handling of the toxic compounds as in the GLC procedure. The detection limit (0·03 ng of ethoxyquin) is comparable to that (0·02 ng) obtained by the GLC method with electron capture detection.[3]

METHODS FOR EVALUATION OF ANTIOXIDANTS

Several procedures[317–326] have been described for measuring the effectiveness of synthetic and natural antioxidants. In fact, many techniques developed for measuring oxidative stability of oils and fatty materials form the basis of methods employed for determining antioxidant activity. The various conventional methods[325,327,328] for measuring the oxidative stability of oils and fats include the Active Oxygen Method (AOM), Schaal oven test, shelf storage, oxygen uptake/absorption, weight gain, oxygen bomb and fully automated Rancimat apparatus. In these tests there is initially little change, the oil effectively resisting oxidation. When this resistance is overcome, perhaps due to exhaustion of the natural and/or added antioxidants, rapid oxidation commences. The onset of this rapid oxidation is quite definite and easily determined. The time during which the oil's resistance to oxidation is effective is called the Induction Period (IP), a time which can in many cases be measured with a fair degree of reproducibility. Some of the techniques mentioned depend on the passage of air through the heated oil. This is the case, for instance with the AOM and the Rancimat test. Under these conditions antioxidants volatile at the temperature of test are swept from the oil. Work carried out in the author's laboratory[329] has shown that for this reason the Rancimat apparatus can give misleading results with, for example,

BHT and BHA. Numerous workers have used IP determinations[330–337] for screening the effectiveness of antioxidants in vegetable oils and animal fats. For example, Olcott & Einset[338] measured the increase in weight of heated oils in the presence of antioxidants, which hindered the oxygen uptake by the oil and thus minimised the weight gain. Palmateer *et al.*[339] absorbed fats and antioxidants on celite and monitored the development of oxidative deterioration by the thiobarbituric acid reaction. Hamilton & Tappel[340] developed a polarographic, haemoglobin-catalysed, oxygen uptake method for measuring antioxidant activity. The evaluation of antioxidant activity in a linoleate emulsion to which haemoglobin was added as pro-oxidant has also been reported.[341]

A rapid, routine method for ranking antioxidant activity has been described by Marco.[342] The procedure is based on minimising β-carotene loss in the coupled oxidation of linoleic acid and β-carotene using an emulsified, aqueous system. Antioxidant activities of several antioxidants such as BHA, PG, BHT, dilauryl thiodipropionate and thiodipropionic acid have been evaluated. The method allows the determination of antioxidant activity residing in oil samples in comparison with any desired antioxidant. Berner *et al.*[343] have also reported a rapid method, involving hemin-catalysed oxygen uptake, for determining antioxidant activity. The increase in the IP by the addition of an antioxidant has been related to antioxidant efficacy which is sometimes expressed as antioxidant index or protection factor. The activities of several antioxidants (viz. TBHQ, PG and BHA) and synergists (ascorbic acid, citric acid and EDTA) using the hemin-catalysed procedure, are compared with those obtained by the classic active oxygen method. Kawashima *et al.*[321] also determined antioxidant activity by the hemin-catalysed oxygen uptake method, using a Gilson differential respirometer. The antioxidant activity was taken as oxygen absorption in 60 min.

A simplified procedure for the evaluation of antioxidants, based on the carotene discoloration method developed by Marco,[342] has been described by Miller.[344] Aqueous emulsions of antioxidant, carotene and lipid are prepared in spectrometer tubes, and the oxidative destruction of carotene in the system is measured colorimetrically at 470 nm. Recently, Taga *et al.*[345] have described the procedure, in detail, involving measurement of carotene bleaching, for determining antioxidant activity. The thiocynate method for the evaluation of natural antioxidants isolated from *Eucalyptus* leaf waxes has been described by Osawa & Namiki.[346,347] The reaction mixture of linoleic acid, ethanol, 0·2 M phosphate buffer and antioxidant is incubated at 40°C and the peroxide value is determined by measuring absorbance at 500 nm after a colour reaction with $FeCl_2$ and thiocyanate. The activity of an antioxidant used in potato flakes has been evaluated by measuring the residual oxygen and pentane in the head space of the packed product.[330] Faria[348] has developed a gas chromatographic reactor for measuring the effectiveness of antioxidants. The procedure involves continuous measurement of the amount of oxygen reacted during the oxidation test. The relative activities, expressed as protective index, of a number of antioxidants (e.g. BHT, PG, α-tocopherol, citric acid and ascorbic acid) on the oxidation of linoleic acid at 85°C and 55 mm Hg have been determined. This dynamic procedure has been described as a fast and reproducible method for the study of the effect of antioxidants on polyunsaturated lipids.

Castro-Martinez *et al.*[349] have used a plate diffusion method for the estimation of antioxidant activity. The agar plates were prepared, incubated at room temperature, and the antioxidant activity was then related to intensity and persistence of the carotene colour. A new procedure for detecting and evaluating antioxidative components from food products has been developed by Chang *et al.*[350] The method involves separation of individual components on a fluorescent TLC plate and visualising by spraying with a tocopherol free soyabean oil solution followed by exposing it to UV light (254 nm). Recently, a simple and rapid method for measuring antioxidant activity by using an oxygen electrode with an automatic recorder of oxygen consumption has been described by Hirayama *et al.*[351]

Finally, it should be mentioned that the relative effectiveness of antioxidants varies with the substrate. For example, the activity of antioxidants on β-carotene will not be the same as on vegetable oil, nor will the activity on vegetable oils be the same as on animal fats. Also, it has been shown[352] that water has a pronounced influence on the efficacy of an antioxidant. Moreover, in our experience, the antioxidant activities of volatile antioxidants such as BHT and BHA when examined at elevated temperature in tests such as the AOM and fully automated Rancimat should be considered with caution. This is so because a continuous stream of air bubbled through the oil system in such methods distills off both the volatile potent anti- and pro-oxidant components.

REFERENCES

1. Miles, R. S., McKeith, F. K., Bechtel, P. J. and Novakofski, J., *J. Food Protect.*, **49** (3) (1986) 222.
2. Gregory, D., *Food,* **6** (6) (June 1984) 18.
3. Winell, B., *Analyst* (*London*), **101** (1976) 883.
4. Olek, M., Declercq, B., Caboche, M., Blanchard, F. & Sudraud, G., *J. Chromatogr.*, **281** (1983) 309.
5. Buck, D. F., *Cereals Foods World,* **29** (5) (1984) 301.
6. Coppen, P. P., in *Rancidity in Foods,* ed. J. C. Allen & R. J. Hamilton. Applied Science Publishers, London, 1983, p. 67.
7. Klaui, H., in *Vitamin C*, ed. G. G. Birch & K. J. Parker. Applied Science Publishers, London, 1974, p. 16.
8. Dziezak, J. D., *Food Technol.*, **40** (9) (September 1986) 94.
9. Sherwin, E. R., *J. Am. Oil. Chem. Soc.*, **53** (1976) 430.
10. Labuza, T., *CRC Crit. Rev. Food Technol.*, **2** (1971) 355.
11. Cort, W. M., *J. Am. Oil Chem. Soc.*, **51** (1974) 321.
12. Klaui, H. & Pongracz, G., in *Vitamin C*, ed. J. N. Counsell & D. H. Hornig. Applied Science Publishers, London, 1982, p. 139.
13. Packer, J. E., Slater, T. F. & Wilson, R. L., *Nature,* **278** (1979) 737.
14. Joint FAO/WHO Expert Committee on Food Additives. FAO Food and Nutrition paper No. 4, Food and Agriculture Organisation of the United Nations, Rome, 1978, p. 179.
15. Taylor, M. J. & Richardson, T., *Adv. Appl. Microbiol.*, **25** (1979) 7.
16. Cillard, J. & Cillard, P., *J. Am. Oil Chem. Soc.*, **63** (9) (1986) 1165.
17. Ahmad, M. M., Al-Hakim, S. & Shehata, Y., *J. Am. Oil Chem. Soc.*, **60** (1983) 837.
18. Marcuse, R., *Rev. Fr. Corps Gras,* **7** (1973) 391.
19. Riisom, T., Sims, R. J. & Fioriti, J. A., *J. Am. Oil Chem. Soc.*, **57** (1980) 354.
20. Farag, R. S., Osman, S. A., Hallabo, S. A. S. & Nasar, A. A., *J. Am. Oil Chem. Soc.*, **55** (1978) 703.
21. Dziedzic, S. Z. & Hudson, B. J. F., *J. Am. Oil Chem. Soc.*, **61** (6) (1984) 1042.
22. Chipault, J. R., Mizuno, G. R., Hawkin, J. M. & Lundberg, W. O., *Food Res.*, **17** (1952) 46.
23. Chipault, J. R., Mizuno, G. R. & Lundberg, W. O., *Food Technol.*, **10** (1956) 209.
24. Houliham, C. M., Ho, C. T. & Chang, S. S., *J. Am. Oil Chem. Soc.*, **62** (1) (1985) 96.
25. Kramer, R. E., *J. Am. Oil Chem. Soc.*, **62** (1) (1985) 111.
26. Fukuda, Y., Nagata, M., Osawa, T. & Namiki, M., *J. Am. Oil Chem. Soc.*, **63** (8) (1986) 1027.
27. Hermann, K. J., *Food Technol.*, **11** (1976) 433.
28. Food Additives and the Consumer, Commission of the European Communities, Catalogue No. CB-25-78-744-EN-C, Office for Official Publications of the European Communities, Luxembourg, 1980.

29. Gray, J., in *Food Intolerance: Fact and Fiction*. Grafton Books, London, 1986, p. 132.
30. Directive 87/55/EEC, *Offic. J. European Communities,* No. L 24/41, 1987.
31. Bunnell, R. H., in *The Vitamins,* ed. P. Gyorgy & W. N. Pearson, 2nd edn, Vol. VI. Academic Press, New York, 1967, p. 261.
32. Desai, I. D., in *Vitamin E: A Comprehensive Treatise*, ed. L. J. Machlin. Marcel Dekker, New York, 1980, p. 67.
33. Endean, M. E., Leatherhead Food Research Association, Science and Technology Survey No. 91, 1976.
34. Nelis, H. J., De Bevere, V. O. R. C. & De Leenheer, A. P., in *Modern Chromatographic Analysis of the Vitamins,* ed. A. P. De Leenheer, W. E. Lambert & M. G. M. De Ruyter. Marcel Dekker, New York, 1985, p. 129.
35. Parrish, D. B., *CRC Crit. Rev. Food Sci. Nutr.,* **13** (1980) 161.
36. Uebersax, P., *Mitt. Geb. Lebensmittel Unters. U. Hyg.,* **63** (4) (1972) 469.
37. Bui-Nguyén, M. H., in *Modern Chromatographic Analysis of the Vitamins,* ed. A. P. De Leenheer, W. E. Lambert & M. G. M. De Ruyter, Marcel Dekker, New York, 1985, p. 267.
38. Cooke, J. R. & Maxon, R. E. D., in *Vitamin C,* ed. J. N. Counsell & D. H. Hornig. Applied Science Publishers, London, 1982, p. 167.
39. Hasselman, C. & Diop, P. A., *Sci. Aliments,* **3** (1983) 161.
40. Gertz, C. & Herrmann, K., *Z. Lebensm.-Unters. Forsch.,* **177** (1983) 186.
41. Pujol Forn, M., *Circ. Farm.,* **39** (270) (1981) 5.
42. Robards, K. & Dilli, S., *Analyst (London),* **112** (1987) 933.
43. Stuckey, B. N. & Osborne, C. E., *J. Am. Oil Chem. Soc.,* **42** (1965) 228.
44. The Association of Public Analysts, Special Rep. No. 1. London, APA, 1963.
45. Ames, S. R., *J. Assoc. Off. Anal. Chem.,* **54** (1971) 1.
46. Taylor, S. L., Lamden, M. P. & Tappel, A. L., *Lipids,* **11** (7) (1976) 530.
47. McBride, H. D. & Evans, D. H., *Analyt. Chem.,* **45** (1973) 446.
48. Candlish, J. K., *J. Agric. Food Chem.,* **31** (1983) 166.
49. Muller-Mulot, W., *J. Am. Oil Chem. Soc.,* **53** (1976) 732.
50. Green, J., Marcinkiewicz, S. & Watt, P. R., *J. Sci. Food Agric.,* **6** (1955) 274.
51. Pokorny', S., Coupek, J. & Pokorny', J., *J. Chromatogr.,* **71** (1972) 576.
52. Mariani, C. & Fedeli, E., *Riv. Ital. Sostanze Grasse,* **59** (1982) 557.
53. Nelson, J. P., Milun, A. J. & Fisher, H. D., *J. Am. Oil Chem. Soc.,* **47** (1970) 259.
54. Cort, W. M., Vincente, T. S., Waysek, E. H. & Williams, B. D., *J. Agric. Food Chem.,* **31** (1983) 1330.
55. Gertz, C. & Herrmann, K., *Z. Lebensm.-Unters. Forsch.,* **174** (1982) 390.
56. Ito, N., Hirose, M., Fukushima, S., Tsuda, H., Shirai, T. & Tatematsu, M., *Food Chem. Toxic.,* **24** (1986) 1071.

57. Conning, D. M. & Phillip, J. C., *Food Chem. Toxic.*, **24** (1986) 1145.
58. Van der Heijden, C. A., Janssen, P. J. C. M. & Strik, J. J. T. W. A., *Food Chem. Toxic.*, **24** (1986) 1067.
59. S.I. No. 105, *The Antioxidants in Food Regulations*. HMSO, London, 1978.
60. FAO/WHO Codex Alimentarius Commission, Alinorm 87/17, Report on 13th session of the Codex Committee on fats and oils, London, 23–27 February 1987, p. 5.
61. Haigh, R., *Food Chem. Toxic.*, **24** (1986) 1031.
62. International Union of Pure and Applied Chemistry, *Standard Methods for the Analysis of Oils, Fats and Derivatives*, 7th edn, ed. C. Paquot & A. Hautfenne. Blackwell, Oxford, 1987.
63. Arens, M., Kroll, S. & Müller-Mullot, W., *Fette, Seifen Anstrichm.*, **86** (4) (1984) 148.
64. Brubacher, G., Müller-Mulot, W. & Southgate, D. A. T., *Methods for the Determination of Vitamins in Foods*. Elsevier Applied Science Publishers, London, 1985, p. 107.
65. Official and Tentative Methods of the American Oil Chemists Society. AOCS, Champaign, IL, 1986.
66. AOAC Official Methods of Analysis, 14th edn. AOAC, Arlington, Virginia, 1984.
67. British Standards Institution, BS standard methods of analysis of fats and fatty oils, Part 2 Other methods. B.S.I., Milton Keynes, 1982.
68. International Organisation for Standardisation, ISO methods, published by the ISO. Available from National Standards Organisation, e.g. B.S.I.
69. Bauernfeind, J. C., *CRC Crit. Rev. Food Sci. Nutr.*, **8** (4) (1977) 337.
70. McLaughlin, P. J. & Weihrauach, J. L., *J. Am. Diet. Assoc.*, **75** (1979) 647.
71. Reynolds, S. L., *Proc. Inst. Food Sci. Technol. (U.K.)*, **18** (1) (1985) 43.
72. Bunnell, R. H., *Lipids*, **6** (1971) 215.
73. Emmerie, A. & Engel, C., *Rec. Trav. Chim.*, **57** (1938) 1351.
74. Stern, M. H. & Baxter, J. G., *Analyt. Chem.*, **19** (1947) 902.
75. Bukovits, G. J. & Lezerovich, A., *J. Am. Oil Chem. Soc.*, **64** (1987) 517.
76. Niederstebruch, A. & Hinsch, I., *Fette, Seifen Anstrichm.*, **69** (1967) 559.
77. Podlaha, O., Eriksson, A. & Toregard, B., *J. Am. Oil Chem. Soc.*, **55** (1978) 530.
78. Waltking, A. E., Kiernau, M. & Bleffort, G. W., *J. Am. Oil Chem. Soc.*, **60** (1977) 890.
79. Knoblock, E., Macha, F. & Mnoucek, K., *Analyt. Chem.* **24** (1952) 19.
80. Wisser, K., Heimann, W. & Fritsche, Ch., *Z. Anal. Chem.*, **230** (1967) 189.
81. Atuma, S. S., *J. Sci. Food Agric.*, **26** (1975) 393.
82. Atuma, S. S. & Lindquist, J., *Analyst (London)*, **98** (1973) 886.
83. Löliger, J. & Saucy, F., *Z. Lebensm.-Unters. Forsch.*, **170** (1980) 413.
84. Chow, C. K., Draper, H. H. & Csallany, A. S., *Analyt. Biochem.*, **32** (1969) 81.

85. Contreras-Guzman, E., Strong III, F. C. & da Silva, W. J., *J. Agric. Food Chem.*, **30** (1982) 1113.
86. Gavind Rao, M. K. & Perkini, E. G., *J. Agric. Food Chem.*, **20** (1972) 240.
87. Slover, H. T., Shelley, L. M. & Burks, T. L., *J. Am. Oil Chem. Soc.*, **44** (1967) 161.
88. Hartman, K. T., *J. Am. Oil Chem. Soc.*, **54** (1977) 421.
89. Meijboom, P. W. & Jongenotter, G. A., *J. Am. Oil Chem. Soc.*, **56** (1979) 33.
90. Slover, H. T., Thompson, Jr R. H. & Merola, G. V., *J. Am. Oil Chem. Soc.*, **60** (1983) 1524.
91. Van Niekerk, P. J., *Analyt. Biochem.*, **52** (1973) 533.
92. Rossell, J. B., King, B. & Downes, M. J., *J. Am. Oil Chem. Soc.*, **60** (2) (1983) 333.
93. Thompson, J. N. & Hatina, G., *J. Liq. Chromatogr.*, **2** (1979) 327.
94. Speek, A. J., Schrijver, J. & Schreurs, W. H. P., *J. Food Sci.*, **50** (1985) 121.
95. Mankel, A., *Dtsch. Lebensm. Rundsch.*, **75** (1979) 77.
96. Barnett, S. A., Frick, L. W. & Baine, H. M., Abstr. 93rd Annual Meeting, AOAC, Washington, Oct. 15–18, 1979, p. 199.
97. Pickston, L., *NZJ Sci.*, **21** (1978) 383.
98. Taylor, P. W. & Barnes, P. J., *Chem. Ind.*, **20** (1981) 722.
99. Thompson, J. N., Erdody, P. & Maxwell, W. B., *Analyt. Biochem.*, **50** (1972) 267.
100. Van Niekerk, P. J., M.Sc. Thesis, University of South Africa, 1975.
101. Rammell, C. G. & Hoogenboom, J. J. L., *J. Liq. Chromatogr.*, **8** (4) (1985) 707.
102. Abe, K., Yuguchi, Y. & Katsui, G., *J. Nutr. Sci. Vitaminol.*, **21** (1975) 183.
103. Van Niekerk, P. J. & du Plessis, L. M., *S. Afr. Food Rev.*, **3** (1976) 167.
104. Devries, J. W., Egberg, D. C. & Heroff, J. C., in *Liquid Chromatographic Analysis of Food and Beverages*, Vol. 2, ed. G. Charalambous. Academic Press, London, 1979, p. 477.
105. Widicus, W. A. & Kirk, J. R., *J. Assoc. Off. Analyt. Chem.*, **62** (1979) 637.
106. Barnett, S. A., Frick, L. W. & Baine, H. M., *Analyt. Chem.*, **52** (1980) 610.
107. Deldime, P., Lefebvre, G., Sadin, Y. & Wybauw, M., *Rev. Franc. des Corps Gras*, **27** (1980) 279.
108. Carpenter, A. P., Jr, *J. Am. Oil Chem. Soc.*, **56** (1979) 668.
109. Cortesi, N. & Fedeli, E., *Riv. Ital. Sost. Grasse*, **57** (1980) 16.
110. Capuano, A. & Daghetta, A., *Riv. Ital. Sost. Grasse*, **57** (1980) 285.
111. Barnes, P. J. & Taylor, P. W., *J. Sci. Food Agric.*, **31** (1980) 997.
112. Manz, U. & Philipp, K., *Int. J. Vitam. Nutr. Res.*, **51** (1981) 342.
113. Zonta, F. & Stancher, B., *Riv. Ital. Sost. Grasse*, **60** (1983) 195.
114. Brubacher, G., Müller-Mulot, W. & Southgate, D. A. T., *Methods for the Determination of Vitamins in Foods*. Elsevier Applied Science Publishers, London, 1985, p. 97.

115. Coors, U. & Montag, A., *Milchwissenschaft,* **40** (8) (1985) 470.
116. Anglin, C., Mahon, J. H. & Chapman, R. A., *J. Agric. Food Chem.*, **4** (1956) 1018.
117. Filipic, V. P. & Ogg, C. L., *J. Assoc. Off. Agric. Chem.*, **43** (1960) 795.
118. Sloman, K. J., Romagnoli, R. J. & Cavagnol, J. C., *J. Assoc. Off. Agric. Chem.*, **45** (1962) 76.
119. Takenori, M., Isao, N. & Masao, I., *Shokuhin Eiseigaku Zasshi,* **18** (3) (1977) 283.
120. Alicino, N. J., Klein, H. C., Quattrone, J. J. & Choy, T. K., *J. Agric. Food Chem.*, **11** (1963) 496.
121. Buttery, R. G. & Stuckey, B. N., *J. Agric. Food Chem.*, **9** (1961) 283.
122. Stoddard, E. E., *J. Assoc. Off. Anal. Chem.*, **55** (1972) 1072.
123. McCaulley, D. F., Fazio, T., Howard, J. W., DiCiurcio, F. M. & Ives, J., *J. Assoc. Off. Anal. Chem.*, **50** (1967) 243.
124. Phipps, A. M., *J. Am. Oil Chem. Soc.*, **50** (1973) 21.
125. Schwien, W. G. & Conroy, H. W., *J. Assoc. Off. Anal. Chem.*, **48** (1965) 489.
126. Laszlo, H. & Dugan, L. R., *J. Am. Oil Chem. Soc.*, **38** (1961) 178.
127. Pujol Forn, M. & Lopez Sabater, M. C., *Circ. Farm.*, **42** (1984) 3.
128. Szalkowski, C. R. & Garber, J. B., *J. Agric. Food Chem.*, **10** (1962) 490.
129. Kobayashi, K., Tsuji, S., Tonogai, Y., Ito, Y. & Tanabe, H., *J. Japanese Soc., Food Sci. Technol.*, **33** (10) (1986) 720.
130. Cassidy, W. & Fisher, A. J., *Analyst,* **85** (1960) 295.
131. Vos, H. J., Wessels, H. & Six, C. W. T., *Analyst,* **82** (1957) 362.
132. Mahon, J. H. & Chapman, R. A., *Analyt. Chem.*, **23** (1951) 1116.
133. Sastry, C. S. P., Ekambareswara Rao, K. & Viplava Prasad, U., *Talanta,* **29** (1982) 917.
134. Viplava Prasad, U., Ekambareswara Rao, K. & Sastry, C. S. P., *Food Chem.*, **17** (1985) 209.
135. Komaitis, M. E. & Kapel, M., *J. Am. Oil Chem. Soc.*, **62** (9) (1985) 1371.
136. Viplava Prasad, U., Divakar, T. E., Hariprasad, K. & Sastry, C. S. P., *Food Chem.*, **25** (1987) 159.
137. Hansen, P. V., Kauffman, F. S. & Wiedermann, L. H., *J. Am. Oil Chem. Soc.*, **36** (1959) 193.
138. Phillips, M. A. & Hinkel, R. D., *J. Agric. Food Chem.*, **5** (1957) 379.
139. Seher, A., *Fette Seifen Anstrichm.*, **60** (1958) 1144.
140. Whetsel, K. B., Roberson, W. E. & Johnson, F. E., *J. Am. Oil Chem. Soc.*, **32** (1955) 493.
141. Wolff, J. P., *Rev. Fr. Corps. Gras.* **5** (1958) 630.
142. Doeden, W. G., Bowers, R. H. & Ingala, A. C., *J. Am. Oil Chem. Soc.*, **56** (1979) 12.
143. Alary, J., Grosset, C. & Coeur, A., *Ann. Pharm. Fr.*, **40** (1982) 301.
144. Guldborg, M., *Fresenius Z. Anal. Chem.*, **309** (1981) 117.
145. Jayaramani, S., Vasundhara, T. S. & Parihar, D. B., *Mikrochim. Acta,* **11** (1976) 365.
146. Pujol Forn, M., *Grasas Aceites* (*Seville*), **31** (1980) 187.
147. Sahasrabudhe, M. R., *J. Assoc. Off. Agric. Chem.*, **47** (1964) 888.

148. Salo, T. & Salminen, K., *Z. Lebensm.-Unters. Forsch.*, **125** (1964) 167.
149. Scheidt, S. A. & Conroy, H. W., *J. Assoc. Off. Anal. Chem.*, **49** (1966) 807.
150. Van Dessel, L. & Clement, J., *Z. Lebensm.-Unters. Forsch.*, **139** (1969) 146.
151. Woggon, H., Uhde, W. J. & Zydek, G., *Z. Lebensm.-Unters. Forsch.*, **138** (1968) 169.
152. Dehority, B. A., *J. Chromatogr.*, **2** (1959) 384.
153. Mitchell, L. C., *J. Assoc. Off. Agric. Chem.*, **40** (1957) 909.
154. Subramanian, S. A., *Curr. Sci.*, **35** (1966) 309.
155. Wheeler, D. A., *Talanta,* **15** (1968) 1315.
156. Rutkowski, A., Koslowska, H. & Szerszynski, J., *J. Rocz. Panstw. Zakl. Hig.*, **14** (1963) 361.
157. Seher, A., *Fette Seifen Anstrichm.*, **61** (1959) 345.
158. Seher, A., *Nahrung,* **4** (1960) 466.
159. Van Peteghem, C. H. & Dekeyser, D. A., *J. Assoc. Off. Anal. Chem.*, **64** (1981) 1331.
160. Hartman, K. T. & Rose, L. C., *J. Am. Oil Chem. Soc.*, **47** (1970) 7.
161. Hayakawa, J., Narafu, T., Takahashi, H., Ishida, Y. & Tsuiki, T., *Shokuhin Eiseigaka Zasshi,* **10** (1969) 190.
162. Mariani, C. & Fedeli, E., *Riv. Ital. Sost. Grasse,* **60** (1983) 667.
163. Min, D. B. & Schweizer, D., *J. Food Sci.*, **48** (1983) 73.
164. Min, D. B., Ticknor, D. & Schweizer, D., *J. Am. Oil Chem. Soc.*, **59** (1982) 378.
165. Pujol Forn, M. & Garrabou Goma, M., *Circ. Farm.*, **34** (1976) 419.
166. Schwien, W. G., Miller, B. J. & Conroy, H. W., *J. Assoc. Off. Agric. Chem.*, **49** (1966) 809.
167. Senten, J. R., Waumans, J. M. & Clements, J. M., *J. Assoc. Off. Anal. Chem.*, **60** (1977) 505.
168. Wyatt, D. M., *J. Am. Oil Chem. Soc.*, **58** (1981) 917.
169. Yu, L. Z., Inoko, M. & Matsuno, T., *J. Agric. Food Chem.*, **32** (1984) 681.
170. Beaulieu, F. & Hadziyev, D., *J. Food Sci.*, **47** (1982) 589.
171. Greenburg, M. J., Hoholick, J., Robinson, R., Kubis, K., Groce, J. & Weber, L., *J. Food Sci.*, **49** (1984) 1622.
172. Anderson, R. H. & Nelson, J. P., *Food Technol.*, **17** (1963) 915.
173. Schwecke, W. M. & Nelsen, J. H., *J. Agric. Food Chem.*, **12** (1964) 86.
174. Takahashi, D. M., *J. Assoc. Off. Agric. Chem.*, **47** (1964) 367.
175. Takahashi, D. M., *J. Assoc. Off. Anal. Chem.*, **50** (1967) 880.
176. Takahashi, D. M., *J. Assoc. Off. Anal. Chem.*, **53** (1970) 39.
177. Toyoda, M., Ogawa, S., Tonogai, Y., Ito, Y. & Iwaida, M., *J. Assoc. Off. Anal. Chem.*, **63** (1980) 1135.
178. Yamamoto, M., Ochi, S., Konogai, I., Komine, S., Ishikawa, M., Narita, H., Masui, T. & Kitada, Y., *Shokuhin Eiseigaku Zasshi,* **26** (1985) 285.
179. Stijve, T. & Diserens, J. M., *Dtsch. Lebensm.-Rundsch.*, **79** (4) (1983) 108.
180. Takeba, K. & Matsumoto, M., *Eisei Kabaku,* **29** (1983) 329.

181. Singh, J. & Lapointe, M. R., *J. Assoc. Off. Anal. Chem.*, **57** (1974) 804.
182. Austin, R. E. & Wyatt, D. M., *J. Am. Oil Chem. Soc.*, **57** (1980) 422.
183. Kline, D. A., Joe, F. L. & Fazio, T., *J. Assoc. Off. Anal. Chem.*, **61** (3) (1978) 513.
184. Garcia, C. D. & Camino, C. P., *Grasas Y Aceites,* **34** (3) (1983) 183.
185. Page, B. D. & Kennedy, B. P., *J. Assoc. Off. Anal. Chem.*, **59** (1976) 1208.
186. Ito, Y., Toyoda, M., Suzuki, H., Ogawa, S. & Iwaida, M., *J. Food Protect.*, **43** (11) (1980) 832.
187. Archer, A. W., *Anal. Chim. Acta.* **128** (1981) 235.
188. Berridge, J. C., Kent, J. & Norcott, K. M., *J. Chromatogr.*, **285** (1984) 389.
189. Bocca, A. & Delise, M., *Riv. Soc. Ital. Sci. Aliment.*, **11** (1982) 345.
190. Constante, E. G., *Grasas Y Aceites,* **26** (3) (1975) 150.
191. Kitada, Y., Ueda, Y., Yamamoto, M., Shinomiya, K. & Nakazawa, H., *J. Liq. Chromatogr.*, **8** (1985) 47.
192. Masoud, A. N. & Cha, Y. N., *J. High Resolut. Chromatogr. Commun.*, **5** (1982) 299.
193. Pellerin, F., Delaveau, P., Dumitiescu, D. & Safta, F., *Ann. Pharm. Fr.*, **40** (3) (1982) 221.
194. Ansari, G. A., *J. Chromatogr.*, **262** (1983) 393.
195. Indyk, H. & Woolard, D. C., *J. Chromatogr.*, **356** (1986) 401.
196. Kitada, Y., Tamase, K., Mizobuchi, M., Sasaki, M., Tanigawa, K., Yamamoto, M. & Nakazawa, H., *Bunseki Kagaku,* **33** (1) (1984) E33.
197. Piironen, V., Varo, P., Syvaoja, E. L., Salminen,, K. & Koivistoinen, P., *Int. J. Vitam. Nutr. Res.*, **54** (1) (1984) 35.
198. Van Niekerk, P. J. & du Plessis, L. M., *J. Chromatogr.*, **187** (1980) 436.
199. Hammond, K., *J. Assoc. Public Anal.*, **16** (1978) 17.
200. Page, B. D., *J. Assoc. Off. Anal. Chem.*, **62** (1979) 1239.
201. Page, B. D., *J. Assoc. Off. Anal. Chem.*, **66** (3) (1983) 727.
202. King, W. P., Joseph, K. T. & Kissinger, P. T., *J. Assoc. Off. Anal. Chem.*, **63** (1980) 137.
203. Anderson, J. & Van Niekerk, P. J., *J. Chromatogr.*, **394** (1987) 400.
204. Cooke, J. R., in *Vitamin C*, ed. G. G. Birch & K. J. Parker. Applied Science Publishers, London, 1974, p. 31.
205. Coustard, J. M. & Sudraud, G., *J. Chromatogr.*, **219** (1981) 338.
206. Matsumoto, K., Yamada, K. & Osajima, Y., *Analyt. Chem.*, **53** (1981) 1974.
207. Nadolna, I., Secomska, B. & Zielinska, Z., *Rocz. Panstw. Zakl. Hig.*, **30** (1) (1979) 51.
208. Rose, R. C. & Nahrwold, D. L., *Analyt. Biochem.*, **114** (1981) 140.
209. Barakat, M. Z., Shehab, S. K., Darwish, N. & El-Zoheiry, A., *Analyt. Biochem.*, **53** (1973) 245.
210. Lau, O. W. & Luk, S. F., *J. Assoc. Off. Anal. Chem.*, **70** (3) (1987) 518.
211. Onishi, I. & Hara, T., *Bull. Chem. Soc. Jpn,* **37** (1964) 1314.
212. Bunton, N. G., Jennings, N. & Crosby, N. T., *J. Assoc. Publ. Anal.*, **17** (1979) 105.

213. Deutsch, M. J. & Weeks, C. E., *J. Assoc. Off. Agric. Chem.*, **48** (1965) 1248.
214. AOAC Official Methods of Analysis, Sec. 43.061-43.067, 13th edn. AOAC, Arlington, VA, 1980.
215. AOAC Official Methods of Analysis, Sec. 43.056–43.060, 13th edn. AOAC, Arlington, VA, 1980.
216. Fung, Y. S. & Luk, S. F., *Analyst,* **110** (1985) 201.
217. Lau, O. W., Luk, S. F. & Wong, K. S., *Analyst,* **111** (1986) 665.
218. Tono, T. & Fujita, S., *Agric. Biol. Chem.*, **45** (1981) 2947.
219. Gerhardt, U. & Windmuller, R., *Fleischwirtschaft,* **61** (9) (1981) 1389.
220. Lindquist, J. & Farroha, S. M., *Analyst,* **100** (1975) 377.
221. Sontag, G. & Kainz, G., *Mikrochim. Acta (Wein),* **1** (1978) 175.
222. Brubacher, G., Müller-Mulot, W. & Southgate, D. A. T., *Methods for the Determination of Vitamins in Foods.* Elsevier Applied Science Publishers, London, 1985, p. 85.
223. Carnevale, J., *Food Technol. Aust.,* **32** (1980) 302.
224. Dennison, D. B., Brawley, T. G. & Hunter, G. L. K, *J. Agric. Food Chem.,* **29** (5) (1981) 927.
225. Finby, J. W. & Duang, E., *J. Chromatogr.*, **207** (1981) 449.
226. Haddad, P. R. & Lau, J., *Food Technol. Aust.,* **36** (1984) 46.
227. Shaw, P. E. & Wilson, C. W. III, *J. Agric. Food Chem.,* **30** (1982) 394.
228. Wills, R. B. H., Wimalasiri, P. & Greenfield, H., *J. Agric. Food Chem.,* **32** (1984) 836.
229. Lloyd, L. L., Warner, F. P., Kennedy, J. F. & White, C. A., *Chromatogr. Intern.,* **17** (6) (1986) 6.
230. Sood, S. P., Sartori, L. E., Wittmer, D. P. & Haney, W. G., *Analyt. Chem.,* **48** (1976) 796.
231. Pachla, L. A. & Kissinger, P. T., *Methods Enzymol.,* **62** (1979) 15.
232. Archer, A. W., Higgins, V. R. & Perryman, D. L., *J. Assoc. Public Anal.,* **18** (3) (1980) 99.
233. Augustin, J., Beck, C. & Marousek, G. I., *J. Food Sci.,* **46** (1981) 312.
234. Miki, N., *Nippon Shokuhin Kogyo Gakkaishi,* **28** (5) (1981) 264.
235. Bui-Nguyên, M. H., *J. Chromatogr.,* **196** (1980) 163.
236. Wilson, C. W. III & Shaw, P. E., *J. Agric. Food Chem.,* **35** (1987) 329.
237. Sachie, I., Yuriko, T. & Hiroyuki, I., *Shokuhin Eiseigaku Zasshi,* **9** (3) (1968) 232.
238. Budslawski, J. & Pogorzelski, K., *Lait,* **44** (1964) 378.
239. Strohecker, R., *Fette Seifen Anstrichm.,* **66** (1964) 787.
240. Sedlácek, B. A. J., *Fette Seifen Anstrichm.,* **63** (1961) 1053.
241. Pongracz, G., *Z. Lebensm.-Unters. Forsch.,* **147** (1971) 83.
242. Alary, J., Grosset, C. & Coeur, A., *Ann. Pharm. Fr.,* **40** (1982) 301.
243. De la Torre Boronat, M., Xirau-verges, M. & Antoja-Ribo, F., *Circ. Farm.,* **27** (1969) 139.
244. Melton, S. L., Harrison, J. & Churchville, D., *J. Am. Oil Chem. Soc.,* **58** (1981) 573A, Abstract 34.
245. Vicente, T. S., Waysek, E. H. & Cort, W. M., *J. Am. Oil Chem. Soc.,* **62** (4) (1985) 745.
246. Woollard, D. C., *NZJ Dairy Sci. Technol.,* **18** (1983) 55.

247. Weihrauch, J. L. & Son, Y. S., *J. Am. Oil Chem. Soc.*, **60** (12) (1983) 1971.
248. Bibby, C. L., *Food Manuf.*, **20** (1945) 441.
249. Brandt, P., Hollstein, E. & Franzke, Cl., *Die Lebensmittel Industrie*, **20** (1973) 31.
250. Kaur, N., Sukhija, P. S. & Bhatia, I. S., *J. Sci. Food Agric.*, **33** (1982) 576.
251. Olcott, H. S. & Van der Veen, J., *J. Food Sci.*, **28** (1963) 313.
252. Privett, O. S. & Quackenbush, F. W., *J. Am. Oil Chem. Soc.*, **31** (1954) 169.
253. Hildebrand, D. H., Terao, J. & Kito, M., *J. Am. Oil Chem. Soc.*, **61** (3) (1984) 552.
254. Smouse, T. H., *J. Am. Oil Chem. Soc.*, **56** (1979) 747A.
255. Gopalakrishna, A. G. & Prabhakar, J. V., *J. Am. Oil Chem. Soc.*, **62** (11) (1985) 1581.
256. Dziedzic, S. Z., Robinson, J. L. & Hudson, B. J. F., *J. Agric. Food Chem.*, **34** (1986) 1027.
257. Hudson, B. J. F. & Ghavami, M., *Lebensm. Wiss & Technol.*, **17** (1984) 191.
258. Hudson, B. J. F. & Lewis, J. I., *Food Chem.*, **10** (1983) 161.
259. Hudson, B. J. F. & Mahgoub, S. E. O., *J. Sci. Food Agric.*, **32** (1981) 208.
260. Kwon, T. W., Snyder, H. E. & Brown, H. G., *J. Am. Oil Chem. Soc.*, **61** (12) (1984) 1843.
261. Cocks, L. V. & van Rede, C., *Laboratory Handbook for Oil and Fat Analysis*. Academic Press, London, 1966, p. 137.
262. Daun, J. K., Davidson, L. D., Blake, J. A. & Yuen, W., *J. Am. Oil Chem. Soc.*, **58** (1981) 914.
263. Pardun, H., *Fette Seifen Anstrichm.*, **66** (6) (1964) 467.
264. Chapman, G. W., *J. Am. Oil Chem. Soc.*, **57** (1980) 299.
265. Pardun, H., *Fette Seifen Anstrichm.*, **83** (1981) 240.
266. Official and Tentative Methods of the American Oil Chemists' Society, Method Ca 12-55. AOCS, Champaign, IL, 1979.
267. AOAC Official Methods of Analysis, Sec. 13.045, 14th edn. AOAC, Arlington, Virginia, 1984.
268. Official and Tentative Methods of the American Oil Chemists' Society, Method Ca 19–86. AOCS, Champaign, IL, 1986.
269. Sinram, R. D., *J. Am. Oil Chem. Soc.*, **63** (5) (1986) 667.
270. Beutler, O. & Henniger, G., *Swiss Food*, **3** (12) (1981) 27.
271. Yuen, W. K. & Kelly, P. C., *J. Am. Oil Chem. Soc.*, **57** (1980) 359.
272. Black, B. C. & Hammond, E. G., *J. Am. Oil Chem. Soc.*, **42** (1965) 1002.
273. Hartman, L., Cardoso-Elias, M. & Esteves, W., *Analyst* (*London*), **105** (1980) 173.
274. Rouser, G., Siakotos, A. N. & Fleischer, S., *Lipids*, **1** (1966) 85.
275. Zhukov, A. V. & Vereshchagin, A. G., *J. Lipid Res.*, **10** (6) (1969) 711.
276. Gente-Jauniaux, M. & Prevot, A., *Rev. Franc. Corps Gras*, **26** (1979) 325.

277. Prevot, A. & Gente-Jauniaux, M., *Rev. Franc. Corps Gras,* **24** (1977) 493.
278. Belcher, R., Bogdanski, S. L. & Townshend, A., *Anal. Chem. Acta,* **67** (1973) 1.
279. Gentner, P. R. & Haasemann, A., *Fette Seifen Anstrichm.,* **81** (9) (1979) 357.
280. Totani, Y., Pretorius, H. E. & du Plessis, L. M., *J. Am. Oil Chem. Soc.,* **59** (4) (1982) 162.
281. Dijkstra, A. J. & Meert, D., *J. Am. Oil Chem. Soc.,* **59** (4) (1982) 199.
282. Erdahl, W. L., Stolyhwo, A. & Privett, O. S., *J. Am. Oil Chem. Soc.,* **50** (1973) 513.
283. Jungawala, F. B., Evans, J. E. & McCluer, R. H., *Biochem. J.,* **155** (1976) 55.
284. Hanson, V. L., Park, J. Y., Osborn, T. W. & Kiral, R. M., *J. Chromatogr.,* **205** (1981) 393.
285. Hox, W. M. A. & Guerts van Kessel, W. S. M., *J. Chromatogr.,* **142** (1977) 735.
286. Kiuchi, K., Ohta, T. & Ebine, H., *J. Chromatogr.,* **133** (1977) 226.
287. Nasner, A. & Kraus, L., *Fette Seifen Anstrichm.,* **83** (1981) 70.
288. Rhee, J. S. & Shin, M. G., *J. Am. Oil Chem. Soc.,* **59** (2) (1982) 98.
289. Hurst, W. J. & Martin, R. A. Jr., *J. Am. Oil Chem. Soc.,* **57** (1980) 307.
290. Kaitaranta, J. K. & Bessman, S. P., *Analyt. Chem.,* **53** (1981) 1232.
291. Buschbeck, R., *Pharmazie,* **20** (1965) 27.
292. Choy, T. K., Quartrone, J. J. Jr & Elefant, M., *Anal. Chim. Acta,* **29** (1963) 114.
293. Masuyama, S., *Kagaku to Kogyo,* **33** (1959) 224.
294. Ohlson, R., Persmark, U. & Podlaha, O., *J. Am. Oil Chem. Soc.,* **45** (1968) 475.
295. Klein, H., *Fleischwirtschaft,* **61** (7) (1981) 1029.
296. Miyakoshi, K. & Komoda, M., *J. Am. Oil Chem. Soc.,* **54** (1977) 331.
297. Tsuji, S., Tonogai, Y. & Ito, Y., *J. Food Prot.,* **49** (11) (1986) 914.
298. Kato, K., Nakaoka, T. & Ito, K., *J. Food Hyg. Soc. Japan,* **27** (6) (1986) 668.
299. Tsuji, S., Tonogai, Y., Ito, Y. & Harada, M., *J. Food Hyg. Soc. Japan,* **26** (4) (1985) 357.
300. Alicino, N. J., Klein, H. C., Quattrone, J. J. & Choy, T. K., *J. Agric. Food Chem.,* **11** (1963) 340.
301. Baldi, M., Zanoni, L., Pietrogrande, M. C. & Maietti, S., *Industrie alimentari,* **21** (5) (1982) 389.
302. Hobson-Frohock, A., *J. Sci. Food Agric.,* **33** (12) (1982) 1269.
303. Perfetti, G. A., Warner, C. R. & Fazio, T., *J. Assoc. Off. Anal. Chem.,* **64** (6) (1981) 1453.
304. Uchiyama, S. & Uchiyama, M., *J. Chromatogr.,* **262** (1983) 340.
305. Dahle, H. K. & Skaare, J. U., *J. Agric. Food Chem.,* **23** (6) (1975) 1093.
306. Spark, A. A., *J. Am. Oil Chem. Soc.,* **59** (4) (1982) 185.
307. Baldi, M., Maietti, S. & Pietrogrande, M. C., *Industrie alimentari,* **21** (10–11) (1982) 771.
308. Bickoff, E. M., Guggolz, J., Livingston, A. L. & Thompson, C. R., *Analyt. Chem.,* **28** (1956) 376.

309. Dunkley, W. L., Franke, A. A. & Low, E., *J. Dairy Sci.*, **51** (1968) 1215.
310. Van Deren, J. M. & Jaworski, E. G., *J. Assoc. Off. Anal. Chem.*, **51** (1968) 537.
311. Weilenmann, H. R., Hurter, J., Stoll, K. & Temperli, A., *Lebensml. Wiss. Technol.*, **5** (1972) 106.
312. Witt, S. C., Bickoff, E. M. & Kohler, G. O., *J. Assoc. Off. Anal. Chem.*, **56** (1973) 167.
313. Bovolenta, A., Baldi, M. & Zanoni, L., *Industrie alimentari*, **21** (7) (1982) 537.
314. Choy, T. K., Quattrone, J. J. Jr & Alicino, N. J., *J. Chromatogr.*, **12** (1963) 171.
315. Perfetti, G. A., Frank, J. L., Jr & Fazio, T., *J. Assoc. Off. Anal. Chem.*, **66** (5) (1983) 1143.
316. AOAC Official Methods of Analysis, 14th edn, Secs. 41.024–41.028. AOAC, Arlington, VA, 1984.
317. Bickoff, E. M., Livingston, A. L. & Thompson, C. R., *J. Am. Oil Chem. Soc.*, **32** (1955) 64.
318. Cort, W. M., *Food Technol.*, **28** (10) (1974) 60.
319. Hadorn, H. & Zuercher, K., *Dtsch. Lebensm. Rundsch.*, **70** (1974) 57.
320. Hammerschmidt, P. A. & Pratt, D. E., *J. Food Sci.*, **43** (1978) 556.
321. Kawashima, K., Itoh, H. & Chibata, I., *Agric. Biol. Chem.*, **43** (1979) 827.
322. Ke, P. J., Cervantes, E. & Robles-Martinez, C., *J. Sci. Food Agric.*, **34** (1983) 1154.
323. Kendrick, J. & Watts, B. M., *Lipids*, **4** (1969) 454.
324. Shearer, G. & Blain, J. A., *J. Sci. Food Agric.*, **17** (1966) 533.
325. Sherwin, E. R., *J. Am. Oil Chem. Soc.*, **45** (1968) 632A.
326. Wade, V. N., Al-Tahiri, R. & Crawford, R. J. M., *Milchwissenschaft*, **41** (8) (1986) 479.
327. Bishov, S. J. & Henick, A. S., *J. Am. Oil Chem. Soc.*, **43** (1966). 477.
328. Rossell, J. B. in *Rancidity in Foods*, ed. J. C. Allen & R. J. Hamilton. Applied Science Publishers, London, 1983, p. 21.
329. Kochhar, S. P. & Rossell, J. B., Unpublished data, 1985.
330. Bracco, U., Löliger, J. & Viret, J. L., *J. Am. Oil Chem. Soc.*, **58** (6) (1981) 686.
331. Boehm, E. E. & Maddox, D. N., *Proc. Inst. Food Sci. Technol. (UK)*, **6** (4) (1973) 210.
332. Fukuzumi, K. & Ikeda, N., *J. Am. Oil Chem. Soc.*, **47** (1970) 369.
333. Gerhardt, U. & Blat, P., *Fleischwirtschaft*, **64** (4) (1984) 484.
334. Kläui, H., *Proc. Inst. Food Sci. Technol. (UK)*, **6** (4) (1973) 195.
335. Moore, R. N. & Bickford, W. G., *J. Am. Oil Chem. Soc.*, **29** (1952) 1.
336. Sherwin, E. R., *J. Am. Oil Chem. Soc.*, **55** (1978) 809.
337. Thompson, J. W. & Sherwin, E. R., *J. Am. Oil Chem. Soc.*, **43** (1966) 683.
338. Olcott, H. S. & Einset, E., *J. Am. Oil Chem. Soc.*, **35** (1958) 161.
339. Palmateer, R. E., Yu, T. C. & Sinnhuber, R. O., *Food Technol.*, **14** (1960) 528.
340. Hamilton, J. W. & Tappel, A. L., *J. Am. Oil Chem. Soc.*, **40** (1963) 52.

341. Taylor, M. J. & Richardson, T., *J. Food Sci.*, **45** (1980) 1223.
342. Marco, G. J., *J. Am. Oil Chem. Soc.*, **45** (1968) 594.
343. Berner, D. L., Conte, J. A. & Jacobson, G. A., *J. Am. Oil Chem. Soc.*, **51** (1974) 292.
344. Miller, H. E., *J. Am. Oil Chem. Soc.*, **48** (1971) 91.
345. Taga, M. S., Miller, E. E. & Pratt, D. E., *J. Am. Oil Chem. Soc.*, **61** (1984) 928.
346. Osawa, T. & Namiki, M., *Agric. Biol. Chem.*, **45** (1981) 735.
347. Osawa, T. & Namiki, M., *J. Agric. Food Chem.*, **33** (5) (1985) 775.
348. Faria, J. A. F., *J. Am. Oil Chem. Soc.*, **59** (12) (1982) 533.
349. Castro-Martinez, R., Pratt, D. E. & Miller, E. E., in *Proceedings World Conference on Emerging Technologies in Fats and Oils Industry*, ed. A. R. Baldwin. AOCS, Champaign, IL, 1986, p. 392.
350. Chang, W. H., Luu, H. X. & Cheng, A. C., *J. Food Sci.*, **48** (1983) 658.
351. Hirayama, T., Yamazaki, M., Watanabe, T., Ono, M. & Fukui, S., *J. Food Hyg. Soc. Japan*, **27** (6) (1986) 615.
352. Labuza, T. P., Silver, M., Cohn, M., Heidelbaugh, M. & Karel, M., *J. Am. Oil Chem. Soc.*, **48** (1971) 527.

Chapter 3

CHEMISTRY AND IMPLICATIONS OF DEGRADATION OF PHENOLIC ANTIOXIDANTS

KIYOMI KIKUGAWA, AKIRA KUNUGI & TSUTAO KURECHI
Tokyo College of Pharmacy, 1432-1 Horinouchi, Hachioji, Tokyo 192-03, Japan

INTRODUCTION

Phenolic antioxidants are used to prevent or delay the autoxidation of fats and oils. Since Bolland[1] showed that autoxidation of fats and oils progresses through the chain reaction of free radicals, many studies on the mechanisms of the action of the antioxidants have been done. It has been shown that the antioxidants scavenge free radicals produced during the propagation steps of autoxidation of fats and oils by donating hydrogen radicals.[2–4] The chemistry of the degradation of antioxidant molecules during the course of autoxidation of fats and oils is important for elucidating the action of the antioxidants and of synergistic effects of compounds that potentiate their action. Degradation products of the antioxidants during the course of their exerting antioxidant activity in fats and oils have been elucidated. This chapter deals with the chemistry of the degradation of antioxidants, the chemistry of the degradation of mixed antioxidants, the activity of the antioxidants during the course of their degradation, and the chemistry of the synergistic effects of compounds that potentiate the activity of antioxidants.

CHEMISTRY OF THE DEGRADATION OF PHENOLIC ANTIOXIDANTS

The degradation of phenolic antioxidants, including monophenolic, diphenolic and triphenolic antioxidants, during the course of oxidation of fats and oils has been demonstrated. The fate of the antioxidants relates to the potency of their antioxidant activities. Most experiments with regard to the degradation of the antioxidants have been done under the conditions for autoxidation of fats and oils, i.e. thermal oxidation, oxidation by active oxygen method and oxidation by ultraviolet and visible light. The formation of antioxidant dimers is the most common feature of the degradation of the antioxidants. The dimers may be produced by the formation of phenoxy radicals followed by radical rearrangement and a coupling reaction.

Butylated Hydroxytoluene (BHT)

Butylated hydroxytoluene (BHT) is one of the most common mono-phenolic antioxidants. Oxidation of BHT by chemical oxidizing agents is well known.[5,6] Anderson & Huntley[7] assumed that BHT added to oils is converted into its oxidation products during the course of autoxidation of oils and they attempted to isolate the oxidation products. Japanese researchers[8] first isolated and identified the de-gradation products of BHT in autoxidized soybean oil. Thus, soybean oil containing BHT was irradiated by ultraviolet light, and five decomposition products of BHT were detected by thin-layer and gas chromatography. The degradation products have been identified as

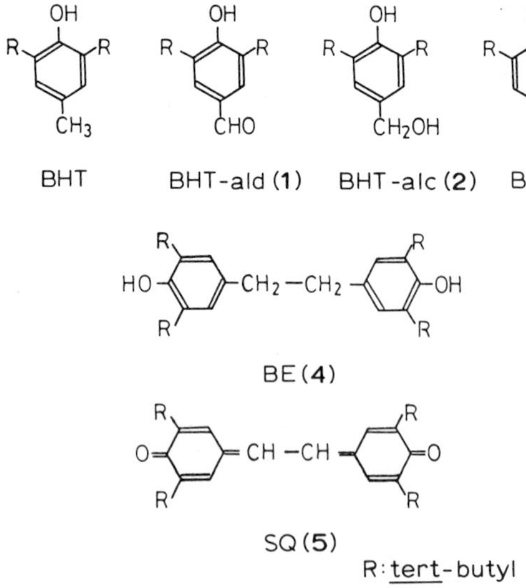

BHT BHT-ald (**1**) BHT-alc (**2**) BQ (**3**)

BE (**4**)

SQ (**5**)

R: *tert*-butyl

CHART 1

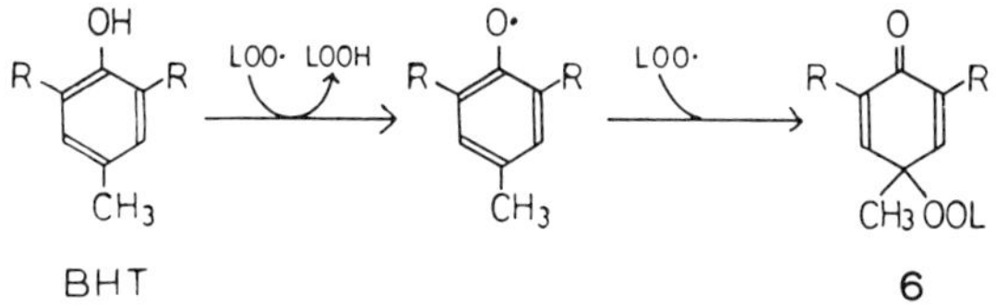

OH R R BHT LOO· LOOH O· R R CH3 LOO· O R R CH3 OOL 6

R: tert-butyl

CHART 2

3,5-di-*tert*-butyl-4-hydroxybenzaldehyde (BHT-ald, **1**), 3,5-di-*tert*-butyl-4-hydroxybenzylalcohol (BHT-alc, **2**), 2,6-di-*tert*-butylbenzoquinone (BQ, **3**), 3,5,3′,5′-tetra-*tert*-butyl-4,4′-dihydroxy-1,2-diphenylethylene (BE, **4**) and 3,5,3′,5′-tetra-*tert*-butylstilbenequinone (SQ, **5**) (Chart 1). The content of BHT decreases with irradiation time of the oil with ultraviolet light. BHT is completely lost after the induction period of the oil has come to an end. Degradation products **1–5** are formed as a result of the loss of BHT. The isolation and identification of SQ (**5**) from a heated edible oil with added BHT have been reported.[9] The initial reaction may be the formation of phenoxy radicals from BHT, which is followed by the rearrangement of the radicals to produce **1–5**.

In the course of oxidation of fats and oils, formation of adduct **6**

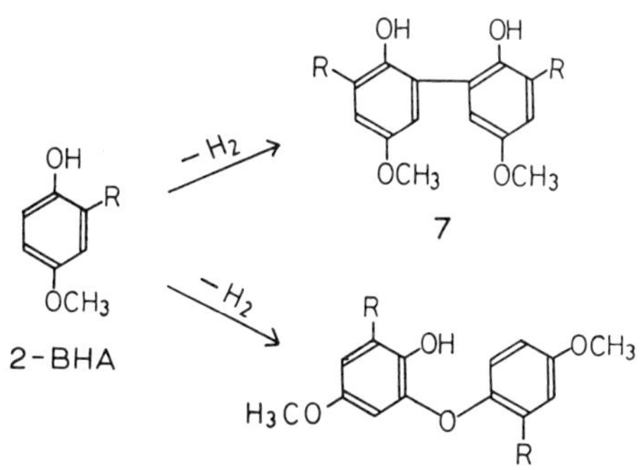

OH R OCH3 2-BHA $-H_2$ OH OH R R OCH3 OCH3 7 $-H_2$ R OH OCH3 H3CO O R 8

R: tert-butyl

CHART 3

between BHT and oxidized lipids is suggested[5,10,11] (Chart 2). Compound **6** is a coupled product between the radicals derived from the phenoxy radicals of BHT and the peroxy radicals of lipids.

Butylated Hydroxyanisole (BHA)

Butylated hydroxyanisole (BHA) used as a food antioxidant is a 9:1 mixture of 2- and 3-isomers (2-BHA and 3-BHA). It has been shown that chemical oxidation of 2-BHA affords a biphenyl type dimer, 2,2′-dihydroxy-5,5′-dimethoxy-3,3′-di-*tert*-butylbiphenyl (**7**) (Chart 3).[12–14] Degradation of BHA by light or heat has been studied by Kurechi and his coworkers.[15–17] Both 2-BHA and 3-BHA in alcohols are readily degraded by heating or by irradiation with visible and ultraviolet light.[15] Ultraviolet irradiation of 2-BHA in benzene gives two oxidation products, **7** and a biphenyl ether type dimer 2′,3-di-*tert*-butyl-2-hydroxy-4′,5-dimethoxybiphenylether (**8**).[16] Both compounds **7** and **8** may be produced by the formation of phenoxy radicals of 2-BHA followed by the radical rearrangement and the coupling reaction. The time course of the loss of 2-BHA and the formation of **7** and **8** in benzene and soybean oil has been followed (Fig. 1).[17] 2-BHA is lost and instead compounds **7** and **8** are produced. The amount of **7** is larger than that of **8** throughout the irradiation of the oils. The sum of the amounts of **7** and **8** produced is much lower than the amount of

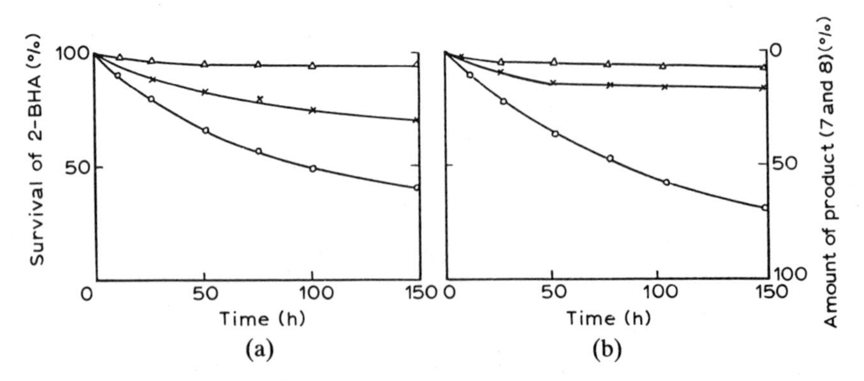

FIG. 1. Time course of the ultraviolet light-induced decrease of 2-BHA and increase of **7** and **8** in benzene (a) and soybean oil (b) containing 2-BHA.[17] 2-BHA (○), **7** (×) and **8** (△).

2-BHA lost throughout the irradiation, which can be explained by the loss of **8**, which is highly sensitive to irradiation. Mihara *et al.*[18] reported similar observations.

Ultraviolet irradiation of 2-BHA in benzene gives another two compounds.[19] However, these two products are adducts of 2-BHA and benzene. Irradiation of BHA in ethanol by γ-rays afforded 6 compounds including **7**, **8**, 2-*tert*-butylhydroquinone, 2-*tert*-butyl-1,4-dimethoxybenzene and two unidentified products.[20]

Tocopherols

α-, β-, γ- and δ-Tocopherols are natural monophenolic antioxidants present in edible oils. A variety of chemical oxidation products of tocopherols have been reported[21–24] Chemical oxidation of tocopherols affords their dimers and trimers.[25] The mechanisms of the formation of their dimers involve initial formation of phenoxy radicals which then react in two different ways: (i) two phenoxy radicals couple through oxygen or the ortho positions on the ring and (ii) phenoxy radicals rearrange to benzyl radicals, followed by coupling. A strong preference is noted for reaction at the 5-position versus the 7-position regardless of whether the substituents in these positions are hydrogen or methyl. The electronic or steric factors underlying this preference are not readily apparent.

Oxidation of tocopherols and formation of their dimers and trimers during the course of oxidation of fats and oils have been extensively studied.[26–39] Csallany *et al.*[26] heated methyl linoleate containing α-tocopherol under aerobic conditions and obtained its dimer and trimer together with α-tocopherylquinone. Fujitani & Ando[27–31] isolated and identified the dimers of α-, γ- and δ-tocopherols in thermally oxidized triglycerides and fatty acid methyl esters oxidized by the active oxygen method. α-Tocopherol in trilinolein under thermal oxidation gives two oxidation products, 5-formyl-γ-tocopherol-3-ene and a biphenylethane type dimer 1,2-bis(γ-tocopherol-5′-yl)ethane (α-TED, **9**) (Chart 4).[27] The time course of the decrease of α-tocopherol and the formation of dimer **9** during the course of thermal oxidation of trilinolein and trilaurin has been followed. While the differences in the rate of decreases of α-tocopherol are not remarkable, formation of **9** in trilaurin is much higher than that in trilinolein. In methyl linoleate by the active oxygen method, α-tocopherol is rapidly lost to produce **9**, but **9** is also degraded.[31]

γ-Tocopherol is more susceptible to oxidation than α-tocopherol.[32]

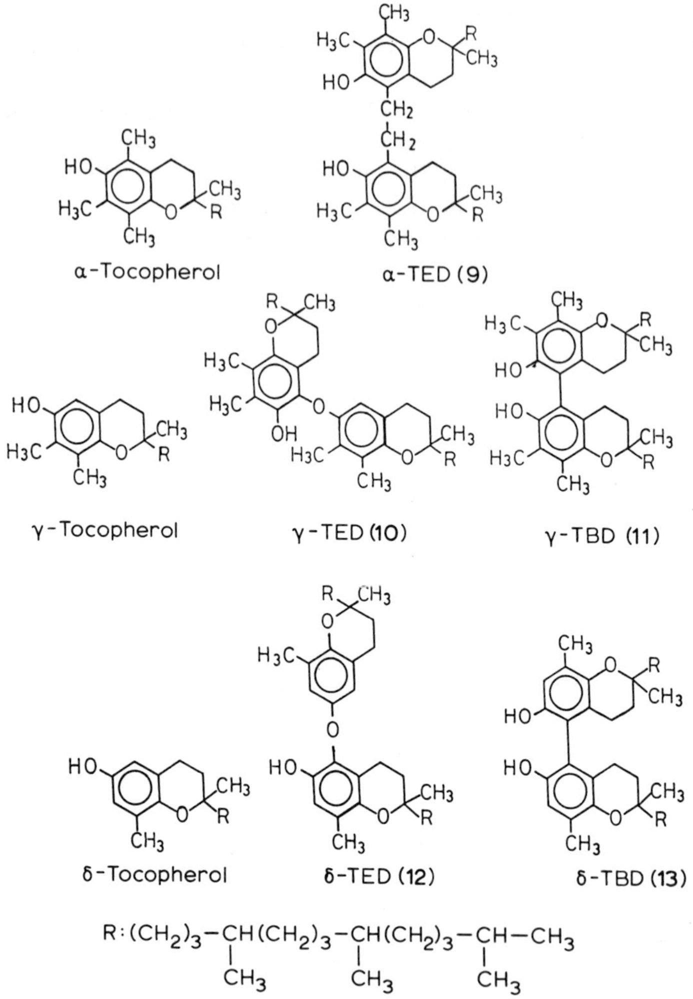

α-Tocopherol α-TED (9)

γ-Tocopherol γ-TED (10) γ-TBD (11)

δ-Tocopherol δ-TED (12) δ-TBD (13)

R: $(CH_2)_3-CH(CH_3)(CH_2)_3-CH(CH_3)(CH_2)_3-CH(CH_3)-CH_3$

CHART 4

γ-Tocopherol affords two dimers, a biphenylether type dimer 5-(γ-tocopheroxy)-γ-tocopherol (γ-TED, **10**) and a biphenyl type dimer 5-(γ-tocopherol-5-yl)-γ-tocopherol (γ-TBD, **11**) with two atropisomers.[28,30] These dimers have been obtained by chemical oxidation[33] and have been found in edible oils.[33–37] Under thermal oxidation of γ-tocopherol at 150°C or 180°C,[28] the amount of γ-TED (**10**) formed is greatly affected by the triglycerides used, and the formation of **10** is favourable in the order of trilaurin, triolein and trilinolein. The amount of γ-TBD (**11**) formed is relatively low and not affected by the triglycerides. This suggests that phenoxy radicals of γ-tocopherol produced during the course of thermal oxidation are more stable in the saturated triglycerides than in the unsaturated triglycerides. The time course of the contents of γ-tocopherol and dimers **10** and **11** in fatty acid methyl esters oxidized by the active oxygen method has been followed[30] (Fig. 2). The loss of γ-tocopherol

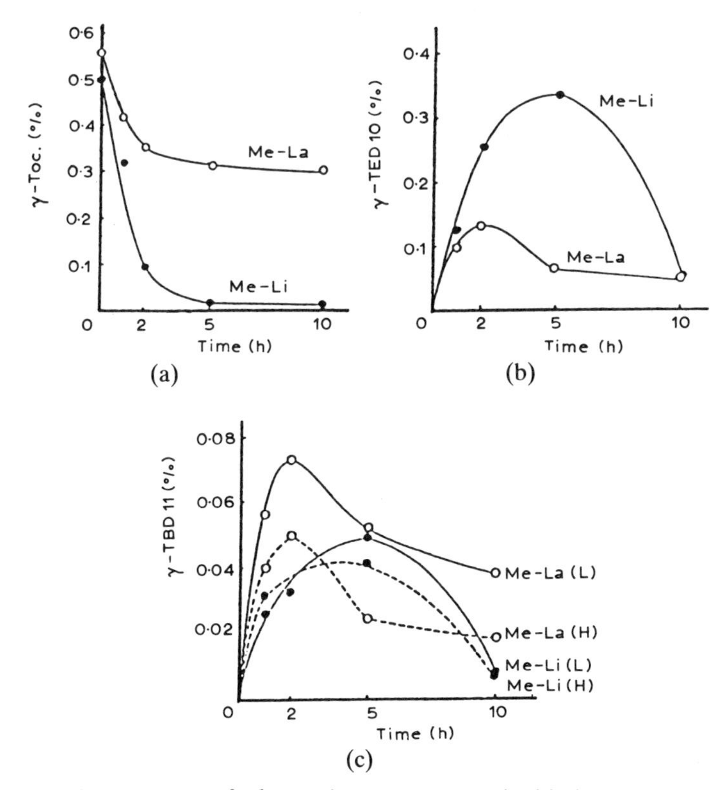

FIG. 2. Time course of the active oxygen method-induced decrease of γ-tocopherol (a), increase of γ-TED (**10**) (b) and increase of L and H atropisomers of γ-TBD (**11**) (c) in methyl linoleate (Me-Li) and methyl laurate (Me-La).[30]

is rapid and complete after 5 h in methyl linoleate, and the content of γ-TED (**10**) increases initially with time but rapidly decreases after 5 h. In contrast, decrease of the content of γ-tocopherol and increase of the content of **10** are much lower in methyl laurate. The formation of two atropisomers of γ-TBD (**11**) is relatively low in both methyl linoleate and methyl laurate.

δ-Tocopherol gives two dimers, 5-(δ-tocopheroxy)-δ-tocopherol (δ-TED, **12**) and 5-(δ-tocopherol-5-yl)-δ-tocopherol (δ-TBD, **13**), by thermal oxidation or by the active oxygen method.[29,31] The formation of these dimers has been proved.[38,39] The formation and degradation of the dimers from δ-tocopherol are similar to those from γ-tocopherol. In general, α-, β- and γ-tocopherols produce the dimers during the course of oxidation of fats and oils. Biphenylether type dimers **10** and **12** are preferentially produced in the oxidation of γ- and δ-tocopherols, indicating that the phenoxy radicals are stable. It is likely

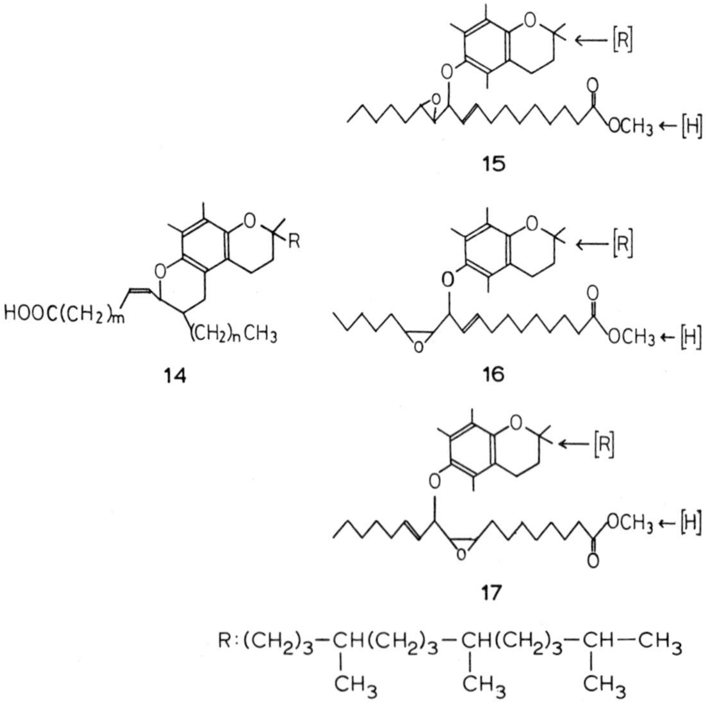
[R]
OCH3 ← [H]
15
R
HOOC(CH2)m
(CH2)nCH3
14
[R]
OCH3 ← [H]
16
[R]
OCH3 ← [H]
17
R: (CH2)3—CH(CH2)3—CH(CH2)3—CH—CH3
CH3 CH3 CH3

CHART 5

that the dimers are more stable than the parent tocopherols and contribute as antioxidants for fats and oils.

Formation of an adduct between α-tocopherol and fatty acid during the course of fatty acid oxidation has been demonstrated.[40–42] α-Tocopherol and linoleic acid form adduct **14** when the mixture is heated at 80°C.[40] In a similar way,[41] methyl linoleate adds under peroxidizing conditions to the 7-methyl of α-tocopherylquinone. Either linoleic acid hydroperoxide or methyl linoleate hydroperoxide reacts anaerobically with α-tocopherol to form principally an addition product (**15–17**) (Chart 5).[42] The mechanism appears to be free radical addition brought about by the catalytic formation of the alkoxy radicals from the hydroperoxides and the phenoxy radicals from α-tocopherol.

Sesamol

Sesamol, one of the constituents of sesame oil, may be responsible for its stability. Certain processing treatments of sesame oil result in the formation of sesamol from its bound form, sesamolin.[43,44] Sesamol is readily oxidized to sesamol dimer (**18**) and then to sesamol dimer quinone (**19**) by treatment with hydrogen peroxide/horseradish peroxidase[45] (Chart 6). It is possible that sesamol is converted into **18** and **19** during the progress of oxidation of sesame oil.[46]

Ethyl Protocatechuate (EP)

Ethyl protocatechuate (EP) is a diphenolic antioxidant which has been used for food. Its photochemical degradation in ethanol and methyl

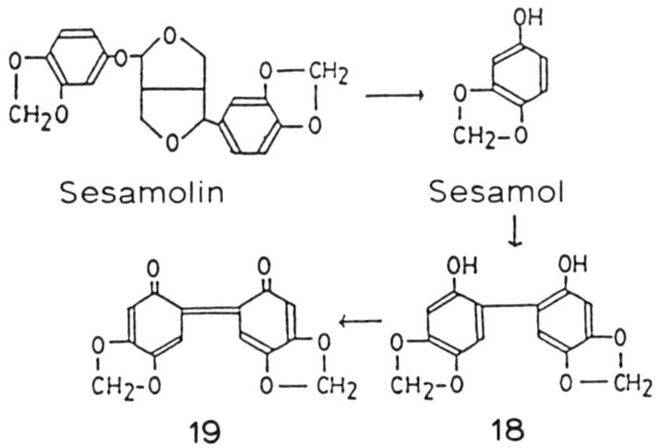

Sesamolin

Sesamol

19

18

CHART 6

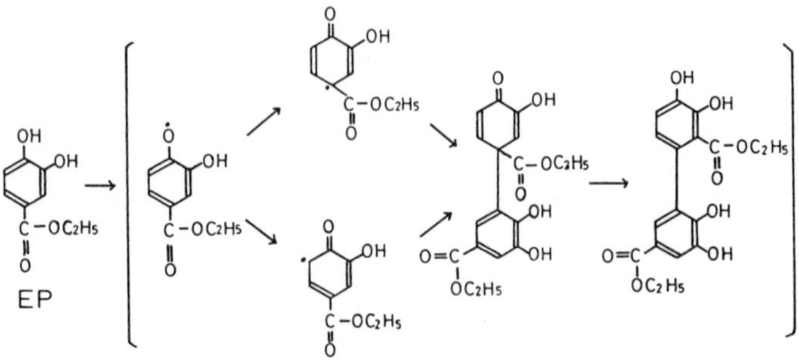

CHART 7

oleate have been studied by Taki & Kurechi.[47] When EP in ethanol or methyl oleate is irradiated by ultraviolet light, its dimer 5′-ethoxycarbonyl-3,3′,4-trihydroxybiphenyl-2,2′-carbolactone (**20**) is produced. This dimer may be formed by the formation of phenoxy radicals of EP, the radical rearrangement and the dienonephenol rearrangement (Chart 7). The time course of the decrease of EP and formation of **20** in methyl oleate oxidized under irradiation by ultraviolet light is shown in Fig. 3. The content of EP progressively decreases with irradiation and it is completely lost after 300 h. Formation of **20** is observed after 60 h of irradiation and reaches a maximum (1% of the EP added to the oil). The compound is unstable

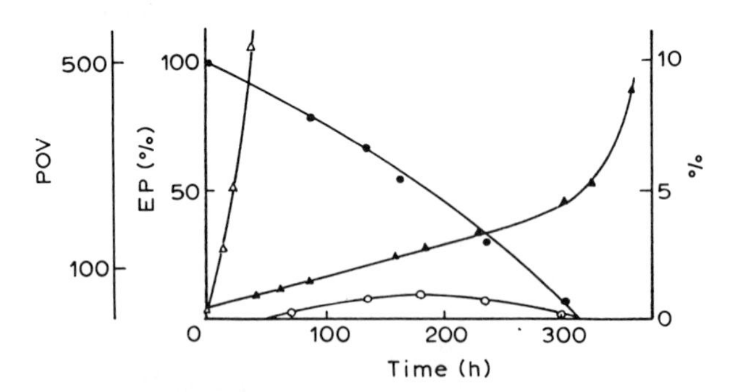

FIG. 3. Time course of the ultraviolet light-induced oxidation of methyl oleate containing 2% EP and the decrease and increase of EP and **20**.[47] Peroxide value of control methyl oleate (△), peroxide value of methyl oleate containing EP (▲), amount of EP in methyl oleate (●), and amount of **20** in methyl oleate (○).

and completely lost after 300 h irradiation. The peroxide value of methyl linoleate progressively increases and eventually both EP and **20** are destroyed completely.

Tert-Butyl Hydroquinone (TBHQ)

Tert-Butyl hydroquinone (TBHQ) is a diphenolic antioxidant useful for food. Oxidation of TBHQ has been studied by Kurechi *et al.*[48,49] Irradiation of TBHQ in benzene by ultraviolet light affords five oxidation products, 2-*tert*-butyl-*p*-benzoquinone (**21**), 2,2-dimethyl-5-hydroxy-2,3-dihydrobenzo[b]furan (**22**), 2-[2-(3′-*tert*-butyl-4′-hydroxyphenoxy)-2-methyl-1-propyl]hydroquinone (**23**), 2-(2-hydroxy-2-methyl-1-propyl)hydroquinone (**24**) and 2-*tert*-butyl-4-ethoxyphenol (**25**). Mechanisms for the formation of these compounds have been proposed (Chart 8).

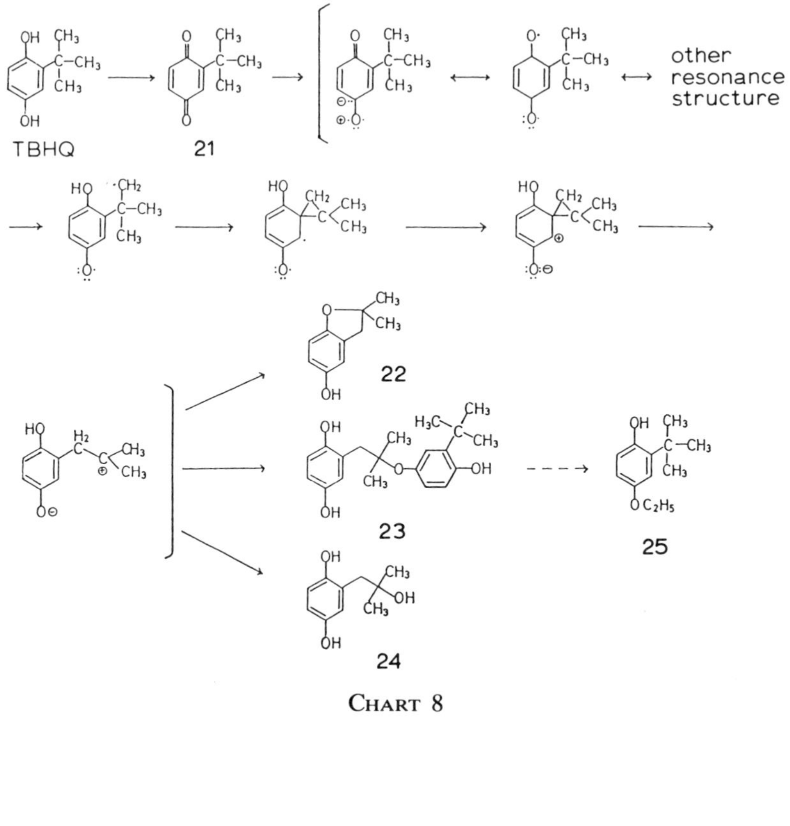

CHART 8

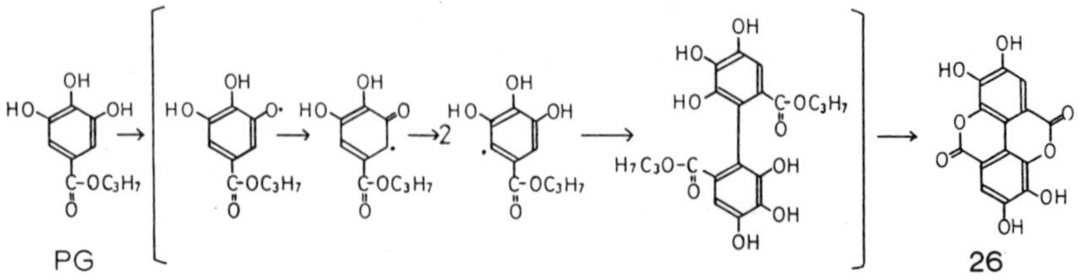

CHART 9

Propyl Gallate (PG)

Propyl gallate (PG) is a representative triphenolic antioxidant of fats and oils. Irradiation of PG in ethanol affords 2,3,7,8-tetrahydroxy-(1)benzopyrano(5,4,3-cde)(1)benzopyran-5,10-dione (ellagic acid, **26**).[50] Compound **26** may be formed by the production of phenoxy radicals from PG, followed by a radical rearrangement and coupling reaction (Chart 9).

CHEMISTRY OF THE DEGRADATION OF MIXED PHENOLIC ANTIOXIDANTS

In practice, two or more phenolic antioxidants are often used concomitantly because synergistic antioxidant effects for fats and oils are expected. The fate of the antioxidant molecules in the mixtures is important for understanding the synergism of the antioxidants.[51–54] As described below, hetero dimers between different antioxidant molecules are formed in the oxidation of mixtures of BHT and BHA, BHA and EP, and BHA and PG. The formation of these dimers may be important for the elucidation of synergistic effects of combined antioxidants. The chemistry of the degradation of mixed phenolic antioxidants has been studied by Kurechi and his coworkers.[50,55,56]

BHT and BHA

Irradiation of a mixture of BHT and 2-BHA in benzene by ultraviolet light affords a new product **27** in addition to the known oxidation products of BHT, **1, 4** and **5**, and 2-BHA, **7** and **8**.[55] Compound **27** is identified as 3,3′,5′-tri-*tert*-butyl-5-methoxy-2,4′-dihydroxydiphenylmethane, a biphenylether type dimer, which is presumably derived from radicals of BHT and 2-BHA (Chart 10). Figure 4 shows the

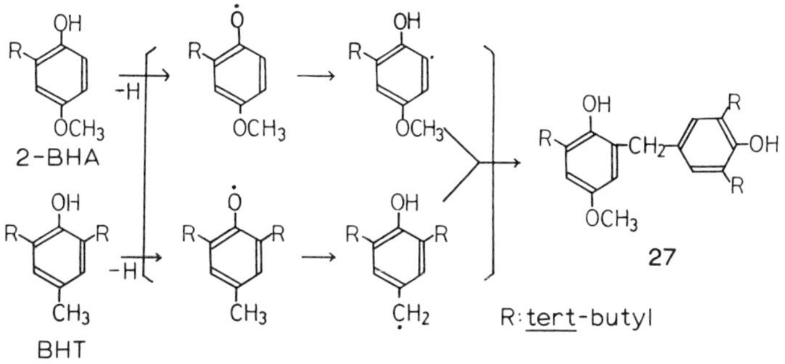

CHART 10

time-dependent changes in the amounts of BHT and 2-BHA, and their oxidation products, in benzene. BHT and 2-BHA gradually decrease with irradiation time, the loss of BHT being more rapid than that of 2-BHA. The amount of BE (**4**), that is probably derived from BHT alone, is the highest. The amount of **4** reaches a maximum of 12% of the original BHT after 144-h irradiation, but it gradually decreases later. SQ (**5**) is increasingly formed with irradiation time and reaches 2% of BHT. Oxidation products **7** and **8** of 2-BHA linearly increase with irradiation time, and reach 6% and 1% of 2-BHA, respectively. The amount of compound **27** increases with time but it is the lowest among the amounts of other oxidation products.

BHA and EP

Irradiation of a mixture of 2-BHA and EP in ethanol by ultraviolet light produces three new products in addition to the known oxidation products of 2-BHA, **7** and **8**, and of EP, **20**.[56] A new major oxidation product (**28**) is identified as 2,2′-dihydroxy-3-*tert*-butyl-5-methoxy-5′-carboethoxydiphenyl ether (Chart 11).

BHA and PG

When a mixture of 2-BHA and PG in ethanol is irradiated by ultraviolet light, new oxidation products are formed in addition to those of 2-BHA, **7** and **8**, and of PG, **26**.[50] The new products have been identified as propyl-3,5-dihydroxy-4-(2′-hydroxy-5′-methoxy-3′-*tert*-butylphenoxy)benzoate (**29**) and propyl-3,4-dihydroxy-5-(2′-hydroxy-5′-methoxy-3′-*tert*-butylphenoxy)benzoate (**30**) (Chart 12).

1·0
0·5
0
Survival (mmol)
(a)
0·05
Amount of products (mmol)
0
5
10
Time (days)
(b)

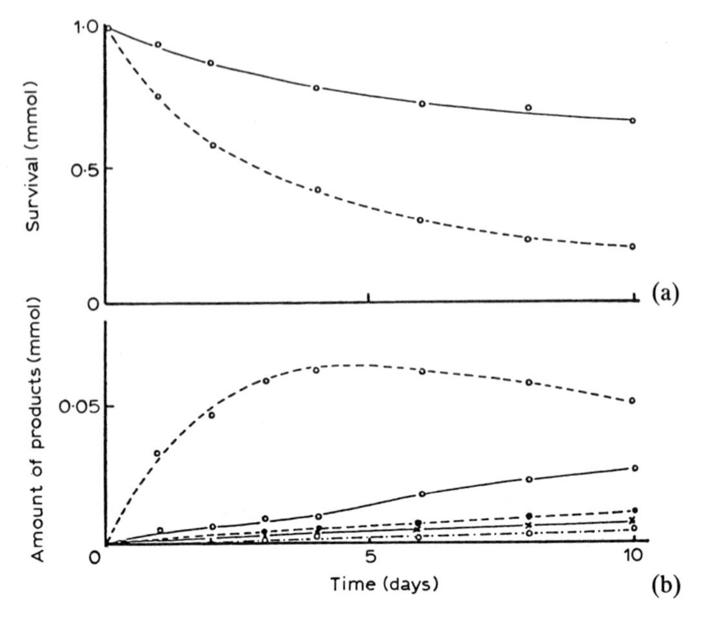

FIG. 4. Time course of the contents of BHT and 2-BHA (a) and their oxidation products (b) in benzene containing 1 mmol BHT and 1 mmol 2-BHA under irradiation by ultraviolet light.[55] (a), BHT (○- - -○) and 2-BHA (○—○); (b), **4** (○ - - - ○), **5** (●- -●), **7** (○—○), **8** (× — ×), and **27** (○-·-○).

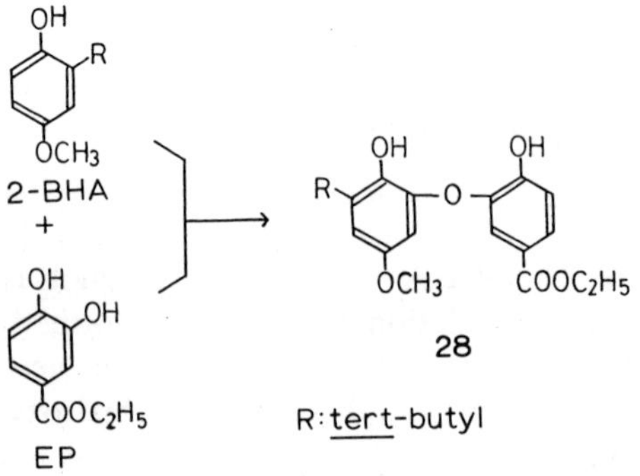

OH
R
OCH3
2-BHA
+
OH
OH
COOC2H5
EP
OH
OH
R
O
OCH3
COOC2H5
28
R: tert-butyl

CHART 11

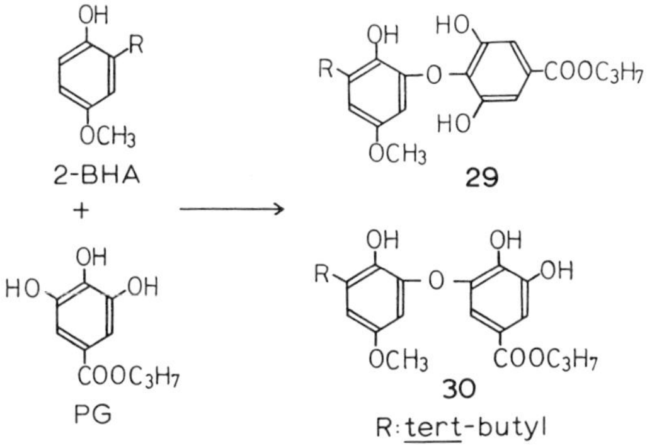

CHART 12

ANTIOXIDANT ACTIVITY OF THE OXIDATION PRODUCTS OF PHENOLIC ANTIOXIDANTS

Most of the oxidation products of the phenolic antioxidants still retain antioxidant activity. This fact will influence the antioxidant activity of the parent antioxidants during the course of their degradation.

The antioxidant activities of five oxidation products (**1**–**5**) of BHT have been studied.[8] Soybean oil containing each of these oxidation products was irradiated by ultraviolet light. The antioxidant activities of these compounds were determined as the ratios of the induction periods of the oil containing the compounds to that of the oil containing the parent antioxidant:

Antioxidant activity $= (t - t_o)/(t_p - t_o)$: where t_o is induction period of the oil without antioxidant, t is induction period of the oil with the new antioxidant, and t_p is induction period of the oil with the parent antioxidant.

The ratios of the activities of these compounds relative to that of BHT are calculated based on the observed induction period of the substrate oil (Table 1). BHT-alc (**2**), BQ (**3**) and BE (**4**) still retain 56, 23 and 64% activity of BHT, respectively, although BHT-ald (**1**) and SQ (**5**) are inactive. BHT-alc (**2**), BQ (**3**) and BE (**4**) may thus protect the oil against autoxidation, together with the remaining BHT.

Oxidation products of 2-BHA also exhibit antioxidant activity. Kurechi[57] evaluated the antioxidant activities of **7** and **8** in several substrate oils by the active oxygen method or irradiation by ultraviolet light or visible light. Antioxidant activities of these compounds

TABLE 1

ANTIOXIDANT ACTIVITIES OF OXIDATION PRODUCTS OF BHT IN SOYBEAN OIL IRRADIATED BY ULTRAVIOLET LIGHT[8]

Compounds (0·1%)	*Antioxidant activity* $\frac{t - t_o}{t_{BHT} - t_o}$
BHT	1·00
BHT-ald (**1**)	0
BHT-alc (**2**)	0·56
BQ (**3**)	0·23
BE (**4**)	0·64
SQ (**5**)	0

evaluated on the basis of the induction periods of the substrate oils under the active oxygen method are shown in Table 2. Oxidation products **7** and **8** show significant antioxidant activity in lard, beef tallow and methyl oleate. The potency of the antioxidant activity is in the order: 2-BHA, **8** and **7**. These compounds also show antioxidant activities for

TABLE 2

ANTIOXIDANT ACTIVITIES OF OXIDATION PRODUCTS OF 2-BHA IN LARD, BEEF TALLOW AND METHYL OLEATE OXIDIZED BY THE ACTIVE OXYGEN METHOD[57]

Compounds	*Concentration (oil)*	*Antioxidant activity* $\frac{t - t_o}{t_{2\text{-}BHA} - t_o}$
2-BHA	0·005% (lard)	1·00
7	0·005% (lard)	0·24
8	0·005% (lard)	0·48
2-BHA	0·02% (lard)	1·00
7	0·02% (lard)	0·33
8	0·02% (lard)	0·70
2-BHA	0·08% (lard)	1·00
7	0.08% (lard)	0·44
8	0.08% (lard)	1·41
2-BHA	0·02% (beef tallow)	1·00
7	0·02% (beef tallow)	0·18
8	0·02% (beef tallow)	0·52
2-BHA	0·02% (methyl oleate)	1·00
7	0·02% (methyl oleate)	0·24
8	0·02% (methyl oleate)	0·71

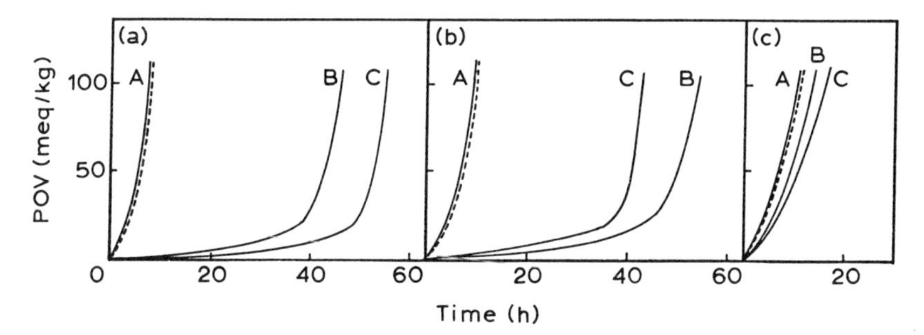

FIG. 5. Antioxidant activities of sesamolin, sesamol and sesamol dimer (**18**) in lard (a), methyl oleate (b) and soybean oil (c).[46] The substrate oils were oxidized by the active oxygen method. Control (– – – –), 0·01% sesamolin (—A—), 0·01% sesamol (—B—) and 0·01% sesamol dimer (**18**) (—C—).

the substrate oils when evaluated on the basis of the induction periods of the oils irradiated by ultraviolet or visible light; showing the potency of the activity in the order of 2-BHA, **7** and **8**. The oxidation products thus still retain antioxidant activity.

Sesamol and sesamol dimer (**18**) exhibit extensive antioxidant activities with clear demonstration of prolonged induction periods, especially in lard and methyl oleate.[46] While the activity of **18** is higher than that of sesamol in lard, it is lower in methyl oleate. In soybean oil, sesamol exhibits low but significant activity, and **18** exhibits slightly higher activity (Fig. 5).

TABLE 3

ANTIOXIDANT ACTIVITIES OF OXIDATION PRODUCTS OF TBHQ IN METHYL OLEATE, SOYBEAN OIL AND LARD OXIDIZED BY ACTIVE OXYGEN METHOD[49]

Compounds (0·01%)	*Substrate and antioxidant activity* $\frac{t - t_o}{t_{TBHQ} - t_o}$		
	Methyl oleate	*Soybean oil*	*Lard*
TBHQ	1·00	1·00	1·00
21	0·05	0·02	0·10
22	1·73	−0·03	0·51
23	1·31	0·68	1·05
24	1·31	0·56	1·39
25	2·53	0·06	0·87

The oxidation product **20** from EP exhibits no significant antioxidant activity in methyl oleate.[47] It is unlikely that the compound can act as an antioxidant during the course of the degradation of EP.

The antioxidant activities of oxidation products **21**–**25** of TBHQ are interesting.[48,49] All these oxidation products still retain antioxidant activity. However, the activity levels are very dependent on the substrate oils (Table 3). The order of the activities is, in **25**, **22**, **23**, **24**, TBHQ and **21** in methyl oleate, but in the order of TBHQ, **23**, **24**, **25**, **21** and **22** in soybean oil, and in the order of **24**, **23**, TBHQ, **25**, **22** and **21** in lard. Compound **21** has little activity in any substrate oils tested

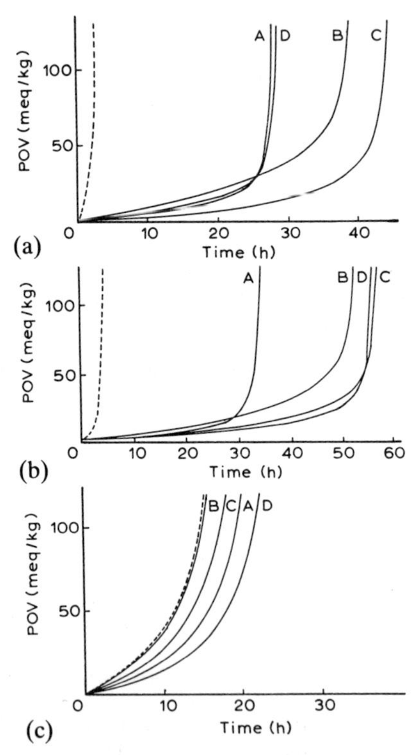

FIG. 6. Antioxidant activities of BHT, 2-BHA, their mixture and hetero dimer **27** in methyl oleate (a), lard (b) and soybean oil (c).[55] The substrate oils were oxidized by the active oxygen method. Control (- - - -), 1 mM BHT (—A—), 1 mM 2-BHA (—B—), 0·5 mM BHT and 0·5 mM 2-BHA (—C—), and 0·5 mM **27** (—D—).

possibly because a large proportion of it is removed by evaporation during the test. It is notable that oxidation products **22–25** exhibit antioxidant activities greater than that of TBHQ, depending on the substrate oils.

The oxidation product of PG (**26**) still retains antioxidant activity when tested on methyl oleate by the active oxygen method.[50]

The antioxidant activities of the hetero dimers **27**, **28**, **29** and **30** derived from the oxidation of the mixtures of BHT and 2-BHA, 2-BHA and EP, and 2-BHA and PG have been evaluated. The antioxidant activity of **27** is compared with those of the parent antioxidants BHT and 2-BHA and the mixtures of these antioxidants in methyl oleate, lard and soybean oil by the active oxygen method.[55] The antioxidant activities in methyl oleate are shown in Fig. 6(a). 2-BHA shows greater antioxidant activity than BHT, but a mixture of BHT and 2-BHA shows much greater activity than BHT or 2-BHA alone owing to a synergistic effect. Compound **27** retains antioxidant activity comparable to that of BHT.

The antioxidant activities in lard are shown in Fig. 6(b). 2-BHA shows greater activity than BHT, but a mixture of BHT and 2-BHA shows much greater activity than either, and compound **27** shows activity comparable to that of the mixture. Thus, compound **27** has greater activity than BHT and 2-BHA in lard.

The antioxidant activities in soybean oil are shown in Fig. 6(c). 2-BHA shows little activity whereas BHT shows significant activity. While the activity of the mixture does not exceed those of the parent antioxidants, compound **27** shows much greater activity than those of

TABLE 4
ANTIOXIDANT ACTIVITIES OF BHT, 2-BHA, THEIR MIXTURE AND HETERO DIMER **27**[55]

Compounds (total 1 mM)	*Substrate and antioxidant activity* $\frac{t - t_o}{t_{BHT} - t_o}$		
	Methyl oleate	*Lard*	*Soybean oil*
BHT	1·00	1·00	1·00
2-BHA	1·42	1·53	0·02
BHT + 2-BHA	1·66	1·67	0·55
27	1·03	1·66	1·47

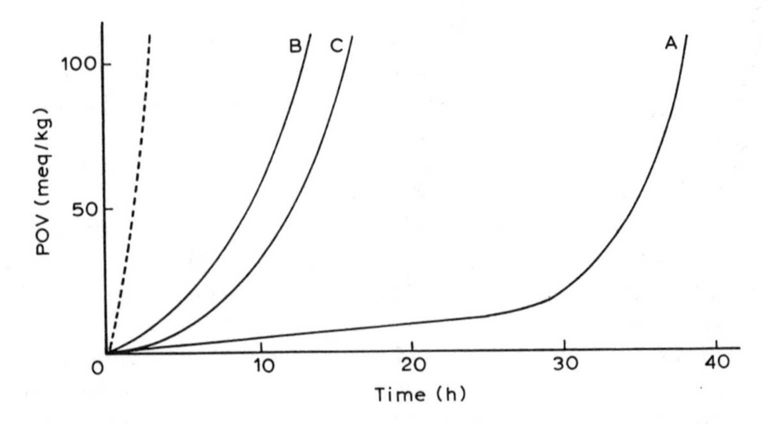

FIG. 7. Antioxidant activities of 2-BHA, EP and hetero dimer **28** in methyl oleate.[56] Methyl oleate was oxidized by the active oxygen method. Control (– – – –), 0·02% 2-BHA (—A—), 0·02% EP (—B—), and 0·02% **28** (—C—).

either of them. The antioxidant activities are tabulated in Table 4. The hetero dimer **27** is thus a significant factor in the evaluation of the antioxidant activity of the mixture of BHT and 2-BHA.

The antioxidant activity of **28** has been compared with those of its parent antioxidants, 2-BHA and EP.[56] The antioxidant activity of **28** in methyl oleate is between those of 2-BHA and EP (Fig. 7). The activity

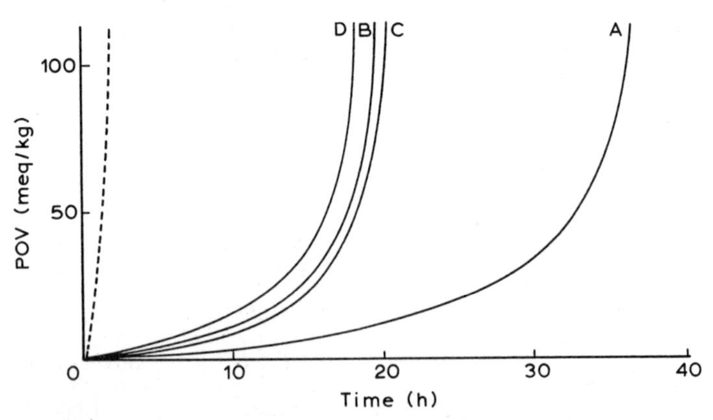

FIG. 8. Antioxidant activities of 2-BHA, PG and hetero dimers **29** and **30** in methyl oleate.[50] Methyl oleate was oxidized by the active oxygen method. Control (– – – –), 0·01% 2-BHA (—A—), 0·01% PG (—B—), 0·01% **29** (—C—), and 0·01% **30** (—D—).

TABLE 5
ANTIOXIDANT ACTIVITIES OF 2-BHA, PG AND HETERO DIMERS **29** AND **30**[50]

Compounds (0·01%)	*Substrate and antioxidant activity* $\frac{t - t_o}{t_{BHA} - t_o}$		
	Methyl oleate	*Lard*	*Soybean oil*
2-BHA	1·00	1·00	1·00
PG	0·52	1·50	12·25
29	0·53	0·88	4·00
30	0·47	0·91	1·50

of **28** in lard is also between those of 2-BHA and EP. These results indicate that hetero dimer **28** has intermediate activity between 2-BHA and EP.

The antioxidant activities of compounds **29** and **30**, both derived from the mixture of 2-BHA and PG, have been compared with those of the parent antioxidants.[50] 2-BHA shows higher activity than PG in methyl oleate, and hetero dimers **29** and **30** exhibit activities comparable to that of PG (Fig. 8 and Table 5). PG has greater activity than 2-BHA in lard and soybean oil, and the hetero dimers have activities greater than that of 2-BHA (Table 5).

CHEMISTRY OF THE SYNERGISM OF PHENOLIC ANTIOXIDANTS WITH NON-PHENOLIC COMPOUNDS

Pronounced synergistic effects between phenolic compounds and certain acidic substances such as ascorbic acid, citric acid and phosphoric acid have been described.[58] These active synergists are also effective metal-chelating agents, which could give rise to the theory that their only activity is that of metal chelation. Considerable evidence has been gathered on the synergistic effects of various compounds in food products.[4] The effects of several synergists such as citric acid, ascorbic acid, phospholipids, amines, amino acids, protein hydrolysates and melanoidin have been elucidated. Kraybill *et al.* showed that BHA exhibits synergism with certain phenolics and phospholipids.[59] As already indicated, BHA, BHT and PG are frequently used in combination since their excellent synergistic effects have been demonstrated.[4,51,52] Several other investigations on the mechanisms of synergism have been made.

Synergistic Effect of Citric Acid

The synergistic effect of citric acid and its esters with tocopherols has been demonstrated.[60–62] Kanematsu *et al.*[63] demonstrated that the antioxidant effect of tocopherols in lard and palm oil is little affected by the addition of citric acid. They concluded that present day fats and oils, due to the progress of oil-refining techniques, contain a smaller amount of metal compounds than formerly. The results indicated that the effect of citric, malic and succinic acids is as would be expected on the basis of the function of the acids as metal deactivators. Monoacylglyceryl citrate and isopropyl citrate show little or no synergistic effects with lard and palm oil, though they may also act as metal deactivators.[64,65]

Synergistic Effect of Ascorbic Acid

Ascorbic acid and ascorbyl palmitate, which function by oxygen scavenging, an entirely different mechanism from that of phenolic antioxidants, are used as antioxidants to remove oxygen in solution. Synergistic effects of ascorbic acid and its derivatives with tocopherols have been demonstrated.[62,66–74]

Packer *et al.*[75] reported direct observations of the interaction of α-tocopherol and ascorbic acid and showed that they act synergistically, α-tocopherol acting as the primary antioxidant and the resulting α-tocopheryl radical then reacting with ascorbic acid to regenerate α-tocopherol. Niki *et al.*[76] demonstrated that the α-tocopheryl radical generated by the stable radical, 2,2-diphenyl-1-picrylhydrazyl (DPPH),

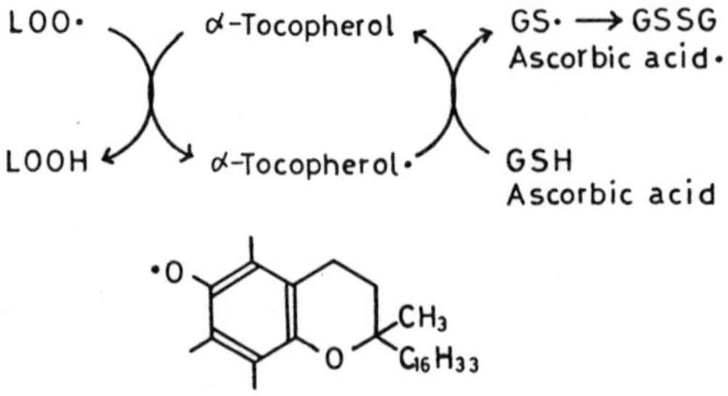

LOO· α-Tocopherol GS· → GSSG
Ascorbic acid·
LOOH α-Tocopherol· GSH
Ascorbic acid

·O CH_3 $C_{16}H_{33}$

CHART 13

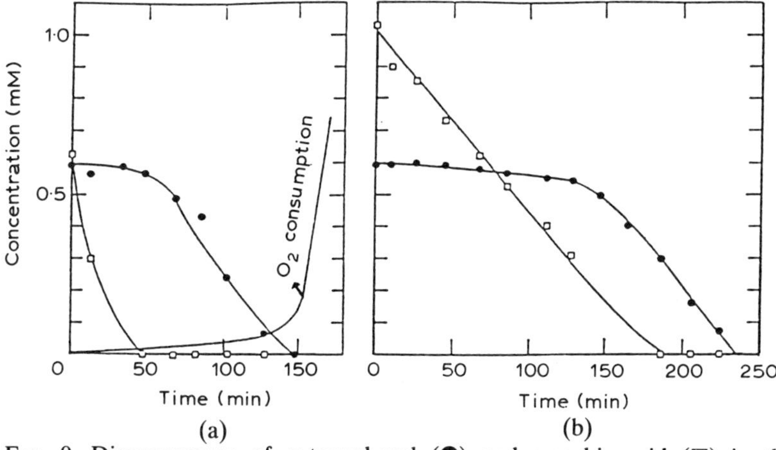

FIG. 9. Disappearance of α-tocopherol (●) and ascorbic acid (□) in the oxidation of methyl linoleate at 37°C in *tert*-butanol/methanol.[77] [Methyl linoleate] = 0·60 M, [2,2′-azabis(2,4-dimethylvaleronitrile)] = 0·010 M, [α-tocopherol] = 0·595 mM, and [ascorbic acid] = 0·620 mM (a) and 1·03 mM (b).

disappears quite rapidly when it is mixed with ascorbic acid. Glutathione similarly acts to destroy the α-tocopheryl radical (Chart 13). Niki *et al.*[77] studied the disappearance of α-tocopherol and ascorbic acid in the oxidation of methyl linoleate initiated with 2,2′-azobis(2,4-dimethylvaleronitrile). Figure 9 shows the consumption of α-tocopherol and ascorbic acid in the oxidation of methyl linoleate. When either α-tocopherol or ascorbic acid is used alone, they disappear linearly with time. However, when both α-tocopherol and ascorbic acid are used, ascorbic acid is consumed first and α-tocopherol is consumed after the ascorbic acid has disappeared. These results strongly suggest that α-tocopherol scavenges the peroxy radical of methyl linoleate more quickly than ascorbic acid, but that the α-tocopheryl radial reacts with ascorbic acid to regenerate α-tocopherol. This sequence may contribute to the synergistic effect of ascorbic acid with α-tocopherol.

Synergistic Effect of Phospholipids

Crude soybean oil exhibits greater oxidative stability than refined oil, and at least part of the increased stability of soybean oil is due to the phospholipids present in crude oil.[78] Linow & Mieth[79] found that the antioxidant activity of α-tocopherol in methyl linoleate is increased by adding phosphatidyl choline or phosphatidyl ethanolamine. The addition of the phospholipids also reduces the decomposition of soybean tocopherols. Hudson & Mahgoub[80] found that phosphatidyl choline and phosphatidyl ethanolamine from eggs act as synergists with α-tocopherol in the autoxidation of lard. The synergistic effects of phosphatidyl choline and phosphatidyl ethanolamine in enhancing the antioxidant properties of polyhydroxy flavonoids in lard have also been demonstrated.[81] Phosphatidyl ethanolamine is very effective but phosphatidyl choline has little synergistic activity. It is suggested that the presence in the synergist molecule of a strongly acid, proton generating function is of importance.

Phosphatidyl ethanolamine exerts synergistic effects with tocopherols in model systems based on both lard and soybean oil.[82] The synergistic effect is specific to the chemical class of phospholipids, being less marked with phosphatidyl choline and negative with phosphatidyl inositol. Dipalmitoyl phosphatidyl ethanolamine is a potent synergist for a wide range of antioxidants such as PG, α-tocopherol, BHA, BHT and TBHQ at elevated temperature, i.e. above 80°C[83] (Table 6). At lower temperatures it has very little synergistic action. At 120°C the synergistic effect increases progressively as the concentration of synergist increases from 0·025% to 0·25%. At a given level of synergist, its effect is proportionately

TABLE 6

EFFECT OF CONCENTRATION OF DIPALMITOYL PHOSPHATIDYL ETHANOLAMINE ON ITS SYNERGISTIC ACTION IN COOPERATION WITH ANTIOXIDANTS ADDED TO LARD AT 120°C[83]

Antioxidant (0·025%)	*Synergistic efficiency*[a]			
	% of dipalmitoyl phosphatidyl ethanolamine			
	0·025	*0·05*	*0·10*	*0·25*
PG	−43	−17	0	+36
α-Tocopherol	+22	+35	+57	+78
BHA	+23	+56	+44	+70
BHT	+8	+25	+37	0
TBHQ	−4	+42	+45	+45

[a] $100[(I_m - I_l) - (I_a - I_l) - (I_s - I_l)]/(I_m - I_l)$: where I_l is induction period of lard, I_a is induction period of lard with antioxidant, I_s is induction period of lard with synergist, and I_m is induction period of lard with antioxidant and synergist.

greater at low rather than high levels of antioxidants. Hildebrand *et al.*[84] have shown that the synergistic effects of phosphatidyl ethanolamine and phosphatidyl inositol with soybean tocopherols in soybean oil are larger than that of phosphatidyl choline. The effects of these phospholipids are not simply a matter of metal inactivation, but rather appear to extend the effectiveness of the tocopherols.

Yuki *et al.*[85] investigated the interaction of lecithin and tocopherols during the thermal oxidation of liquid paraffin. While the oxidation of tocopherols at 180°C gives only a trace of tocopherols and their dimers, oxidation in the presence of lecithin causes a much sharper increase in the amounts of intact tocopherols and their dimers, suggesting that lecithin inhibits the degradation of tocopherols and their dimers. Fujitani & Ando[86–89] reported a similar protective activity of lecithin toward tocopherols in triglyceride oxidation. Dziedzic *et al.*[90] described the protective effect of dipalmitoyl phosphatidyl ethanolamine toward PG in lard oxidation. When dipalmitoyl phosphatidyl ethanolamine is simultaneously present, the life of PG is extended and longer induction periods of lard are observed (Fig. 10).

Synergistic Effect of Amines

Yuki *et al.*[91] demonstrated that trimethylamine oxide (TMAO) has a remarkable synergistic effect with tocopherols in inhibiting the oxidation of fats and oils. The synergistic effect of TMAO is closely related to the kind of oil, temperature and concentration of the antioxidants. Ishikawa *et al.*[92–94] investigated the mechanisms of the synergistic effect of TMAO with tocopherols. The amounts of the dimers of tocopherols formed are larger in the presence of TMAO than in the absence of TMAO, and the dimers are likely to play an important role in synergism with tocopherols and TMAO. From the reaction of methyl linoleate hydroperoxide with TMAO, TMAO was found to act as a peroxide decomposer. TMAO produces a keto acid, which is formed only in the presence of tocopherols and TMAO. From the results, Ishikawa *et al.*[94] tentatively postulated the possible mechanisms of synergistic action of TMAO (Chart 14). Synergism between tocopherols and TMAO is mainly due to that between the dimers of tocopherols (AAH) and TMAO. AAH is consumed by offering $H^{\cdot}$ to the $LOO^{\cdot}$ radical to give the $AA^{\cdot}$ radical and LOOH. TMAO then reacts with LOOH to produce a keto acid and an active intermediate (TMAO-H), which offers $H^{\cdot}$ to the $AA^{\cdot}$ radical, thus regenerating the AAH.

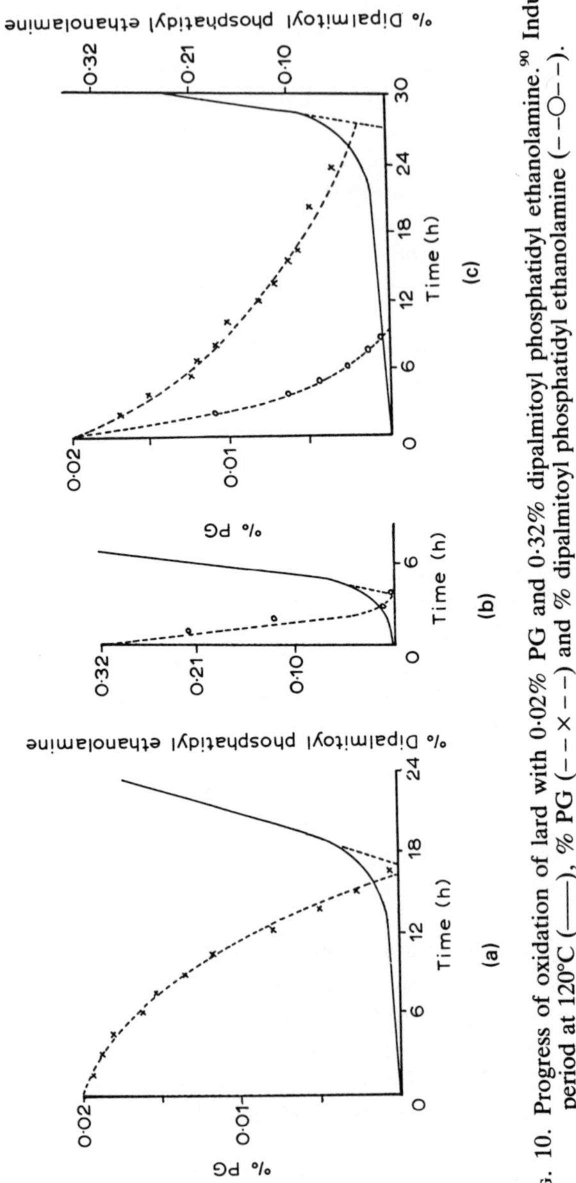

FIG. 10. Progress of oxidation of lard with 0·02% PG and 0·32% dipalmitoyl phosphatidyl ethanolamine.[90] Induction period at 120°C (——), % PG (– – × – –) and % dipalmitoyl phosphatidyl ethanolamine (– –○– –).

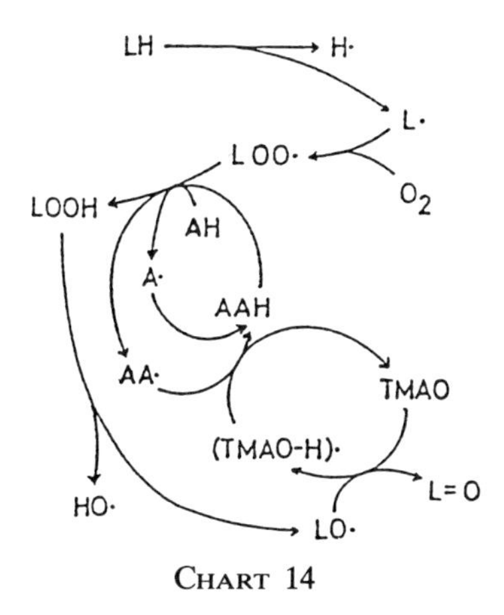

CHART 14

Trioctylamine has been found to have antioxidant activity[95] and to act as a strong peroxide decomposer.[94,96] Harris & Olcott[96] reported on the mechanism of antioxidation of trioctylamine, and they proposed a pathway in which the dimer reacts with LOOH to form trioctylamine oxide and LOH. Remarkable synergism has been observed in the mixture of γ-tocopherol and trioctylamine.[97] The amounts of the dimers of γ-tocopherol markedly increase in the presence of trioctylamine. Keto acid (L=O) is formed as a main product of methyl linoleate oxidation in the presence of γ-tocopherol and trioctylamine. Trioctylamine is much superior to trioctylamine oxide in synergism with tocopherols.[98] The course of the reactions from trioctylamine to trioctylamine oxide may be important in the remarkable synergism. Comparative studies of amino compounds on the formation of γ-tocopherol dimers in autoxidizing linoleate have been reported.[99]

Synergistic Effect of Other Food Constituents

Amino acids are known to exert synergistic effect with antioxidants.[100–106] Trolox-C, a derivative of α-tocopherol, has substantial antioxidant activity.[107,108] Various amino acids were covalently attached to Trolox-C to produce Troloxyl-amino acids (**31**) (Chart 15), and in some oils Troloxyl-amino acids exert even higher antioxidant activities.[109]

Proteins and protein hydrolysates have synergistic effects with antioxidants.[110–115] Melanoidins, reaction products of amino acids and

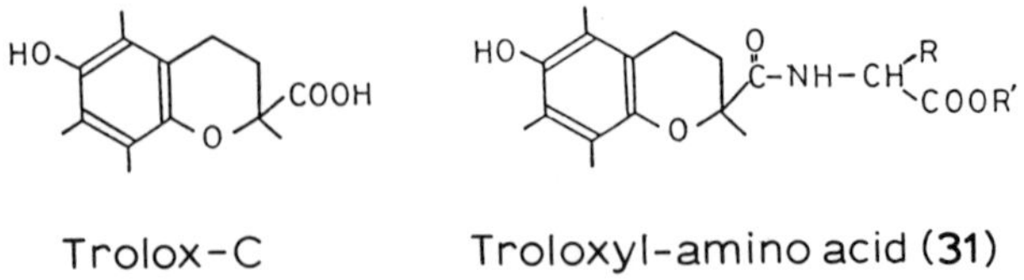

Trolox-C Troloxyl-amino acid (**31**)

CHART 15

sugars, exhibit antioxidant activity and synergistic effects with antioxidants.[116,117] Melanoidins can act as hydroperoxide decomposers.[117]

MECHANISMS OF SYNERGISTIC EFFECTS OF COMBINED ANTIOXIDANTS

As has been shown, BHA, BHT and PG are frequently used in combination since they exhibit marked synergistic antioxidative effects.[4,51,52] The mechanism of the synergism between BHT and BHA have been investigated by Kurechi and his coworkers.[118–122]

The hydrogen donating capabilities of BHT, 2-BHA and mixtures of them to the stable free radical, DPPH, have been compared.[118] 2-BHA donates 0·75–1·2 hydrogen atoms, and BHT donates much less hydrogen. The synergistic effect of hydrogen donation to DPPH

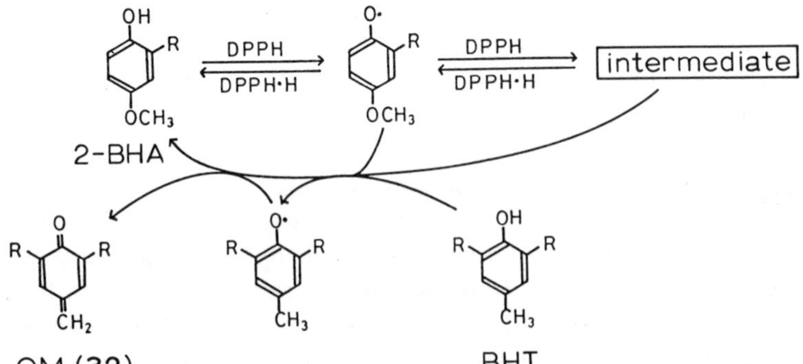

2-BHA DPPH DPPH·H DPPH DPPH·H intermediate

QM (**32**) BHT

R: tert-butyl

CHART 16

(2·78 hydrogen atoms) is observed with a combination of 2-BHA and BHT. The results concur with the observations of Boguth *et al.*[123] who demonstrated the synergistic effect of tocopherols and BHT as a result of catalytic activation of the hydrogen donating capability of inactive BHT by tocopherols. Similar synergistic effects have been observed when 2,6-di-*tert*-butylphenols are combined with 4-methoxyphenol derivatives.[119] The possible reaction processes of 2-BHA and BHT with DPPH are shown in Chart 16.[120] In the reaction of 2-BHA with an excess amount of DPPH, DPPH and 2-BHA are regenerated in a molar ratio of 2:1 by addition of DPPH H, and 2-BHA and 2,6-di-*tert*-butylquinone methide (QM, **32**) are formed in a ratio of 1:1 by addition of BHT. 2-BHA reacts with DPPH to form a stable intermediate via its phenoxy radicals. This intermediate may react with BHT to regenerate 2-BHA with the enhanced oxidation of BHT to QM (**32**) via the phenoxy radicals of BHT. These processes may be associated with the synergism between 2-BHA and BHT in hydrogen donation to DPPH, and thus to peroxy radicals of fats and oils.

2-BHA and BHT can reduce 3 and 0·6 equivalents of ferric ion, respectively.[121] Combination of 2-BHA and BHT produces 1·5 times

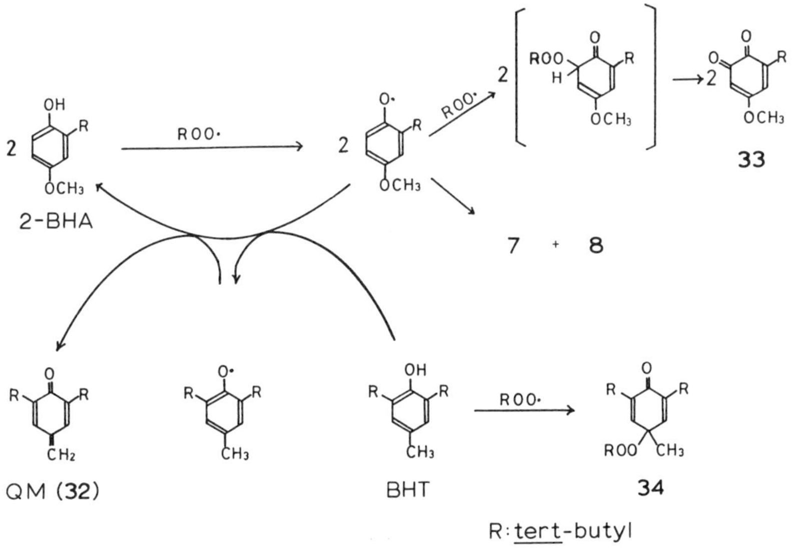

R: *tert*-butyl

CHART 17

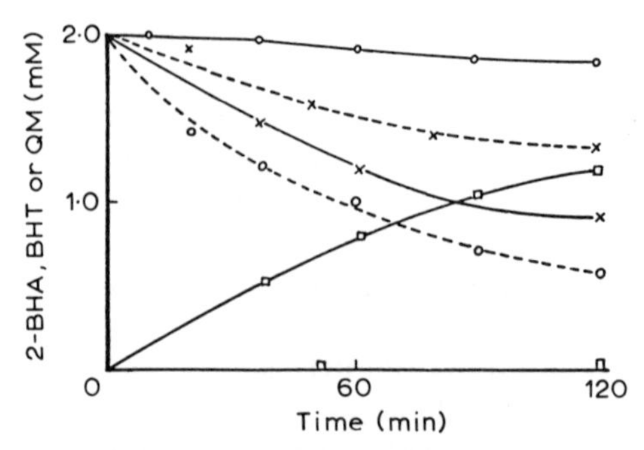

FIG. 11. Time course of loss of 2-BHA and BHT and formation of QM (**32**) in the reaction mixture of 2-BHA and/or BHT with *tert*-butylhydroperoxide/Co^+ (Ref. 122). 2-BHA (2 mM) and BHT (2 mM) were treated with *tert*-butylhydroperoxide (2 mM) and Co^+ (0·1 mM) at 20°C. 2-BHA and BHT (——), and 2-BHA or BHT alone (– – – –). 2-BHA (○), BHT (×) and QM (**32**) (□).

more ferrous ions than the calculated amount. It is evident from product analysis that loss of BHT is enhanced by 2-BHA.

A kinetic explanation for the reaction of the peroxy radicals with a mixture of 2-BHA and BHT, which shows synergism in the protection of lard from peroxidation, has been proposed.[124] Hydrogen donation from 2-BHA and/or BHT to the peroxy radicals prepared by cobalt-catalysed cleavage of *tert*-butylhydroperoxide, and the relation of this process to synergism in the antioxidative effect, have been studied (Chart 17).[122] 2-BHA donates one hydrogen atom to the *tert*-butyl peroxy radicals to form phenoxy radicals, and these react with the peroxy radicals to produce **33**, or with the phenoxy radicals of 2-BHA to produce **7** and **8**. BHT reacts with the peroxy radicals to form **34**. In the combination of 2-BHA and BHT, 2-BHA first donates one hydrogen to the peroxy radicals to form phenoxy radicals. Two hydrogen atoms are transferred from BHT to the phenoxy radicals to regenerate 2-BHA with enhanced oxidation of BHT to QM (**32**). The reaction process is consistent with the finding that the oxidation of 2-BHA is depressed and that of BHT is enhanced in the combination of 2-BHA and BHT (Fig. 11). Hydrogen donation of BHT to 2-BHA may thus be closely correlated with synergism in the antioxidative effect of the mixtures of 2-BHA and BHT.

REFERENCES

1. Bolland, J. L. *Proc. R. Soc., A,* **186** (1946) 218.
2. Bolland, J. L., & Have, P. T. *Trans. Faraday Soc.,* **43** (1946) 201.
3. Shelton, J. R. & Vincent, D. N., *J. Appl. Polym. Sci.,* **85** (1963) 2433.
4. Stucky, B. N., In *Handbook of Food Additives,* ed. T. E. Furia. Chemical Rubber Co., Cleveland, Ohio, 1968, p. 209.
5. Campbell, T. W., & Coppinger, G. M., *J. Am. Chem. Soc.,* **74** (1952) 1469.
6. Moore, R. F., & Walters, W. A., *J. Chem. Soc.* (1954) 243.
7. Anderson, R. H., & Huntley, T. E., *J. Am. Oil Chem. Soc.,* **40** (1963) 349.
8. Harano, Y., Hoshino, O., & Ukita, T., *J. Hyg. Chem. (Japan),* **13** (1967) 197.
9. Leventhal, B., Daun, H., & Gilbert, S. G., *J. Food Sci.,* **41** (1976) 467.
10. Walling, C. & Hodgdon, R. B. Jr, *J. Am. Chem. Soc.,* **80** (1958) 228.
11. Horswill, E. C., Howard, J. A. & Ingold, K. U., *Can. J. Chem.,* **44** (1966) 985.
12. Rosenwald, R. H. & Chenicek, J. A., *J. Am. Oil Chem. Soc.,* **28** (1951) 185.
13. Balts, J. & Volbert, F. *Fette Seifen Anstrichmittel,* **57** (1950) 660.
14. Hewgill, F. R., Kennedy, B. R. & Kilpin, D., *J. Chem. Soc.* (1965) 2904.
15. Kurechi, T. & Ogino, Y., *J. Food. Hyg. Soc. (Japan),* **6** (1965) 453.
16. Kurechi, T., *J. Hyg. Chem. (Japan),* **13** (1967) 191.
17. Kurechi, T. & Senda, H., *J. Hyg. Chem. (Japan),* **23** (1977) 267.
18. Mihara, M., Kondo, T. & Tanabe, H., *J. Food Hyg. Soc. (Japan),* **15** (1974) 276.
19. Mihara, M., Kondo, T. & Tanabe, H., *J. Food Hyg. Soc. (Japan),* **15** (1974) 270.
20. Ishizaki, M., Ueno, S., Kataoka, F., Oyamada, N., Murakami, R., Kubota, K., Katsumura, K. & Hosogai, Y., *J. Food Hyg. Soc. (Japan),* **16** (1975) 230.
21. Nilsson, J. L. G., *Acta Pharm. Suecica,* **6** (1969) 1.
22. Skinner, W. A. & Parkhurst, R. M., *Lipids,* **6** (1971) 240.
23. Boguth, W., In *Vitamins and Hormones,* ed. R. D. Harries. Academic Press, New York, 1970, p. 1.
24. Gallo-Tores, H. E., In *Vitamin E,* ed. L. J. Machlin *et al.* Marcel Dekker, New York, 1980, p. 193.
25. Nilsson, J. L. G., Daves, G. D., Jr & Folkers, K., *Acta Chem. Scand.,* **22** (1968) 207.
26. Csallany, A. S., Chiu, M. & Draper, H. H., *Lipids,* **5** (1970) 63.
27. Fujitani, T. & Ando, H., *J. Japan Oil Chem. Soc.,* **26** (1977) 337.
28. Fujitani, T. & Ando, H., *J. Japan Oil Chem. Soc.,* **26** (1977) 768.
29. Fujitani, T. & Ando, H., *J. Japan Oil Chem. Soc.,* **28** (1979) 896.
30. Fujitani, T. & Ando, H., *J. Japan Oil Chem. Soc.,* **30** (1981) 145.
31. Fujitani, T. & Ando, H., *J. Japan Oil Chem. Soc.,* **33** (1984) 356.

32. Lehmann, J. & Slover, H. T., *Lipids,* **11** (1976) 853.
33. McHale, D. & Green, J., *Chem. Ind.* (*London*) (1963) 982.
34. Shone, G., *Chem. Ind.* (*London*) (1963) 335.
35. Komoda, M. & Harada, I., *J. Am. Oil Chem. Soc.,* **46** (1969) 18.
36. Nilsson, J. L. G., Daves, G. D., Jr & Folkers, K., *Acta Chem. Scand.,* **22** (1968) 200.
37. Seino, H., Watanabe, S. & Abe, Y., *J. Japan Oil Chem. Soc.,* **20** (1971) 218.
38. Nakamura, T. & Kijima, S., *Chem. Pharm. Bull.,* **20** (1972) 1297.
39. Seino, H., Sugata, S., Watanabe, S. & Abe, Y., *J. Japan Oil Chem. Soc.,* **22** (1973) 145.
40. Porter, W. L., Levasseur, L. A. & Henick, A. S., *Lipids,* **6** (1971) 1.
41. Komoda, M. & Harada, I., *J. Am. Oil Chem. Soc.,* **47** (1970) 249.
42. Gardner, H. W., Eskins, K., Grams, G. W. & Inglett, G. E., *Lipids,* **7,** (1972) 324.
43. Budowski, P., Menezes, F. G. T. & Dollear, F. G., *J. Am. Oil Chem. Soc.,* **27** (1950) 377.
44. Beroza, M., *J. Am. Oil Chem. Soc.,* **32** (1955) 348.
45. Kurechi, T., Kikugawa, K. & Aoshima, S., *Chem. Pharm. Bull.,* **29,** (1981) 2351.
46. Kikugawa, K., Arai, M. & Kurechi, T., *J. Am. Oil Chem. Soc.,* **60** (1983) 1528.
47. Taki, T. & Kurechi, T., *Yakugaku Zasshi* (*Japan*), **97** (1977) 1174.
48. Kurechi, T., Aizawa, M. & Kunugi, A., *J. Am. Oil Chem. Soc.,* **60** (1983) 1878.
49. Kurechi, T. & Kunugi, A., *J. Am. Oil Chem. Soc.,* **60** (1983) 1882.
50. Kurechi, T. & Kunugi, A., *J. Am. Oil Chem. Soc.,* **60** (1983) 33.
51. Kraybill, H. R. & Dugan, J. R., *J. Agric. Food Chem.,* **2** (1954) 81.
52. Gearhart, W. H. & Stuckey, B. N., *J. Am. Oil Chem. Soc.,* **32** (1955) 386.
53. Ingold, K. U., *Chem. Rev.,* **61** (1961) 563.
54. Ingold, K. U., *Inst. Petrol.,* **47** (1961) 375.
55. Kurechi, T. & Kato, T., *J. Am. Oil Chem. Soc.,* **57** (1980) 220.
56. Kurechi, T. & Yamaguchi, T., *J. Am. Oil Chem. Soc.,* **57** (1980) 216.
57. Kurechi, T., *J. Hyg. Chem.* (*Japan*), **15** (1969) 301.
58. Lundberg, W. O., *Autoxidation and Antioxidants,* Vol. 1. Interscience, New York, 1961, p. 133.
59. Kraybill, H. R., Dugan, L. R., Beadle, B. W., Vibrans, F. C. *et al., J. Am. Oil Chem. Soc.,* **26** (1949) 449.
60. Cooney, P. M., Evans, C. D., Schwab, A. W. & Cowan, J. C., *J. Am. Oil Chem. Soc.,* **35** (1958) 152.
61. Evans, C. D., Frankel, E. N. & Cooney, P. M., *J. Am. Oil Chem. Soc.,* **36** (1959) 73.
62. Masuyama, S., *J. Japan Oil Chem. Soc.,* **14** (1965) 692.
63. Kanematsu, H., Aoyama, M., Maruyama, T., Niiya, I., Tsukamoto, M., Tokairin, S. & Matsumoto, T., *J. Japan Oil Chem. Soc.,* **32** (1983) 695.
64. Kanematsu, H., Aoyama, M., Maruyama, T., Niiya, I., Tsukamoto, M., Tokairin, S. & Matsumoto, T., *J. Japan Oil Chem. Soc.,* **32** (1983) 731.

65. Aoyama, M., Maruyama, T., Kanematsu, H., Niiya, I., Tsukamoto, M., Tokairin, S. & Matsumoto, T., *J. Japan Oil Chem. Soc.*, **34** (1985) 558.
66. Tappel, A. L., Brown, W. D., Zalkin, H. & Maier, V. P., *J. Am. Oil Chem. Soc.*, **38** (1961) 5.
67. Pongracz, G., *Int. J. Vit. Nutr. Res.*, **43** (1973) 517.
68. Cort, W. M., *J. Am. Oil Chem. Soc.*, **51** (1974) 321.
69. Cort, W. M., *Adv. Chem. Ser.*, **200** (1982) 533.
70. Reinton, R. & Rogstad, A., *J. Food Sci.*, **46** (1981) 970.
71. Barclay, L. R. C., Locke, S. J. & MacNeil, J. M., *Can. J. Chem.*, **61** (1983) 1288.
72. Kanematsu, H., Aoyama, M., Maruyama, T., Niiya, I., Tsukamoto, M., Tokairin, S. & Matsumoto, T., *J. Japan Oil Chem. Soc.*, **33** (1984) 361.
73. Aoyama, M., Maruyama, T., Kanematsu, H., Niiya, I., Tsukamoto, M., Tokairin, S. & Matsumoto, T., *J. Japan Oil Chem. Soc.*, **34** (1985) 48.
74. Aoyama, M., Maruyama, T., Kanematsu, H., Niiya, I., Tsukamoto, M., Tokairin, S. & Matsumoto, T., *J. Japan Oil Chem. Soc.*, **34** (1985) 123.
75. Packer, J. E., Slater, T. F. & Willson, R. L., *Nature (London)*, **278** (1979) 738.
76. Niki, E., Tsuchiya, J., Tanimura, R. & Kamiya, Y., *Chem. Lett.* (1982) 789.
77. Niki, E., Saito, T., Kawakami, A. & Kamiya, Y., *J. Biol. Chem.*, **259** (1984) 4177.
78. Smouse, T. H., *J. Am. Oil Chem. Soc.*, **56** (1979) 747A.
79. Linow, F. & Mieth, G., *Nahrung*, **20** (1976) 19.
80. Hudson, B. J. F. & Mahgoub, S. E. O., *J. Sci. Food Agric.*, **32** (1981) 208.
81. Hudson, B. J. F. & Lewis, J. I., *Food Chem.*, **10** (1983) 111.
82. Hudson, B. J. F. & Ghavami, M., *Lebensm. Wiss. u. Technol.*, **17** (1984) 191.
83. Dziedzic, S. Z. & Hudson, B. J. F., *J. Am. Oil Chem. Soc.*, **61** (1984) 1042.
84. Hildebrand, D. H., Terao, J. & Kito, M., *J. Am. Oil Chem. Soc.*, **61** (1984) 552.
85. Yuki, E., Morimoto, K., Ishikawa, Y. & Noguchi, H., *J. Japan Oil Chem. Soc.*, **27** (1978) 425.
86. Fujitani, T. & Ando, H., *J. Japan Oil Chem. Soc.*, **30** (1981) 140.
87. Fujitani, T. & Ando, H., *J. Japan Oil Chem. Soc.*, **31** (1982) 427.
88. Fujitani, T. & Ando, H., *J. Japan Oil Chem. Soc.*, **33** (1984) 277.
89. Fujitani, T. & Ando, H., *J. Japan Oil Chem. Soc.*, **34** (1985) 271.
90. Dziedzic, S. Z., Robinson, J. L. & Hudson, B. J. F., *J. Agric. Food Chem.*, **34** (1986) 1027.
91. Yuki, E., Ishikawa, Y., Yamaoka, I. & Yoshiwa, T., *J. Food Sci. Technol. (Japan)*, **20** (1973) 411.
92. Ishikawa, Y. & Yuki, E., *Agric. Biol. Chem.*, **39** (1975) 851.
93. Ishikawa, Y., Yuki, E., Kato, H. & Fujimaki, M., *Agric. Biol. Chem.*, **42** (1978) 703.
94. Ishikawa, Y., Yuki, E., Kato, H. & Fujimaki, M., *Agric. Biol. Chem.*, **42** (1978) 711.

95. Olcott, H. S., *J. Food Sci. Technol. (Japan)*, **11** (1964) 544.
96. Harris, L. A. & Olcott, H. S., *J. Am. Oil Chem. Soc.*, **43** (1966) 11.
97. Ishikawa, Y., *J. Japan Oil Chem. Soc.*, **29** (1980) 844.
98. Ishikawa, Y., Yuki, E., Kato, H. & Fujimaki, M., *J. Japan Oil Chem. Soc.*, **26** (1977) 765.
99. Ishikawa, Y., *J. Am. Oil Chem. Soc.*, **59** (1982) 505.
100. Olcott, H. S. & Kuda, E. J., *Nature*, **183** (1959) 1812.
101. Marcuse, R., *J. Am. Oil Chem. Soc.*, **39** (1962) 97.
102. Karel, M., Tannenbaum, S. R., Wallace, D. H. & Maloney, H., *J. Food Sci.*, **31** (1966) 892.
103. Watanabe, Y. & Ayano, A., *J. Japan Soc. Food Nutr.*, **25** (1972) 219.
104. Yuki, E., Ishikawa, Y. & Yoshiwa, T., *J. Japan Oil Chem. Soc.*, **23** (1974) 497.
105. Yuki, E., Ishikawa, Y. & Yoshiwa, T., *J. Japan Oil Chem. Soc.*, **23** (1974) 714.
106. Kawashima, K., Itoh, H. & Chibata, I., *J. Agric. Food Chem.*, **26** (1977) 202.
107. Cort, W. M., Scott, J. W. & Harley, J. H., *Food Technol.*, **29** (1975) 46.
108. Cort, W. M., Scott, J. W., Araujo, M., Mergens, W. J., Cannalonga, M. A., Osadca, M., Harlen, H., Parrish, D. R. & Pool, W. R., *J. Am. Oil Chem. Soc.*, **52** (1975) 174.
109. Taylor, M. J., Richardson, T. & Jasensky, R. D., *J. Am. Oil Chem. Soc.*, **58** (1981) 622.
110. Bishov, S. J., Masuoka, Y. & Henick, A. S., *Food Technol.*, **21** (1967) 148A.
111. Bishov, S. J. & Henick, A. S., *J. Food Sci.*, **37** (1972) 873.
112. Bishov, S. J. & Henick, A. S., *J. Food Sci.*, **40** (1975) 345.
113. Yamaguchi, N., Yokoo, Y. & Fujimaki, M., *J. Food Sci. Technol. (Japan)*, **22** (1975) 425.
114. Yamaguchi, N., Yokoo, Y. & Fujimaki, M., *J. Food Sci. Technol. (Japan)*, **22** (1975) 431.
115. Kawashima, K., Ito, H. & Chibata, I., *Agric. Biol. Chem.*, **43** (1979) 827.
116. Yamaguchi, N. & Okada, Y., *J. Food Sci. Technol. (Japan)*, **15** (1968) 187.
117. Yamaguchi, N. & Fujimaki, H., *J. Food Sci. Technol. (Japan)*, **21** (1974) 13.
118. Kurechi, T., Kikugawa, K. & Kato, T., *Chem. Pharm. Bull.*, **28** (1980) 2089.
119. Kurechi, T. & Kato, T., *Chem. Pharm. Bull.*, **29** (1981) 3012.
120. Kurechi, T. & Kato, T., *Chem. Pharm. Bull.*, **30** (1982) 2964.
121. Kurechi, T., Kikugawa, K., Kato, T. & Numasato, T., *Chem. Pharm. Bull.*, **28** (1980) 2228.
122. Kurechi, T. & Kato, T., *Chem. Pharm. Bull.*, **31** (1983) 1772.
123. Boguth, W., Repges, R. & Zell, R., *Int. Z. Vit. Forshung*, **40** (1970) 323.
124. Ivanova, R. A., Pimenova, N. S., Kozlov, E. I. & Tsepalov, V. F., *Kinet. Katal.*, **20** (1979) 1423.

Chapter 4

NATURAL ANTIOXIDANTS EXPLOITED COMMERCIALLY

PETER SCHULER
Department VM/H, F. Hoffmann-La Roche & Co., 4002-Basle, Switzerland

GENERAL INTRODUCTION

The object of this chapter is to provide information on natural antioxidants commercially exploited. It outlines their manufacture, some physical and chemical properties, methods of application and their efficacy.

The chapter is organized into three parts, namely a general introduction, a second part discussing individually the various natural antioxidants available, and a third part which draws attention to some practical aspects of protecting food with natural antioxidants.

Antioxidants and Synergists

According to a very general definition antioxidants are 'substances capable of delaying, retarding or preventing oxidation processes'. Synergists are 'substances enhancing the activity of antioxidants, without being antioxidants themselves'.

The problem of oxidation is one aspect of food preservation, especially when the oxidised products develop an unpleasant odour or taste. The flavour threshold of oxidation products may be much below 1 ppm (Table 1) and economical losses due to such traces of undesired substances may be considerable.[1,2]

Lipids and lipid-soluble substances which may be susceptible to oxidation are present in almost all foods. They include edible fats and oils, triglycerides as well as mono- and diglycerides (emulsifiers), sterols, fat soluble vitamins, phospholipids, flavours and aromas, carotenoids (colorants), and others.

In aqueous food systems oxidation may also be a severe problem

TABLE 1
FLAVOUR THRESHOLD OF SOME OXIDATION PRODUCTS OF SOYA BEAN OIL

Substance	*Type of flavour*	*Threshold (ppm)*
Oct-1-en-3-one	Metallic	0·001
2-Pentylfuran	Liquorice	2
Oct-1-en-3-ol	Mushrooms	0·007
Pent-1-en-3-one	Metallic	0·0001

Source: Berger.[1]

and the protection against oxidative deterioration of a range of constituents has to be ensured.

General Rules for the Use of Antioxidants

The use of antioxidants can never exempt food manufacturers from the scrupulous observance of good manufacturing practice. Antioxidants can inhibit or delay oxidation, but they cannot improve the quality of an already oxidized product. Especially in lipid systems the degradation reactions are irreversible.

Oxidation needs oxygen, and the reactions have to be initiated by energy input. Consequently it is important to:

—Minimize exogenous oxidation activators: store cool and avoid unnecessary light, especially UV radiation.
—Eliminate endogenous oxidation activators: avoid or reduce metal traces (Cu, Fe), phytin pigments (chlorophylls, hemins) or peroxides. Use good quality raw material.
—Eliminate oxygen as far as possible and keep oxygen uptake low during processing and storage.
—Utilize appropriate containers and packaging materials.

It also should not be forgotten that the usual triplet oxygen in the substrate can be activated to the 1000-fold more active singlet oxygen

$$^{3}O_2 \xrightarrow[\text{Photosensitizers}]{\text{Energy (155 kJ/mol)}} {}^{1}O_2$$

$$RH + {}^{1}O_2 \longrightarrow ROOH$$

FIG. 1. Activation of oxygen. $^{3}O_2$, Triplet oxygen; $^{1}O_2$, singlet oxygen; RH, substrate; ROOH, peroxide of the substrate.

in the presence of photosensitizers (e.g. phytin pigments, FD&C red No. 3). Singlet oxygen is able to produce peroxides directly from suitable organic compounds[3] according to Fig. 1.

Features of Food Antioxidants

An antioxidant for food use should meet several essential requirements:

—effectiveness at low concentrations;
—compatibility with the substrate;
—absence of sensory influence on the food product (off-odour, off-taste, off-colour);
—absence of toxicity to consumers.

Food antioxidants also have to fulfill additional requirements.

Legislation in most countries closely regulates the use of antioxidants by means of 'positive lists'. Such lists are usually based on the one hand on sanitary considerations and toxicological data, and on the other hand on the needs of the food industry, the intention being to keep the number of admissible food additives as low as possible.

In many countries the list of ingredients has to be declared on pre-packed foods. Consequently, the manufacturers prefer antioxidants which are more easily accepted by critical consumers. Antioxidants should be easy and safe to handle, and finally, they should be cost effective.

ANTIOXIDANTS FOR FOOD USE

The list of antioxidants used in industry is of impressive length. In an overview Kurze[4] ranges more than 150 compounds according to their chemical structure and their mode of action (radical scavengers, oxygen scavengers, peroxide decomposers, UV-absorbers, etc.).

The list of antioxidants permitted for food use is much shorter (Table 2). This list is also subjected to occasional modifications. New toxicological findings or alterations in legislation may especially provoke the elimination of substances. Today the approval for a new antioxidant demands extensive toxicological studies (including mutagenicity, teratogenicity and carcinogenicity) and, in most countries, a proof for its superiority over existing ones.

The most important natural antioxidants commercially exploited are

TABLE 2
MOST USUAL FOOD ANTIOXIDANTS COMMERCIALLY EXPLOITED—DISREGARDING LOCAL LEGAL RESTRICTIONS

Ascorbic acid and derivatives:	Calcium ascorbate
	Sodium ascorbate
	Ascorbyl palmitate
	Ascorbyl stearate
Butylated hydroxyanisole (BHA)	
Butylated hydroxytoluene (BHT)	
Tert. Butylhydroquinone (TBHQ)	
Erythorbic acid	Sodium erythorbate
Derivatives of gallic acid:	Propyl gallate
	Octyl gallate
	Dodecyl gallate
Gum Guaiac	
Tocopherols	
Sulphur dioxide and various sulphites	

the *tocopherols* and *ascorbic acid.* Furthermore vegetable extracts have achieved a certain importance (especially spice extracts containing carnosic acid and rosmaric acid). Various soya and oat products are also included in food formulations to oppose the oxidation of the product.

Tocopherols

Eight different substances having vitamin E activity occur in nature. They belong to two families with the generic names *tocols* and *tocotrienols*. The members of each family are designated with α, β, γ or δ, depending on number and position of the methyl groups attached to a chromane ring. The side chain is saturated in the tocols and unsaturated in the tocotrienols. Commonly the tocols are also called *tocopherols*. Structures are shown in Table 3.

The structures of the tocopherols (Table 3) indicate three centres of asymmetry at C-2, C-4′ and C-8′ respectively. The tocotrienols possess one centre of asymmetry at C-2 in addition to the sites of geometrical isomerism at C-3′ and at C-7′. Thus, of each tocopherol and tocotrienol four stereoisomers exist.

With regard to vitamin E activity α-tocopherol is the most potent form. Vitamin E preparation for the food and pharmaceutical industries as well as for feeds are based on this compound. The determination of the biological activity is complicated by the fact that

TABLE 3
STRUCTURE AND PROPERTIES OF VITAMIN E AND RELATED COMPOUNDS

Compound	*Formula mol wt.*	*Structure*
Tocols		
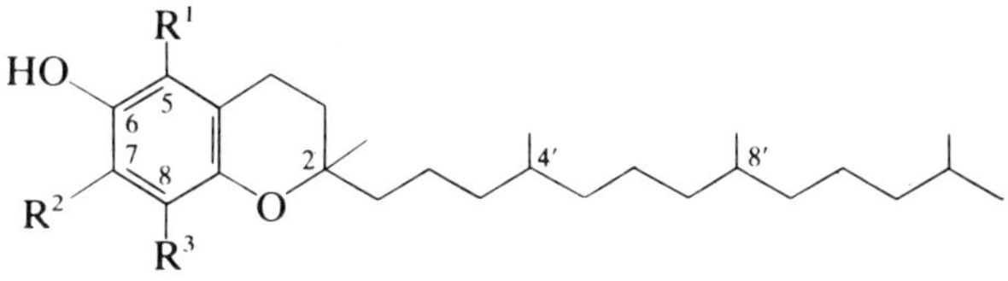		
Tocol	$C_{26}H_{44}O_2$ 388·64	R^1: H R^2: H R^3: H
8-Methyltocol (δ-Tocopherol)	$C_{27}H_{46}O_2$ 402·67	R^1: H R^2: H R^3: CH_3
5,8-Dimethyltocol (β-Tocopherol)	$C_{28}H_{48}O_2$ 416·69	R^1: CH_3 R^2: H R^3: CH_3
7,8-Dimethyltocol (γ-Tocopherol)	$C_{28}H_{48}O_2$ 416·69	R^1: H R^2: CH_3 R^3: CH_3
5,7,8-Trimethyltocol (α-Tocopherol)	$C_{29}H_{50}O_2$ 430·72	R^1: CH_3 R^2: CH_3 R^3: CH_3
Tocotrienols		
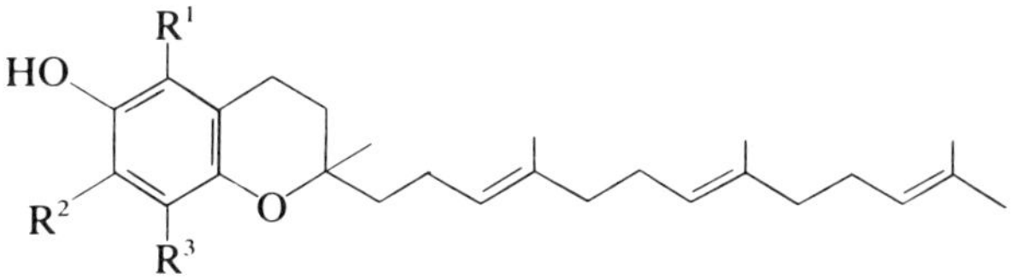		
8-Methyltocotrienol (δ-Tocotrienol)	$C_{27}H_{40}O_2$ 396·62	R^1: H R^2: H R^3: CH_3
5,8-Dimethyltocotrienol (β-Tocotrienol)	$C_{28}H_{42}O_2$ 410·65	R^1: CH_3 R^2: H R^3: CH_3
7,8-Dimethyltocotrienol (γ-Tocotrienol)	$C_{28}H_{42}O_2$ 410·65	R^1: H R^2: CH_3 R^3: CH_3
5,7,8-Trimethyltocotrienol (α-Tocotrienol)	$C_{29}H_{44}O_2$ 424·67	R^1: CH_3 R^2: CH_3 R^3: CH_3

TABLE 4
BIOLOGICAL ACTIVITY OF VITAMIN E COMPOUNDS

α-tocopherol 100%	α-tocotrienol 15–30%
β-tocopherol 15–40%	β-tocotrienol 1–5%
γ-tocopherol 1–20%	γ-tocotrienol 1%
δ-tocopherol 1%	δ-tocotrienol 1%
tocol is inactive	

different types of bioassay result in varying potencies. Average values are given in Table 4.

In addition, the configurations of the compounds influence their biological activities. So the naturally occurring 2R, 4′R, 8′R-α-tocopherol has shown to be approximately 35% more active than the synthetic all-rac-α-tocopherol. These facts have to be considered in the calculation of the vitamin E content of foods.[5]

Natural Occurrence of Tocopherol and Tocotrienols in Foods

In foods of animal origin in practice only α-tocopherol occurs in substantial amounts, depending on the feeding of the animals (Table 5). Vegetable foods contain in their lipid important quantities of the different tocopherols and tocotrienols. Cereals and cereal products, oil-seeds, nuts and vegetables such as peas, beans and carrots are rich sources. The contents may depend very much on the variety and growing conditions of the plant and on the processing and storage of the oil. Under good manufacturing practice the total tocopherol content of oils is reduced by no more than 30–40% during refining (deacidifying, bleaching, and deodorizing). Table 6 shows the range of contents in vegetable oils.

TABLE 5
APPROXIMATE α-TOCOPHEROL CONTENT OF SELECTED FOODS OF ANIMAL ORIGIN (mg/kg)

Beef	6	Cod	2	Milk spring	0·2
Chicken	4	Halibut	9	Milk autumn	1·1
Pork	5	Shrimp	7	Butter	10–33
Lard	12			Eggs	5–11

TABLE 6
APPROXIMATE CONTENTS OF TOCOPHEROL AND TOCOTRIENOL FOUND IN VEGETABLE OILS (mg/kg)

	Tocopherols				*Tocotrienols*			
	α	β	γ	δ	α	β	γ	δ
Coconut	5–10	—	5	5	5	Trace	1–20	—
Cottonseed	40–560	—	270–410	0	—	—	—	—
Maize, grain	60–260	0	400–900	1–50	—	0	0–240	0
Maize, germ	300–430	1–20	450–790	5–60	—	—	—	—
Olive	1–240	0	0	0	—	—	—	—
Palm	180–260	Trace	320	70	120–150	20–40	260–300	70
Peanut	80–330	—	130–590	10–20	—	—	—	—
Rapeseed	180–280	—	380–590	10–20	—	—	—	—
Safflower	340–450	—	70–190	230–240	—	—	—	—
Soyabean	30–120	0–20	250–930	50–450	0	0	0	—
Sunflower	350–700	20–40	10–50	1–10	—	—	—	—
Walnut	560	—	590	450	—	—	—	—
Wheat germ	560–1 200	660–810	260	270	20–90	80–190	—	—

Properties of Tocopherols

For antioxidant purposes α-, γ- and δ-tocopherols are of commercial interest.

Formula:	α: $C_{29}H_{50}O_2$		
	γ: $C_{28}H_{48}O_2$		
	δ: $C_{27}H_{46}O_2$		
Molecular weight:	α: 430·72		
	γ: 416·69		
	δ: 402·67		
Description:	Tocopherols are clear, viscous oily substances of pale yellow colour, nearly odourless. They oxidize and darken in air and on exposure to light.		
Solubility:	Insoluble in water. Miscible at any ratio with vegetable oils, ethanol, ether, chloroform, acetone.		
UV absorption:	In ethanol $E^{1\%}_{1\,cm}$		
	dl-α:	292 nm:	72–76 (maximum)
		255 nm:	6·0–8·0 (minimum)
	dl-γ:	298 nm:	91–97 (maximum)
		257–258 nm:	5·0–8·0 (minimum)
	dl-δ:	297–298 nm:	89–95 (maximum)
		257 nm:	3·0–6·0 (minimum)
Refraction:	N^{20}_{D}		
	dl-α: 1·503–1·507		
	dl-γ: 1·503–1·507		
	dl-δ: 1·500–1·504		
Viscosity:	*dl*-α: 5000–6000 cP (20°)		

Antioxidant Activity of Tocopherols

The antioxidant activity of the tocopherols is mainly based on the 'tocopherol-tocopherylquinone redox system' as shown in Fig. 2.

Tocopherols (AH_2) are so called radical scavengers, i.e. they quench free radicals ($R^{\cdot}$) according to the general scheme.

$$R^{\cdot} + AH_2 \rightarrow RH + AH^{\cdot}$$

$$2AH^{\cdot} \rightarrow A + AH_2$$

The result of the action is a regenerated molecule (RH). After the further reaction of two tocopheryl semiquinone radical molecules

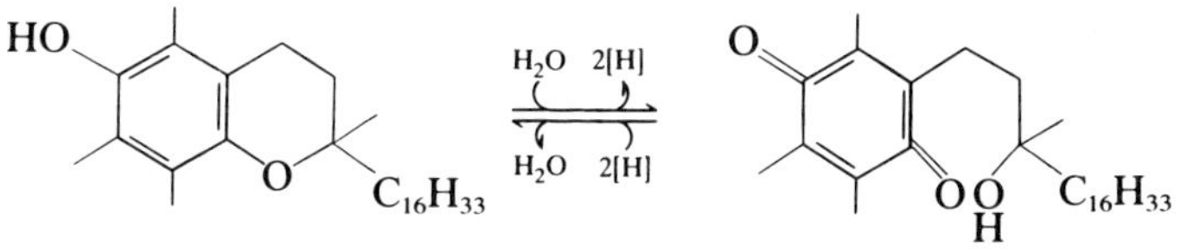

α-tocopherol

α-tocopherylquinone (stable)

FIG. 2. α-Tocopherol—α-tocopherylquinone redox system.

(AH˙) (Fig. 3(a)) one molecule of tocopherylquinone(A), and a regenerated molecule of tocopherol is formed. The elimination of peroxy radicals may happen by the same reaction.

The mechanism of oxidation of α-tocopherol with linoleic acid hydroperoxides has been studied in detail:[6] the sequence includes various steps. The first one is, after the release of one hydrogen atom, the formation of an α-tocopheryl free radical (Fig. 3(a)) followed by the release of a second hydrogen atom to give methyl-tocopherylquinone (Fig. 3(b)). The latter compound is unstable and gives rise finally to α-tocopherylquinone (Fig. 2) as a main product.

However, the oxidation pathways of tocopherols depend on the conditions. The reaction between two radicals may also lead to α-tocopherol dimers, again possessing antioxidant properties. Also, disproportionation reactions based on radicals and yielding methyl-tocopherylquinone and tocopherol have been suggested.

Further antioxidant reactions of tocopherols have been described.

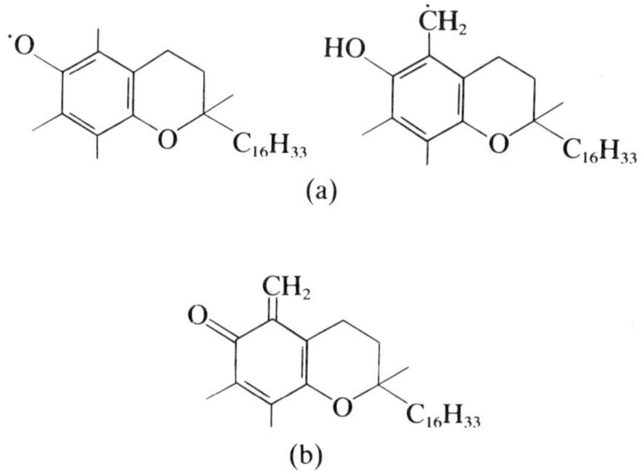

(a)

(b)

FIG. 3. (a) α-Tocopheryl semiquinone radical. (b) Methyl-tocopherylquinone.

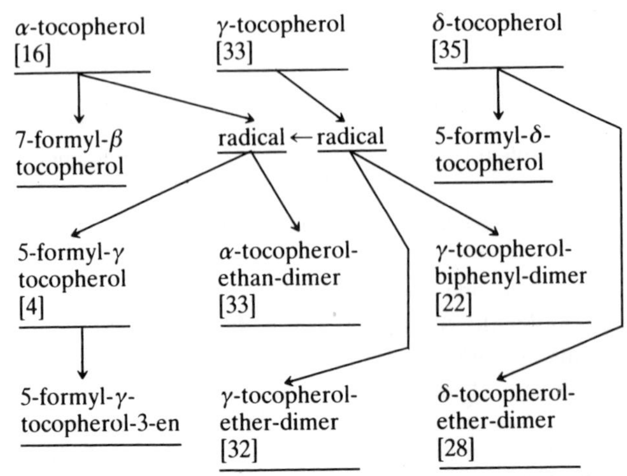

FIG. 4. Antioxidant activity of some tocopherols and their decomposition products (number in brackets: length of induction period in days (control without antioxidants added: 2 days)). (Source: Ishikawa & Yuki.[7])

Ishikawa & Yuki[7] for instance oxidized α-, γ- and δ-tocopherol with trimethylamine oxide. Some of the oxidation products were isolated and then tested for their antioxidant activity (Schaal oven test with tocopherol free lard, 400 ppm of antioxidants, 60°C, evaluation of peroxide values). Figure 4 shows some of the identified decomposition products and their antioxidant efficacies.

Looking at the antioxidant reaction, it is obvious that α-tocopheryl acetate (the commercially available stable form of vitamin E) is not an antioxidant. The active hydroxyl group is protected. However, under certain conditions, e.g. acidic aqueous systems, a slow hydrolysis of the tocopheryl acetate can be observed. The released tocopherol then acts as an antioxidant.

The application of tocopheryl acetate to avoid rancidity in cured raw sausages has been reported by Schriener.[8] The pH value of such products is about 4·8–5·0.

A second example (Table 7) show the antioxidant effect of α-tocopherol acetate on vitamin A-palmitate in aqueous Tween 80 dispersions at low pH.[9]

It seems to be evident that the speed of tocopherol release corresponds to the amount of tocopherol required as antioxidant: in practice this may be achieved in various emulsions, such as sauces, soft drinks and some dairy products.

TABLE 7

INFLUENCE OF pH VALUE ON THE ANTIOXIDANT EFFECT OF α-TOCOPHERYL ACETATE (substrate: vitamin A palmitate in Tween 80 dispersions)

	Retention after 15 days (%)	
pH of dispersion	*Control (no antioxidant)*	*With antioxidant 0·3% dl-α-tocopheryl acetate*
2·5	2·5	68·5
4·7	19·9	78·7
6·5	28·3	65·8
8·3	32·2	32·0

Source: Fatterpekar & Ramasarma.[9]

The antioxidant activity of the various tocopherol stereoisomers is equal, i.e. pure all-rac-α-tocopherol has the same effect as pure natural 2R, 4′R, 8′R-α-tocopherol.

Commercial Products

For antioxidant applications mainly oily product forms are marketed: pure all-rac-α-tocopherol, mixed tocopherols having various contents of α-, γ- and/or δ-tocopherols (usually diluted in a vegetable oil) and synergistic mixtures composed of tocopherols, ascorbyl palmitate or other antioxidants, synergists such as lecithin, citric acid and carriers. The latter products usually are pastes, or, rarely, also viscous liquids. The pastes have to be dissolved at a temperature of 50–60°C in the fat phase of the food product to be stabilized.

Biosynthesis of Tocopherols

The biosynthesis of tocopherols follows the usual path of terpenoid synthesis as far as homogentisic acid.[10] From that point, two routes are possible (Fig. 5). In the ‘tocopherol route’, typically in leaves and in most tissues of algae, phytyl pyrophosphate condenses with homogentisic acid to give δ-tocopherol. After methylation with S-adenosyl methionine, β- and γ-tocopherols, and finally α-tocopherol are formed.

In the ‘tocotrienol route’ homogentisic acid condenses with geranyl geranyl pyrophosphate to give δ-tocotrienol. Again, after methylation with S-adenosyl methionine, β-, γ- and finally α-tocotrienol are formed. A last possible step in this route is the hydrogenation of α-tocotrienol to give α-tocopherol.

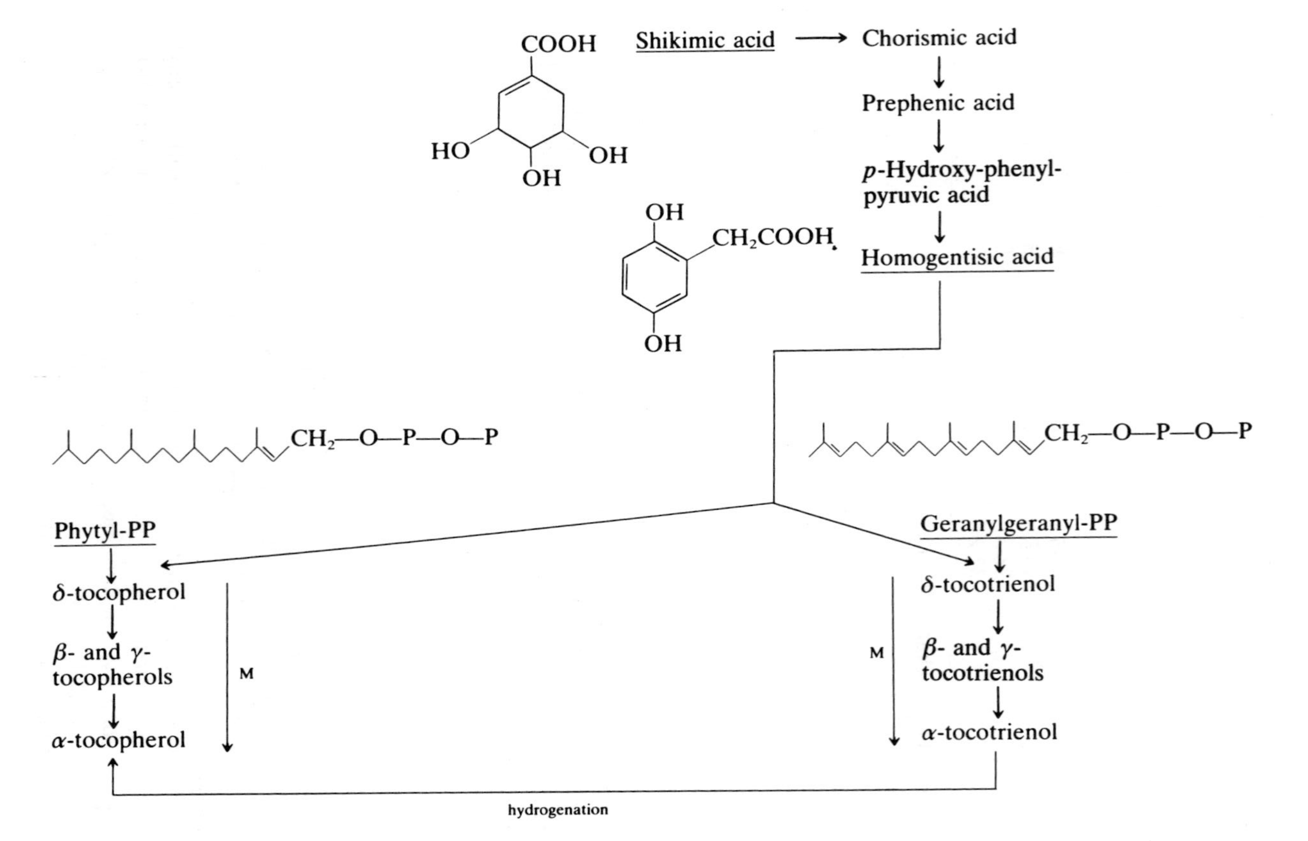

FIG. 5. The biosynthesis of tocopherols.[10] M = methylation with S-adenosyl methionine.

Commercial Production of Tocopherols

Tocopherols are either obtained by extraction from natural sources or by chemical synthesis.

Extraction from natural sources. The most important raw materials are deodorizer sludges obtained in the deodorization of vegetable oils and fats. Besides the various tocopherols and tocotrienols (Table 6, Table 3) such distillates may also contain sterols, esters of sterols and free fatty acids as well as triglycerides. The choice of the processing parameters (temperature, vacuum, quantity of injected steam) is critical for the yield of tocopherols.[11] The separation of the tocopherols from the other extracted compounds (e.g. soya oil gives approx. 20% sterols, 8% tocopherols, 20% free fatty acids and triglycerides) is possible by several methods:[12] by esterification with a lower alcohol, washing and vacuum distillation,[13] by saponification,[14] or by fractional liquid–liquid extraction.[15] Further purification steps may be molecular distillation, extraction, crystallization or a combination of these processes.

The tocopherol concentrates intended for use as antioxidants are mixtures with relatively high contents of γ- and δ-tocopherols, but α-tocopherol is also present. The total tocopherol content usually lies between 30 and 80%, the rest being triglycerides.

The greater part of extracted tocopherol concentrates is, however, not marketed as such but as vitamin E: after chemical methylation (β-, γ- and δ-tocopherol are converted into α-tocopherol) and subsequent acetylation the relatively stable D-α-tocopheryl acetate is obtained.

Chemical synthesis. The various possibilities for total chemical synthesis, as well as partial synthesis have been described extensively in the literature. A good introduction to this field, for example, has been provided by Kasparek.[16] Commonly, the total synthesis of α-tocopherol is based on the condensation of 2,3,5-trimethylhydroquinone with phytol, isophytol or phytylhalogenides (Fig. 6). The crude tocopherol is purified by vacuum distillation. Today the use of synthetic isophytol is preferred. Therefore, the resulting α-tocopherol is a racemic mixture of all eight possible stereoisomers (= all-rac-α-tocopherol, or dl-α-tocopherol).

Earlier, d-phytol or isophytol from natural sources were used. Consequently, the products were mixtures of two isomers, namely 2R,4′R,8′R and 2S,4′R,8′R-α-tocopherol.

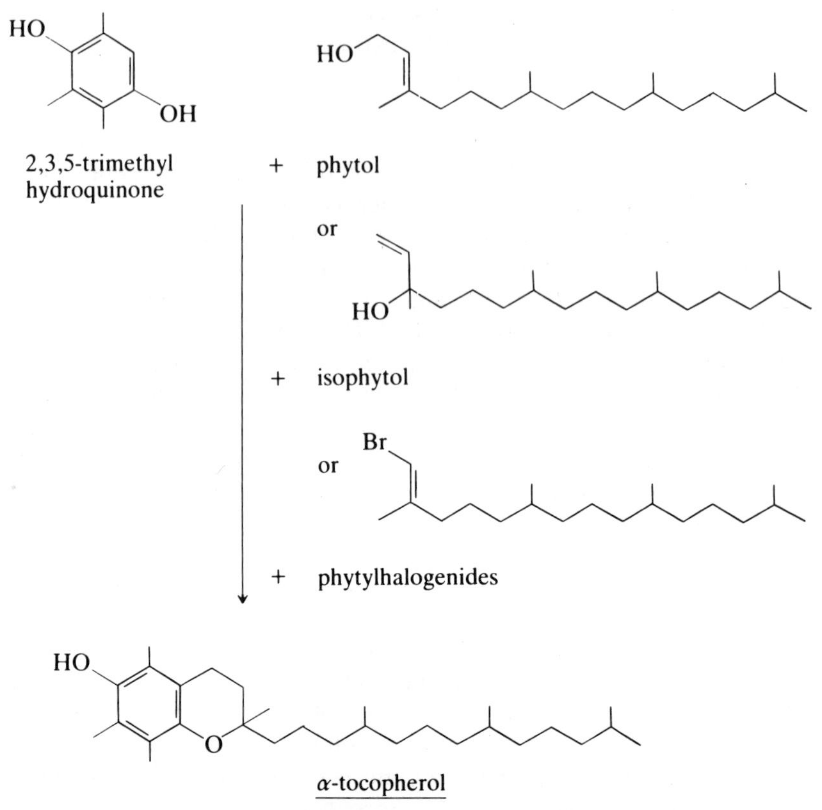

FIG. 6. Total synthesis of α-tocophenol.

The starting materials for the final condensation step, 2,3,5-trimethylhydroquinone and phytol or isophytol, are obtainable by various methods. Synthetic isophytol is obtained from acetone by a stepwise synthesis whereby, at each step, two or three carbon atoms are added.[10,16]

The total synthesis of the other tocopherols is possible by the same principle. So γ-tocopherol can be prepared by condensation of 2,3-dimethylhydroquinone with isophytol (Fig. 7).

Various other possible total syntheses for tocopherols and tocotrienols, the partial syntheses and also the methods of obtaining 'natural' tocopherols (2R, 4′R, 8′R-tocopherol) by chemical syntheses or biosynthesis are discussed in the literature.[12,16]

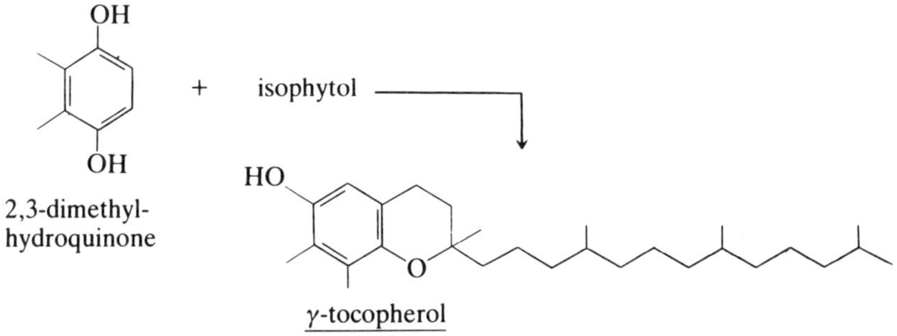

OH

OH

2,3-dimethyl-hydroquinone + isophytol

HO

O

γ-tocopherol

FIG. 7. Synthesis of γ-tocopherol.

Ascorbic Acid

L-Ascorbic acid, or 'vitamin C' is a substance very widespread in nature. Important quantities are also manufactured by chemical synthesis. It can be assumed that half of the totally manufactured quantity is used in the food industry, of which probably only a minor part serves for food enrichment. Ascorbic acid is constantly gaining importance as a versatile natural food additive: it helps to improve the quality and increases the shelf-life of many food products.

Figure 8 shows the structure of L-ascorbic acid and its three other possible stereochemical isomers. Of the three isomers only D-iso-ascorbic acid has found some importance in food manufacturing, but this substance is not completely interchangeable with L-ascorbic acid. To avoid any confusion and to make a clear distinction from L-ascorbic acid (vitamin C) the designation erythorbic acid was introduced, and sodium erythorbate for sodium iso-ascorbate.

The two isomers D-ascorbic acid and L-iso-ascorbic acid have not found commercial interest up to now, and do not show any vitamin C activity.[37]

As well as free ascorbic acid, sodium ascorbate, calcium ascorbate, ascorbyl palmitate and ascorbyl stearate are used as antioxidants in food manufacturing.

Properties of Ascorbic Acid

Table 8 summarizes some physical and chemical properties of ascorbic acid and its derivatives.

Ascorbic Acid versus Erythorbic Acid

In many technological applications in the food field ascorbic acid and erythorbic acid have frequently been used interchangeably on the

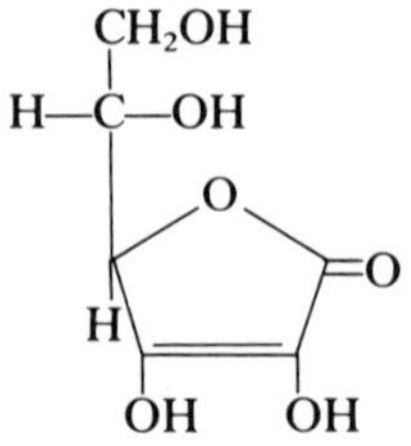

L-Ascorbic acid
(Vitamin C)

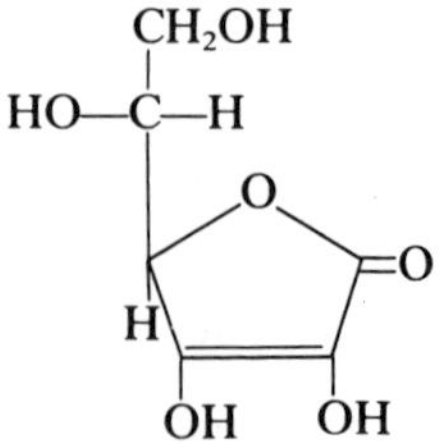

D-Isoascorbic acid
(Erythorbic acid)

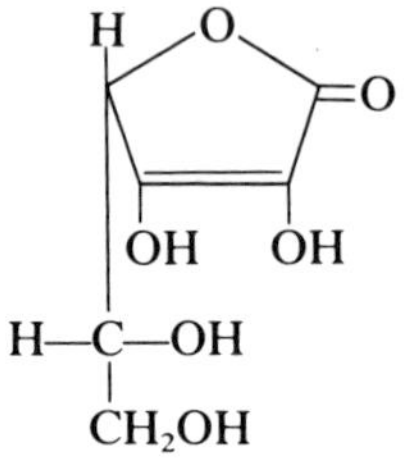

D-Ascorbic acid

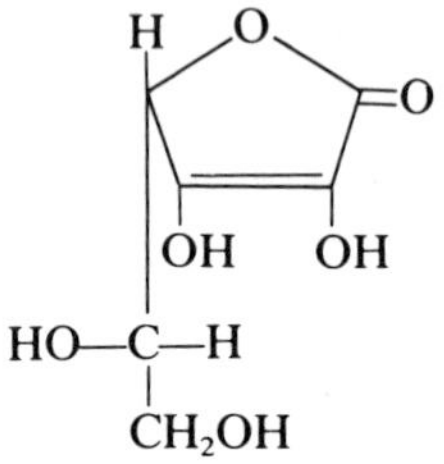

L-Isoascorbic acid

FIG. 8. Stereochemical configuration of ascorbic acid and erythorbic acid.

assumption that the compounds have identical properties. Apart from the stereochemical configuration (Fig. 8) further differences between the two acids should be noted:

1. Erythorbic acid does not occur naturally in foodstuffs.
2. Its biological activity (antiscorbutic) is only 5% of that of ascorbic acid.
3. Due to its close chemical relationship to ascorbic acid it interferes with a number of biological functions of ascorbic acid, e.g. by decreasing the tissue uptake of ascorbic acid and thus lowering the vitamin C stores in the tissue (liver, spleen, adrenal glands, kidney, etc).[24–27]
4. In many countries it is not permitted as a food additive (especially in Europe).
5. Various physical properties are different (Table 9).
6. Differences in chemical behaviour have been observed: in model solutions erythorbic acid was more rapidly oxidized than ascorbic acid.[17] Its faster oxidation has been confirmed by determinations of the redox potential.[18] In the presence of a copper

TABLE 8

PHYSICAL AND CHEMICAL PROPERTIES OF ASCORBIC ACID AND ITS DERIVATIVES

	AA	*Na ascorbate*	*Ca ascorbate*	*A palmitate*	*A stearate*
Formula	$C_6H_8O_6$	$C_6H_7O_6Na$	$(C_6H_7O_6)_2$ Ca $2H_2O$	$C_{22}H_{38}O_7$	$C_{24}H_{42}O_7$
Mol. wt	176·13	198·11	426·36	414·55	442·60
MP (°C)	190–192 (+)	Approx. 200 (+)	Approx. 200 (+)	~113°C (+)[a] ~180°C (+)[b]	~117° (+)[a] ~180° (+)[b]
Appearance	White cryst. powder	White to slightly yellowish cryst. powder	White to slightly greyish cryst. powder	White powder	White powder
Taste	Acidic	Soapy	Bitter	Soapy	Fatty
Solubility (approx.), 25°C					
g/100 ml H_2O	33	89	55	0·0002	0·0002
g/100 ml ethanol (100%)	2	Insoluble	Insoluble	12·5	4
g/100 ml veget. oil	Insoluble	Insoluble	Insoluble	0·01–0·12	0·01–0·10
g/100 ml glycerol	1			0·01	0·01
Ether	Insoluble	Insoluble	Insoluble	0·7	0·6
Chloroform RT	Insoluble	Insoluble	Insoluble	0·03	0·02
pH of 2% aqueous solution	2·4–2·8	6–7·5	6–7·5	—	—
Specific rotation	589 nm, 20°C C = 1 in water	589 nm, 20°C C = 10 in water:	589 nm, 20°C C = 5 in water:	589 nm, 20°C C = 10 in methanol:	589 nm, 20°C C = 5 in Methanol
(on dry material)	+ 20·5° to 21·5°	+ 103° to + 108°	+ 95° to + 97°	+ 21·5° to + 24·0°	+ 19·5–22·5°
Biological activity[c]	100%	89%	83%	42%	40%

Insoluble = practically insoluble,
[a] In presence of oxygen.
[b] In absence of oxygen.
[c] depends on mol. wt of the compound.
(+) with decomposition.

TABLE 9

COMPARISON OF PHYSICAL PROPERTIES OF ASCORBIC AND ERYTHORBIC ACIDS

	Ascorbic acid	*Erythorbic acid*
Molecular weight		
Free acid	176·13	176·13
Sodium salt	198·11	216·12 (monohydrate)
Melting range		
Free acid	190–192°C(+)	164–169°C(+)
Sodium salt	approx. 200°C(+)	approx. 200°C(+)
Solubility (g/100 ml water, 25°C)		
Free acid	30	40
Sodium salt	77	16
Optical rotation $[\alpha]_D^{20}$ (H_2O, C = 1)		
Free acid	+21°	−17°

(+) Decomposition.

catalyst erythorbic acid is oxidized faster than ascorbic acid at pH < 7·5.[19] Furthermore, the antioxidative decomposition of erythorbic acid in comparison to ascorbic acid was shown to be faster in acid solutions containing Fe^{3+} ions.[20]

7. It has been confirmed that the different chemical behaviour found in model experiments also influences the practical use. As a result of its more rapid decomposition, erythorbic acid has a weaker protective effect on foodstuffs than has ascorbic acid. Investigations of the antioxidant activity of the two substances in fruits and meat products have been reported.[21–23]

Reactions of Ascorbic Acid

Ascorbic acid resembles the sugars in structure, and under some conditions, it reacts like sugars (e.g. non-enzymatic browning reactions). The molecule shows some unusual and interesting chemical properties due to the enediol grouping. The dissociation constants are $pK_1 = 4{\cdot}17$ (at the 3-OH) and $pK_2 = 11{\cdot}57$ (at the 2-OH). Ascorbic acid is a moderately strong reducing agent. It forms stable monobasic salts: sodium ascorbate and calcium ascorbate especially are of industrial importance.

In aqueous solutions ascorbic acid has a high affinity for oxygen. It is readily oxidized to give dehydroascorbic acid, a reaction which is catalysed by heavy metal ions. Dehydroascorbic acid can be reduced back to ascorbic acid. It is relatively unstable but still exerts full vitamin C activity. In the presence of oxygen dehydroascorbic acid is irreversibly degraded to diketogulonic acid. Final decomposition products are oxalic acid and threonic acid, but various other pathways have also been identified.[28,29]

In the absence of oxygen a different series of final products have been identified: at lower pH values 2-hydroxy-furfural and carbon dioxide predominate, at a pH near the pK_1 of ascorbic acid other 5-carbon compounds plus carbon dioxide have been found.[30] Figure 9 gives an overview on some decomposition pathways of ascorbic acid frequently encountered.

With regard to the antioxidant activity of ascorbates the first step, the formation of dehydroascorbic acid, is of special interest. In fact dehydroascorbic acid if formed via the intermediate semidehydroascorbic acid which is also called monodehydroascorbic acid or ascorbate free radical.

In nature the compounds represent a buffer system (Fig. 10). The semidehydroascorbic acid concentration in such systems is very low since it is either reduced to give ascorbic acid again or oxidized to give dehydroascorbic acid. The redox cycle is completed in living tissues by enzymatic reduction of dehydroascorbic acid to ascorbic acid.

A review on the oxidation of L-ascorbic acid and other reactions of the molecule at its various positions has been published by Seib.[31]

In food systems in which ascorbic acid is added as an antioxidant, the following reactions are those of most importance:

- —Quenching of various forms of oxygen (singlet oxygen, hydroxyl radicals as well as superoxide), ascorbic acid being oxidized to the ascorbate free radical.
- —Reductions of free radicals, thus terminating radical reactions and preventing damage to food components.
- —Reduction of primary antioxidant radicals, thus acting as a synergist.
- —Oxidation of ascorbate by molecular oxygen in the presence of metal ions. This reaction should be prevented by chelating agents such as EDTA, oxalate, or substances such as flavonoids (e.g. quercetin, rutin, p. 130) and amino acids such as L-cysteine.

Occurrence of Ascorbic Acid in Nature

Ascorbic acid occurs in all tissues of living organisms where it is responsible for the normal functioning of important metabolic processes.

Table 10 lists the approximate ascorbic acid content of a selection of foodstuffs.

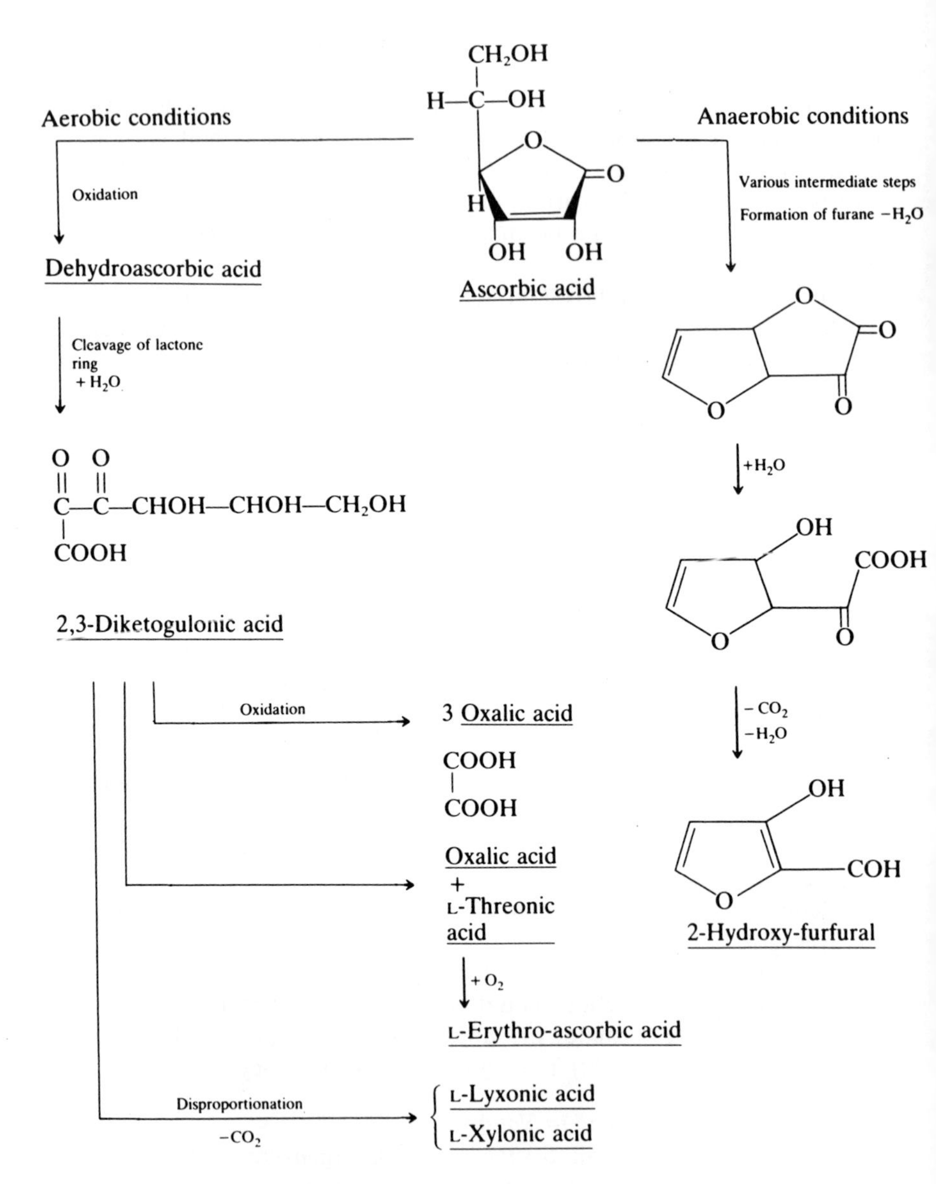

FIG. 9. Decomposition pathways of ascorbic acid.

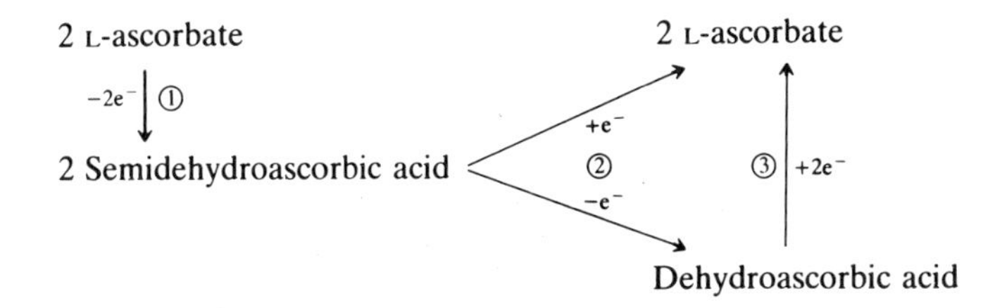

① Oxidation of ascorbate to semidehydroascorbic acid

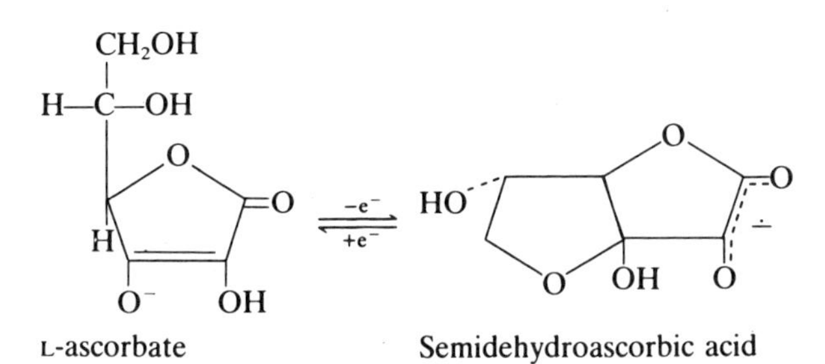

② Disproportionation of semidehydroascorbic acid

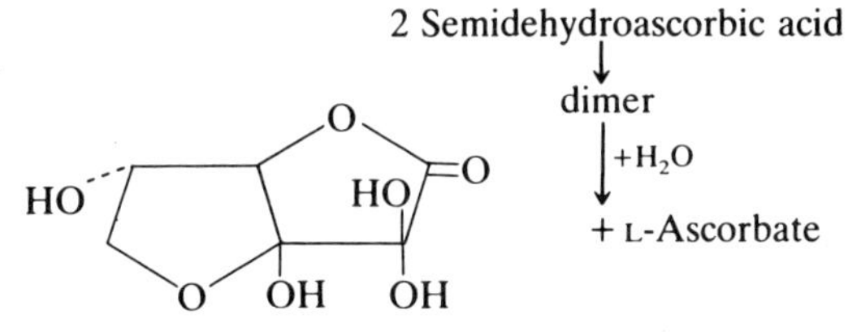

③ Reduction of dehydroascorbic acid

Dehydroascorbic acid $\xrightarrow{+2e^-}$ L-ascorbic acid

in living tissue: $NADPH_2$ → $NADP$

in vitro, e.g. $2H_2SO_3$ → $2H_2SO_4$

FIG. 10. Ascorbic acid–dehydroascorbic acid redox system.

TABLE 10
ASCORBIC ACID CONTENTS OF FOODS (mg/kg)

Vegetable foodstuffs	
Acerola cherries	20 000
Rose-hips	10 000
Blackcurrants	1 400
Lemons	450
Oranges	360
Tomatoes	230
Apples	40
Parsley	1 700
Green pepper (paprika)	1 300
Horseradish	1 200
Green kale	1 100
Potatoes	160
Sauerkraut	140
Mushrooms	50
Animal foodstuffs	
Beef liver	310
Chicken	25
Lean pork	20
Cow's milk	10

The ascorbic acid levels normally found in human organs and in breast milk are given in Table 11.

Biosynthesis of Ascorbic Acid
Ascorbic acid is synthesised in most animals and plants and only a few of them are not able to produce this vital compound for their own

TABLE 11
VITAMIN C LEVELS FOUND IN HUMAN ORGANS AND BREAST MILK

	Approximate content (mg/kg)
Pituitary	450
Leucocytes	350
Eye lens	300
Liver	140
Brain	140
Muscle	25
Breast milk	50
Average body content	20

needs. One of these few exceptions is man. He needs an exogenous source for his vitamin C. In animals the syntheses may either start with D-glucose or D-galactose. The principle of synthesis is based on the reversion of the C-chain: the C-1 atom of the sugar will become the C-6 atom of ascorbic acid. Figure 11 illustrates this synthetic pathway.

In humans the enzyme L-gulono-lactono-oxidase is missing and therefore the pathway is interrupted. In plant tissues other means of synthesis have been identified.[32–35]

Production of Ascorbic Acid

Ascorbic acid can be extracted from natural sources or obtained by chemical synthesis. Products from both sources are identical.

Extraction from natural sources. Today the extraction from natural sources is of little significance, and all ascorbic acid used for food technological purposes is chemically synthesized. However, at the beginning of the 1930s extraction from cabbage, citrus fruits, paprika and leaves of iris was practised.[35]

The chief problems involved in the extraction of such vegetable materials were their susceptibility to deterioration and the variabilities in the content of ascorbic acid. Large quantities of raw materials had to be processed within the short period of harvest. After the establishment of chemical syntheses, and the sudden fall of price, extraction became obsolete.

Today, some ascorbic acid concentrates from acerola are available. They are used for the enrichment of health food products.

Chemical synthesis. Ascorbic acid is chemically synthesised on a large scale, i.e. over 35 000 metric tons/year. The principle of the actual chemical synthesis is to twist the C-chain of a glucose molecule in such a way that the C-1 atom of glucose becomes the C-6 atom of ascorbic acid, and the C-6 atom of glucose the C-1 atom of ascorbic acid. For this purpose the C-1 atom of glucose has to be reduced, whereas the C-5 and C-6 atoms have to be oxidized. The method industrially exploited is based on the work of Reichstein *et al.*[36] and includes the following steps, which, in the meantime, have been further developed to improve process economics.[12,31,35,37]

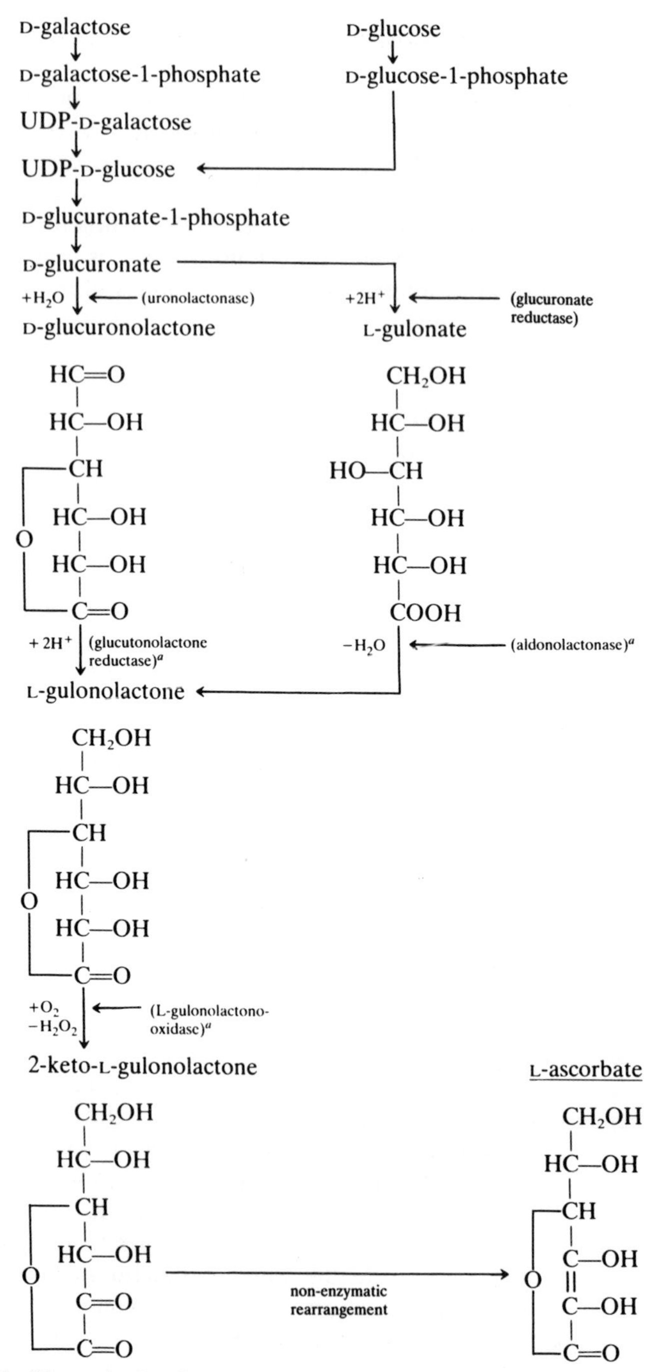

FIG. 11. Biosynthesis of ascorbic acid in animals.[a] Enzymes not present in some species.

1. Reduction of the C-1 atom of D-glucose under pressure with hydrogen to D-sorbitol in the presence of RANEY nickel. Elimination of the nickel catalyst from the aqueous sorbitol solution.

D-Glucose ⟶ D-Sorbitol

2. Microbiological oxidation of the C-5 atom of D-sorbitol to L-sorbose, using *acetobacter suboxidans*. The sorbose is isolated by crystallization.

D-Sorbitol ⟶ L-Sorbose

3. The introduction of acetone to protect hydroxyl groups during the following oxidation steps:

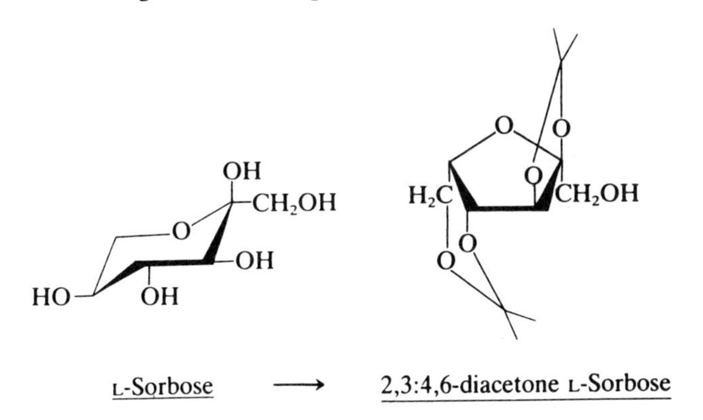

L-Sorbose ⟶ 2,3:4,6-diacetone L-Sorbose

Two methods are used for producing 2,3:4,6-diacetone L-sorbose:

(a) The sorbose is converted in solution containing acetone and concentrated sulphuric acid. The equilibrium between the various reaction products depends on temperature, water content and the amounts of starting materials. The solution is neutralized with soda and diacetone L-sorbose is obtained by elimination of the acetone by distillation and subsequent extraction with an organic solvent such as toluene. Extracted diacetone L-sorbose can be directly used in the next step without additional purification.

(b) A method avoiding sulphuric acid requires catalysts, such as perchloric acid, ferric chloride or ferric bromide. The reaction water formed is continuously eliminated by azeotropic distillation with a water-immiscible organic solvent.

4. Oxidation of 2,3:4,6-diacetone L-sorbose to 2,3:4,6 diacetone keto-L-gulonic acid.

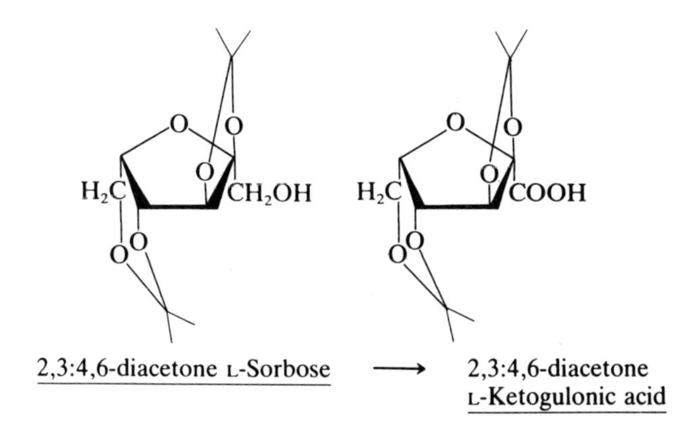

Industrially three methods are applied:

(a) Oxidation with sodium hypochlorite in presence of nickel salt catalysts. Sodium hypochlorite is manufactured on site by electrolysis of sodium chloride. The resulting hydrogen from both (a) and (b) may be used in step No. 1. The resulting sodium salt of diacetone L-ketogulonic acid is neutralized with hydrochloric acid.

(b) Direct electrochemical oxidation. The direct electrochemical oxidation is carried out in alkaline solution.

(c) Catalytic oxidation with oxygen. This oxidation is also carried out in aqueous alkaline solution using oxygen/air or oxygen/nitrogen mixtures. The oxidation is accelerated by the use of precious metal catalysts (Pt or Pd).

5. Conversion of 2,3:4,6 diacetone L-ketogulonic acid to L-ascorbic acid. Several methods are known, and may include intermediate steps. On an industrial scale the direct cyclization of 2,3:4,6 diacetone L-ketogulonic acid with hydrochloric acid gas in a water-free chloroform/ethanol system proceeds to a good yield.

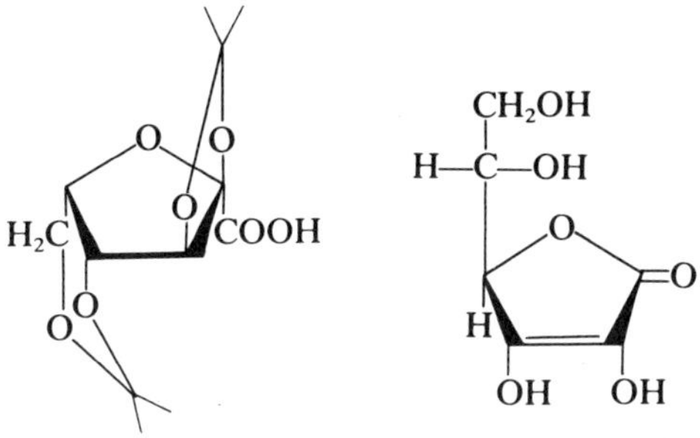

2,3:4,6-diacetone L-Ketogulonic acid ⟶ L-Ascorbic acid

6. Recrystallization. The aqueous solution of the crude product is treated with active charcoal and crystallized, yielding pure L-ascorbic acid.

Synthesis and Properties of Ascorbyl Palmitate

Ascorbyl palmitate is prepared synthetically by reacting ascorbic acid with sulphuric acid followed by re-esterification with palmitic acid. Ascorbyl palmitate is subsequently purified by recrystallization.

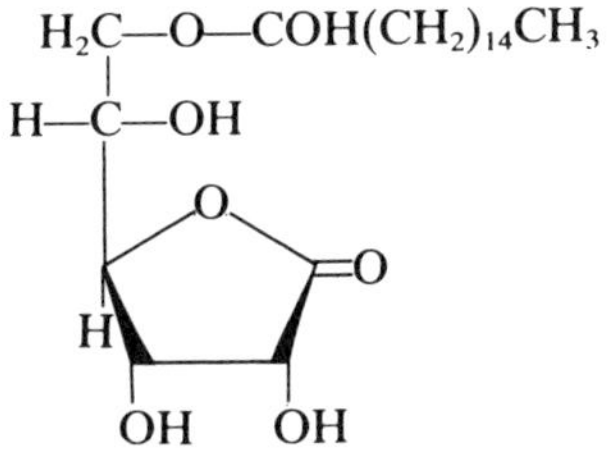

$H_2C—O—COH(CH_2)_{14}CH_3$

Ascorbyl palmitate (6-O-palmitoyl-L-ascorbate)

TABLE 12

SOLUBILITY OF ASCORBYL PALMITATE

Substrate	*Temp. (°C)*	*Approx. solubility (g/kg substrate)*
Glycerides		
Peanut oil	25	0·300
	70	1·600
	100	9
Sunflower oil	25	0·280
Olive oil	25	0·300
Coconut oil	25	1·2
	100	50
Paraffin oil	25	<0·050
Middle chain triglyceride	25	1·7
(Miglyol 812®)	70	20
	100	100
Mono-palmitate	100	480
Di-palmitate	100	350
Tri-palmitate	100	<40
Mono-oleate	100	520
Mono-di-oleate	100	150
Water		
Deionized	25	0·002
	50	0·07
	70	2
	100	10
pH = 8·1	25	0·01
	60	100
Acetone	25	69
Methanol	25	183
	60	600
Ethanol		
100%	0	45
100%	25	125
95%	25	108
50%	25	0·4
Isopropyl alcohol	25	50
	70	200
Ethyl ether	25	7·6
Chloroform	25	0·3
	60	89
Glycerol	25	0·1
Propylene glycol	25	48
Triacetin	25	2·6

Other 6-fatty acid esters of L-ascorbic acid, e.g. ascorbyl stearate,[31,37] are prepared in the same manner.

Ascorbyl palmitate is a light powder used for the stabilization of fats, oils and fatty products. From Table 8 it is obvious that its poor solubility in edible fats and oils, and its insolubility in cold water cause problems.

Table 12 gives an overview of the solubility of ascorbyl palmitate in various solvents, which helps in finding a suitable method of addition to a substrate. When dissolving ascorbyl palmitate it should be remembered that its decomposition in the presence of oxygen starts at approx. 113°C. Its stability also strongly depends on pH value (unstable in alkalis).

Other Natural Antioxidants

Food ingredients or extracted food constituents which act as natural antioxidants have been used from earliest times. Several traditional processing techniques which increase resistance to oxidation are also known. The following overview gives some examples of such natural antioxidants. It also shows that the distinction between 'foodstuff' and 'food-additive' is not clear-cut. Beside the widespread occurrence of the natural antioxidants, tocopherols, ascorbic acid and their derivatives, the activity of various other compounds is (or has been) commercially utilized or will, perhaps, gain importance in the future.

Certainly, there is no assurance that a compound or a fraction isolated from natural food is safe. Many substances in our daily food are known to be toxic. This leads to the demand that food extracts also have to be subjected to toxicological evaluation when their application as food additives leads to substantially increased consumption.

Soya Bean Products

Soybean products are applied in various foods. In soybean oil, however, only the tocopherols, especially γ-tocopherol, but also δ- and, to a minor extent α-tocopherol, are active antioxidants. The sterols, such as campesterol, stigmasterol or β-sitosterol, do not show antioxidant activity.[38]

Soybean flour and other soybean derivatives are sources of a large range of antioxidant compounds: tocopherols, flavonoids, isoflavone glycosides and their derivatives and, as synergists, phospholipids, amino acids and peptides.

Various qualities of soybean flour, aqueous extracts as well as organic solvent extract, from soy flour or soybeans, soy protein concentrates, soybean isolates and soy protein hydrolysates have been suggested as antioxidants. In practice, mainly soy flours, protein isolates and protein hydrolysates are utilized in a range of food products such as fats and oils, bakery products and meat products at levels of 0·1–20%. A useful review on this subject has been published by Hayes *et al.*[39]

Oat Products

Oat products have also been suggested as antioxidants for foods. Commercially germinated oat grains or aqueous extracts are available and used in some countries at levels of 1·5% in fatty products and to impregnate wrapping papers. Antioxidant efficacy is supposedly based on their contents of dihydrocaffeic acid and phospholipids.[40]

Components of Crude Vegetable Oils

Crude oils are generally more resistant to oxidation than refined oils. Refining eliminates various antioxidant compounds, but only tocopherols are recovered from deodorizer sludge. Cantarelli & Montedoro[41] have elucidated this disappearance of antioxidants in olive oil: olives contain a range of natural antioxidants such as cyanidin glycosides, flavonols, flavans and phenolic acids. Some of them are also present in crude olive oil, but in refined olive oils α-tocopherol is practically the only natural antioxidant left.

Amino Acids, Peptides and Proteins

The antioxidant activity of some amino acids and their synergism with other antioxidants is known, and patents have been applied for.[42] For example, skim milk powder in margarine increases the antioxidant activity of the antioxidants added.

The meat of antarctic krill has been proposed for increasing the resistance of canned and preserved foods (especially fish products) against oxidation.[43] The antioxidant principle of krill has been isolated and identified as a tocopherol and a mixture of free amino acids. Furthermore, the same authors showed that quantitative as well as qualitative aspects of combinations of amino acids is of decisive importance for an optimum synergistic effect.

Guaiac Gum

Guaiac gum is an antioxidant and preservative that, for many years, has been used mainly to stabilize refined animal fats in combination with phosphoric acid derivatives as synergists. Guaiac resin is obtained from *guajacum officinale L,* a tree originating in the West Indies. Guaiac gum cannot therefore be regarded as a common foodstuff, but as an additive. In most legislations it is handled as such, and its use is limited to a few countries only.

The antioxidant principles of guaiac gum are the α- and β-guaiaconic acids.[40]

Nordihydroguaiaretic Acid (NDGA)

Nordihydroguaiaretic acid is a greyish-white crystalline substance which was widely used as an antioxidant in the 1950s and 1960s.

NDGA is a major constituent of the resinous exudate of the creosote bush, *Larrea divaricata,* in South-Western USA. Besides the isolation of natural material, NDGA has also been obtained by chemical synthesis.

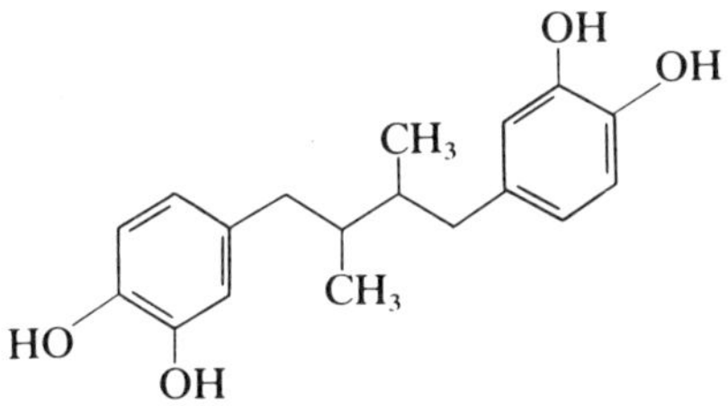

Nordihydroguaiaretic acid (NDGA)

A review of NDGA, its manufacture, efficacy, application, pharmacological and toxicological properties has been published by Oliveto.[44] Due to unfavourable toxicological findings, NDGA has been removed by the US Food and Drug Administration from its GRAS (generally recognized as safe) list. Today NDGA is no longer of practical importance in the food field.

Flavonoids

Derivatives of flavones are quite effective antioxidants, e.g. in milk, lard, butter, especially in combination with synergists such as citric acid, ascorbic acid or phosphoric acid.

The most effective flavonoids (see below) are quercetin, rhamnetin, kämpferol, rutin or vitamin P, and quercitrin. The antioxidant activity is mainly based on the hydroxl group at the C-3 and the double bond between C-2 and C-3 atoms. Metal complexing capacity is centred around the C-3′ and C-4′ atoms. The above-mentioned and various other vegetable flavonoids have been tested as antioxidants in aqueous food systems (to protect vitamin C) and in fatty systems.

Quercetin especially proved to be effective at 150 mg/kg in lard and butter or at 3 mg/kg in methyl linoleate. Nevertheless the substances did not find real commercial exploitation. Some of the reasons might be their physical and chemical properties (solubility) and the necessity for toxicological evaluation. For example, Stan & Huni[45] reported on the mutagenic effects of flavonoids.

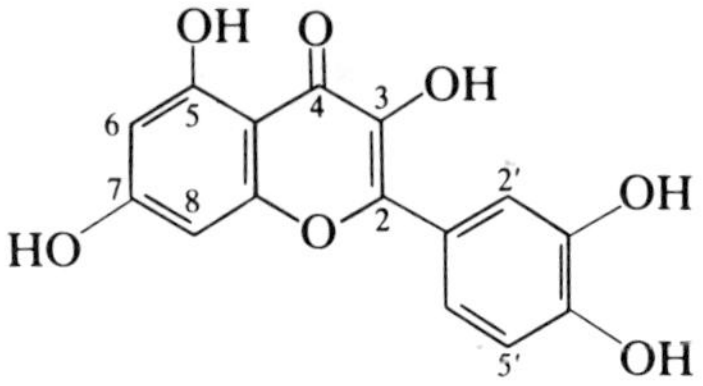

Quercetin
soluble in alcohol, aqueous
alkaline solutions, etc.,
practically insoluble in water

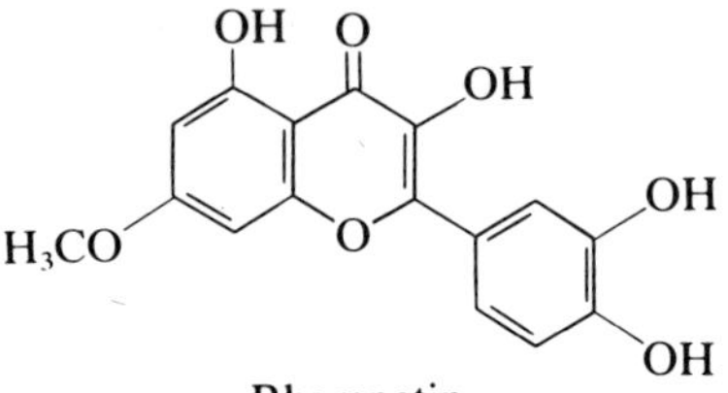

Rhamnetin

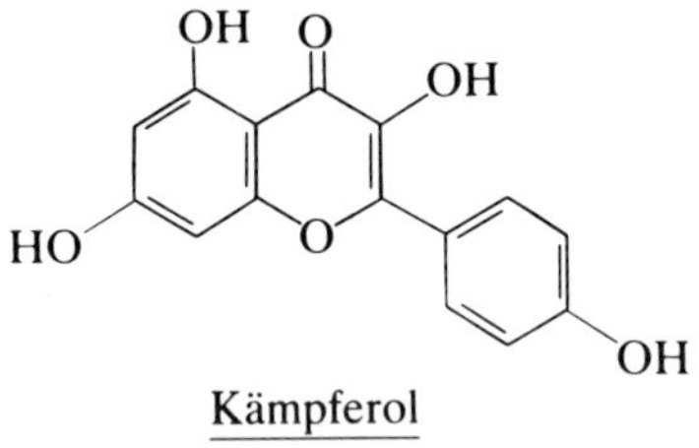

Kämpferol

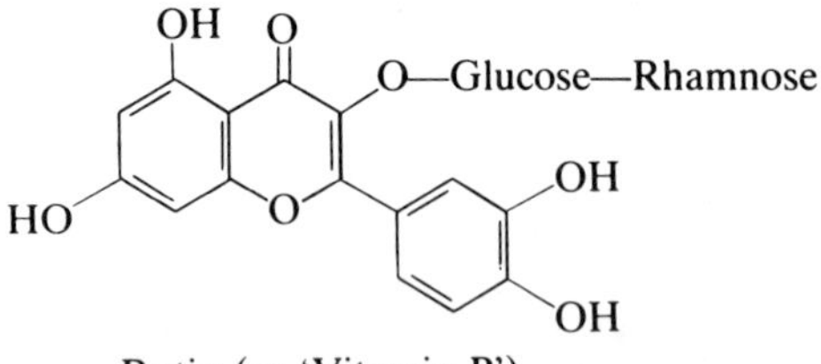

Rutin (or 'Vitamin P')

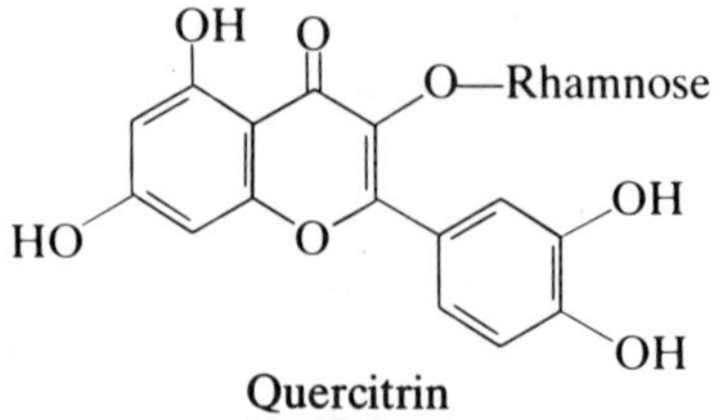

Quercitrin

Spices and Herbs

Spices and herbs have achieved some commercial importance as antioxidants. For many years the beneficial influence of certain ground herbs and spices on fat stability has been known. They have been added to meat products and baked goods as well as to various other preserved foods. Various comprehensive studies on this topic have been published. Herrmann[46] gives an overview on the most important ones up to 1980.

A systematic investigation during 4 years and covering 32 species of herbs and spices has been carried out by Chipault *et al.*[47,48] Several food models have been tested under various conditions (lard: AOM; emulsion: Warburg apparatus; ground pork: frozen storage; mayonnaise: storage at room temperature; pie crust: storage at 63°C) and the peroxide values determined. The antioxidant efficacy was expressed as a protection factor indicating the ratio of the stability of a sample containing the spice to the one of a control sample without spice (Table 13).

From Chipault's figures,[47,48] as well as from other reports, it can be concluded that the antioxidant activity of spices and ground herbs varies over a very large range. Especially in the presence of light, many of them can show a pro-oxidant effect.[46]

The most potent antioxidant spices are generally recognized as

TABLE 13

PROTECTION FACTOR OF SPICES IN VARIOUS FOODS

Substrate:	*Lard*	*Pie crust*	*O/W emulsion*	*Ground pork*		*Sauce mayonnaise*	
Test temp. °C:	99	63	40	−5	−15	20	20
g spice per 100 g fat:	0·2	0·2	0·1	0·25	0·25	0·20	0·56
Allspice	1·8	1·1	16·7	>5·3	>10·0	1·4	3·1
Cloves	1·8	1·3	85·8	>5·3	>10·0	2·0	4·6
Oregano	3·8	2·7	7·9	>7·2	3·7	8·5	9·1
Rosemary	17·6	4·1	10·2	>5·3	>10·0	2·2	—
Sage	14·2	2·7	7·8	>5·3	>10·0	2·4	3·4
Thyme	3·0	1·9	6·8	6·0	3·2	1·8	—

Source: Chipault *et al.*[48]

rosemary and sage. The effectiveness of spices and herbs depends not only on variety and quality but also on the substrate and storage conditions. Only a practical experiment using well defined qualities of spices and herbs can really determine success or failure.

Spice Extracts

Spice extracts have attracted interest since the addition of powdered herbs to liquid fats, delivered in bulk to the food industry, is obviously impracticable. Also, the taste and odour of spices cannot be tolerated in most applications.

Rosemary and sage have been found to constitute the most potent antioxidant spices, and various extracts have been investigated in theory and practice.[49–53,55] However, many of the extracts have a strong odour and bitter taste, and therefore are unsuitable for use in many food products.

Manufacture of spice extract. In 1977 S. Chang *et al.*[54] reported on a process for the extraction of rosemary and sage, yielding a purified antioxidant, which has little of the odour of the original spice. To produce a bland odourless and tasteless antioxidant this product was subjected to molecular distillation or vacuum steam distillation.

Bracco *et al.*[55] reported on a new industrial process to obtain natural antioxidants from spices based on the following steps: micronization of the spice in an edible oil to obtain a transfer of the antioxidant to the lipid phase, cleaning of the lipid phase and molecular distillation. The two processes[54,55] are outlined in Fig. 12.

Antioxidant activity of spice extracts. The antioxidant components of the spices have been investigated. Extracts of spices contain a great number of substances having some antioxidant efficacy. In rosemary carnosic acid (below) has been described as the most active antioxidant constituent by Brieskorn & Dömling.[56] Other phenolic compounds have been investigated, e.g. rosmaric acid (below), which has in model systems an activity comparable to that of caffeic acid. But Gerhardt & Schröter[57] could not find a correlation between the antioxidative efficacy of the tested spices and their rosmaric acid content. There are indications that the antioxidant activities of

(1)

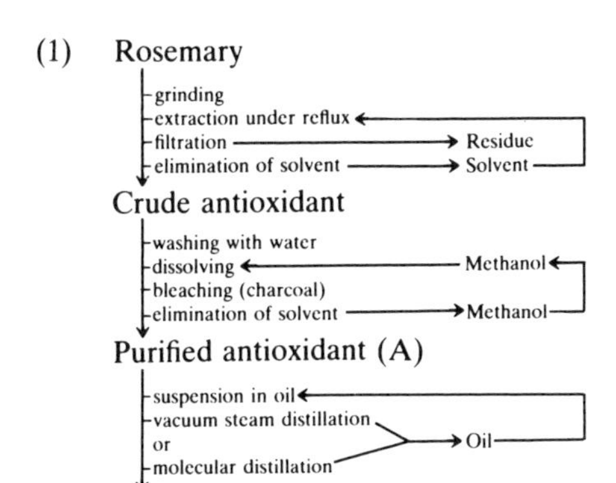

(2)

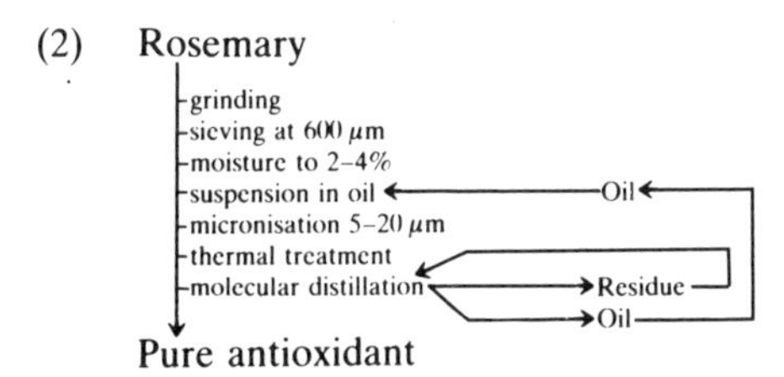

(A) Slightly coloured powder with some odour of the original spice.
(B) Bland in flavour and odour.

FIG. 12. Processes for recovery of rosemary antioxidants: (1) according to Chang *et al.*[54]; (2) according to Bracco *et al.*[55]

rosemary extracts mainly depend on their carnosic acid content.

Carnosic acid
m.p.: 185–190°C

Rosmaric acid
m.p.: 168–169°C

Table 14 compares the antioxidant activities of two different

TABLE 14

COMPARISON OF ANTIOXIDANT ACTIVITIES OF ROSEMARY EXTRACTS, CARNOSIC ACID AND SOME OTHER ANTIOXIDANTS

Antioxidant	*Concentration (mg per kg fat)*	*Protection factor*	
		Lard	*Peanut oil*
Commercial rosemary extract (~16% carnosic acid)	200	3·9	1·3
Experimental rosemary extract (24·8% carnosic acid)	200	5·3	1·4
Carnosic acid	100	9·3	—
(*synthetic*)	200	17·6	2·2
	332 (= 1 mmol)	23·9	—
α-tocopherol	100	5·1	—
	200	7·9	1·1
	421 (= 1 mmol)	8·4	—
γ-tocopherol	100	9·0	—
	200	17·2	1·3
	416 (= 1 mmol)	28·1	—
BHA	200	6·4	1·2
BHT	200	3·3	1·2
Octylgallate	200	13·7	2·5

Testing method: Rancimat test.

$$\text{Protection factor: } \frac{\text{Induction period with antioxidant (min)}}{\text{Induction period without antioxidant (min)}}$$

Source: Pongracz *et al.*[99]

rosemary extracts with those of pure canosic acid[99] as well as BHA, BHT and α- and γ-tocopherols in peanut oil and lard respectively.

As is also true of the tocopherols and the synthetic antioxidants BHA and BHT, carnosic acid is mainly a potent antioxidant for animal fats. Up to a concentration of 332 mg/kg lard (1 mmol carnosic acid), carnosic acid showed a direct correlation between concentration and the protection factor measured in the Rancimat test.

Commercial products. Antioxidant extracts from spices (usually rosemary) commercially available are usually fine powders. Depending on their content of active substances it is recommended to use them at levels between 200 and 1000 mg/kg of finished product to be stabilized. Generally the powders are dispersible in oils or fats, insoluble in water, but soluble in organic solvents. Due to their powder characteristics they can also be used by dry mixing in powdered food.

One of the features mentioned for versatile food antioxidants is freedom from odour and taste: obviously this requirement is of high importance for spice extracts, otherwise their application would be limited to products compatible with the specific spice flavour. Pure carnosic acid is not available as a food antioxidant.

'Maillard Products'

According to many observations various 'Maillard' products formed through the interaction of proteins and carbohydrates during the heat-processing of foods exert antioxidant activity. Two food systems subjected to oxidation may be mentioned: Griffith & Johnson[58] reported on findings in sugar cookies. According to Kessler[59] the storage stability of condensed and dried milk is increased by a heat treatment prior to the removal of water. In food manufacturing the utilization of 'Maillard' products is limited to such applications: extracts have not been commercialized.

Evans *et al.*[60] isolated amino-hexose reductones from the reaction mixtures of aldohexoses and secondary amines, and demonstrated that these reductones acted as antioxidants. Later, alcohol extracts of Maillard-browning systems were also investigated. Rhee & Kim[61] found that the major antioxidant compounds of such systems are probably colourless or almost colourless intermediates such as reductones and dehydro-reductones produced in the earlier stages of the browning reaction.

SYNERGISM

As a general rule antioxidant synergism can be expected between substances with differing modes of action. Among natural substances, the following ones are of interest:

- —Radical scavengers: tocopherols, spice extracts (carnosic acid), flavonoids.
- —Oxygen scavengers: ascorbates.
- —Synergists (mostly acting as metal complexing agents, peroxide decomposers or reactivators of antioxidants): citric acid, phosphoric acid, tartaric acid, lecithin, amino acids, ascorbic acid.

The synergistic effect permits the reduction of the level of antioxidants added to food formulations considerably. As well as reduced costs, less undesirable side effects on the food product itself may be expected. For practical reasons most food manufacturers prefer finished synergistic mixtures to individual substances. Consequently, a number of such mixtures are available on the market, adapted to various customers' need. The following are the most important synergistic systems involving natural antioxidants.

Tocopherol and Ascorbate

In living tissues it is supposed that vitamin E acts as a primary antioxidant. The resulting tocopheryl radical is regenerated by vitamin C. The resulting ascorbate free radical is reduced back to vitamin C by an NADH-dependent system. This mechanism has also been identified under experimental conditions *in vitro*.[62] The interaction between ascorbic acid and tocopherol in food model systems has also been investigated. Lambelet *et al.*[63] observed the interaction of the two substances in the oxidized liquid fraction of subcutaneous chicken fat. The authors showed that the determined α-tocopheroxyl radicals were quenched by ascorbyl palmitate and that radicals of ascorbyl palmitate were subsequently formed. Determination of the concentrations of ascorbyl palmitate and dl-α-tocopherol during the oxidation experiments showed that ascorbyl palmitate was consumed first. This implied that dl-α-tocopherol, which was the antioxidant the most readily oxidized, was regenerated by ascorbyl palmitate.

Table 15[99] shows the behaviour of ascorbyl palmitate (500 mg/kg)

TABLE 15

ANTIOXIDANT TESTS IN LARD AT 100°C: CORRELATION BETWEEN AMOUNTS OF ANTIOXIDANTS SURVIVING AND PEROXIDE VALUE

Storage (h)	*Retention (%)*		*POV (meq/kg)*
	Ascorbyl palmitate	*dl-α-tocopherol*	
0	100	100	0·5
2	24	98	0·8
4	7	95	1·0
8	0	88	3·2
12	0	70	6·0
16	0	52	8·0
20	0	26	13·0
24	0	0	>40

Source: Pongracz *et al.*[99]

and dl-α-tocopherol (100 mg/kg) during the oxidation of lard at 100°C: again ascorbyl palmitate was consumed first, followed by α-tocopherol. After 20 h, the increase of peroxide value accelerated, and the tocopherol was used up.

Figure 13[99] shows the effect of a commercial antioxidant mixture composed of 25% ascorbyl palmitate, 5% dl-α-tocopherol and 70% lecithin. The latter functions as a solvent for ascorbyl palmitate as well as a synergist. Lard was tested at 120°C in the Rancimat test, and the columns represent the prolongation of the induction period (0·5 h for the control sample without antioxidant added). Apparently the anti-

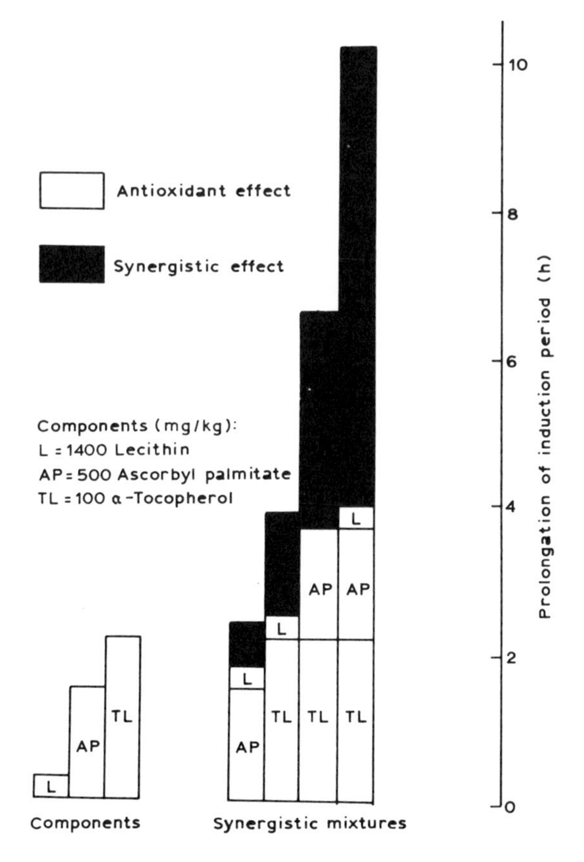

FIG. 13. Antioxidant effect of components and synergism of a commercial antioxidant mixture. (Source: Pongracz *et al.*[99])

TABLE 16
SYNERGISM OF CARNOSIC ACID WITH SOME OTHER NATURAL ANTIOXIDANTS IN LARD (RANCIMAT TEST 120°C)[99]

Antioxidant	*Level of antioxidant addition (mg/kg)*	*Induction period (min)*	*Protection factor*
Control (pure lard)	—	40	—
α-Tocopherol (α-TL)	100	265	6·6
Ascorbyl palmitate (AP)	500	180	4·5
Carnosic acid (CA)	100	440	11·0
α-TL + CA	100 + 100	445	11·1
AP + CA	500 + 100	990	24·8
α-TL + CA + AP	100 + 100 + 500	1070	26·8

oxidant effect of all mixtures is greater than the sum of the effect of the components. γ- or δ-tocopherol show a synergism similar to that of α-tocopherol with ascorbyl palmitate, lecithin or citric acid.[99]

Carnosic Acid, Ascorbates and Tocopherols

According to Chang *et al.*[54] the antioxidant activity of rosemary extract in prime steam lard could further be enhanced by the addition of ascorbic acid (200 mg rosemary extract plus 500 mg ascorbic acid per kg of lard).

The synergism between carnosic acid and ascorbyl palmitate was, confirmed by Pongracz *et al.*[99], but they also observed that carnosic acid suppressed the antioxidant activity of α-tocopherol (Table 16) in lard.

ANTIOXIDANT-PRO-OXIDANT PROPERTIES

Tocopherols

It has been observed that the activity of some phenolic antioxidants does not increase linearly with increasing amounts of additive (Fig. 14 gives examples). At sufficiently high levels of addition it may even become a pro-oxidant effect. Among the natural antioxidants this fact has to be considered mainly with α- and β-tocopherol, whereas γ- and δ-tocopherol do not produce a pro-oxidant effect in the range of practically realistic levels of addition.

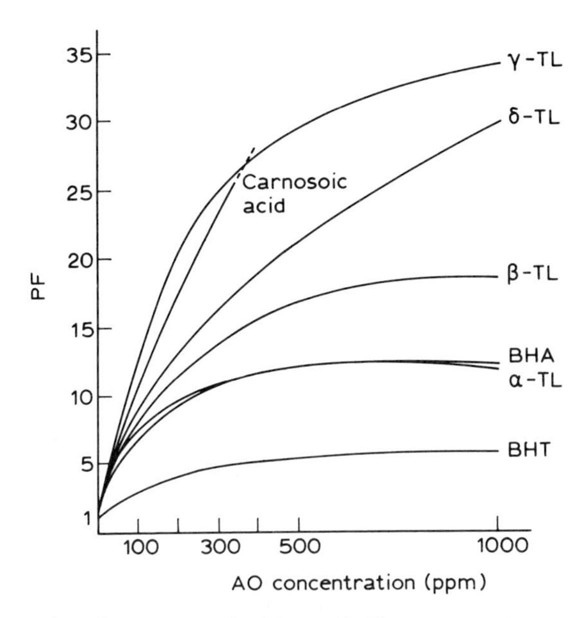

FIG. 14. Relation between antioxidant (AO) concentration and activity. Substrate: lard; testing method: Rancimat, 120°C PF: ratio duration induction period with AO: duration induction period without AO; TL: tocopherol. (Source: Pongracz *et al.*[99])

The reactions can be summarized as follows:

Antioxidant activity (low concentration of tocopherol):

$$ROO^{\cdot} + AH_2 \rightarrow ROOH + AH^{\cdot}$$

Inactivation of radicals:

$$ROO^{\cdot} + AH^{\cdot} \rightarrow ROOH + A \quad \text{(oxidation)}$$

$$AH^{\cdot} + AH^{\cdot} \rightarrow AA \quad \text{(dimerization)}$$

$$AH^{\cdot} + AH^{\cdot} \rightarrow AH_2 + A \quad \text{(dismutation)}$$

$$ROO^{\cdot} + AH^{\cdot} \rightarrow ROOA \quad \text{(addition)}$$

Pro-oxidant activity (high concentration of tocopherol):

$$ROOH + AH^{\cdot} \rightarrow ROO^{\cdot} + AH_2$$

ROOH: hydroperoxides
$ROO^{\cdot}$: peroxide radicals
AH_2: antioxidant (tocopherol)
$AH^{\cdot}$: antioxidant radical

The pro-oxidant effect was demonstrated by Lundberg *et al.*[64] In their experiment they oxidized lard at 100°C with 0·5% hydroquinone added. Initially there was a rapid increase of the peroxide value up to 40. After 15 days the hydroquinone concentration had dropped below 0·2% and the peroxide value dropped to 20. After 35 days all hydroquinone was oxidized, and the peroxide value again sharply increased. In this experiment hydroquinone showed a pro-oxidant effect when added at a level higher than 0·2% of lard, and an antioxidant effect below 0·2%.

Heimann & von Pezold[65] investigated the influence of various concentrations of phenolic antioxidants in vegetable and animal fats. For α-tocopherol they concluded that low amounts (below approx. 600–700 mg/kg fat) do not show pro-oxidant activity at room temperature. When the temperature is increased in such a system the formation of antioxidant radicals accelerates more rapidly than the autoxidation of the substrate. Consequently, the antioxidant becomes a pro-oxidant through the increased decomposition of hydroperoxides. When the antioxidant concentration is increased the number of antioxidant radicals formed exceeds the number of available free and peroxide radicals. Again, the antioxidant radicals may decompose hydroperoxides.

In aqueous model systems the pro-oxidant effect of α-tocopherol during the autoxidation of linoleic acid has been investigated by J. Cillard *et al.*[66,67] It depended on two factors, namely

—the solvent: only in aqueous systems a considerable pro-oxidant effect was observed;
—the concentration: below 5 mmol α-tocopherol/litre mol linoleic acid no pro-oxidant effect was found.

In linoleic acid dispersions with 0·5% Tween 20 at pH 6·9 mainly α-tocopherolquinone (Fig. 2) and very little α-tocopherol dimer had been detected as oxidation products.[67]

In food manufacturing practice it is recommended to keep the amount of total α-tocopherol (natural and added) at levels between 50 and 500 mg/kg, depending on the kind of food product.

Ascorbic Acid

Under certain conditions ascorbic acid may also act as a pro-oxidant, especially in aqueous fat systems. According to some authors it accelerates the oxidation of linoleate in the presence of non-enzymatic catalysts, but inhibits the same reaction under other experimental conditions. Kanner *et al.*[68] investigated a β-carotene–linoleate system and found that ascorbic acid was a pro-oxidant at low concentrations. Fe^{3+} and Co^{2+} promoted synergistically the pro-oxidant effect. Low Cu^{2+} ion concentration also promoted the oxidation of ascorbate. According to Grosch[69] volatile aldehydes can be formed in emulsions of hydroperoxides of linoleic and linolenic acids, ascorbic acid and trace metals. In the absence of ascorbic acid the formation of volatile aldehydes was much lower, as it was also when oxygen was excluded, or when the metal traces were inactivated by EDTA. It was postulated that an ascorbic acid–Fe^{2+}–oxygen complex is responsible for the decomposition of hydroperoxides and the increased autoxidation of unsaturated fatty acids in such systems.

For practical work it is not yet possible to give a generally valid recommendation. The stabilization of food systems with an aqueous and a lipid phase (especially with polyunsaturated fatty acids and their hydroperoxides), metals and oxygen needs in any case special care to avoid pro-oxidant effects. The optimum ascorbic acid treatment and the necessary amount of metal complexing agents or other antioxidants (tocopherols), processing parameters and packing materials have to be determined for each specific product.

The problem of possible pro-oxidant effects with ascorbic acid or its derivatives has mainly to be considered in systems such as milk and other dairy products, oil-in-water emulsions such as sauces or mayonnaise, or water-in-oil emulsions such as margarine or butter.

PROTECTION OF FOODS WITH NATURAL ANTIOXIDANTS

Most food systems are rather complex and more than one component may be susceptible to oxidation. The following sections demonstrate a range of applications by describing the overall problem and giving some details that could be of interest. The natural antioxidants discussed are of striking versatility, therefore also uses of minor importance are mentioned, and others of more practical importance are only casually discussed.

In terms of tonnage, ascorbic acid is predominantly used in meat and meat products, followed by beverages, fruits and vegetables. For tocopherols and synergistic mixtures the uses are in fats, oils and fatty products.

Literature on the mode of application and results has been published in abundance (e.g. Refs 35, 37, 79, 84, 100), and of course the manufacturers of the antioxidants mentioned willingly supply such background information.

Fats and Oils

The autoxidation of fatty acids follows well known pathways. In an *initiation phase* free radicals are formed. During the subsequent *propagation phase* free radicals react with oxygen to produce peroxide radicals which have the ability to attack other fatty acids. This chain reaction ends in the *termination phase* in which stable deterioration products are formed. With ongoing autoxidation the palatability of the fats or oils decreases due to rancid off-flavours.

A comparison of the various natural antioxidants has already been given in Fig. 14. As a model system lard has been chosen, and for comparison, some commonly used synthetic antioxidants have been included.[99] Figure 14 illustrates some facts generally valid for the use of antioxidants in mono-, di- and triglycerides, viz.:

—The duration of the induction period cannot be lengthened to any extent by increasing the amount of added antioxidant.

—Maximum activity is approximately at 300–600 ppm for BHA, BHT and α- and β-tocopherol. For the other substances the maximum activity is at a much higher level than the usual levels of application which are at 100–300 ppm.

—In vegetable fats and oils the content of natural tocopherols should be included in the calculation (Table 6). Since there is no synergism between the shown phenolic antioxidants, the 'stronger' one covers the activity of the weaker one, i.e. it is not very useful to add α-tocopherol (or one of the other 'weaker' antioxidants) to a product that already contains a considerable amount of a 'stronger' one such as γ- or δ-tocopherols, or carnosic acid from a spice extract. Table 17[99] demonstrates this effect with peanut oil (cf. also Table 16).

The excellent synergism between the various tocopherols (respectively spice extracts) and ascorbyl palmitate (respectively ascorbyl stearate) and citric acid or lecithin, amino acids, etc., can be observed in all fatty systems (cf. Synergism).

A special problem in food manufacturing is the stabilization of deep

TABLE 17
ABSENCE OF SYNERGISM BETWEEN TOCOPHEROLS IN PEANUT OIL (Rancimat Test, 120°C)

Natural contents (ppm)	*Antioxidant added (ppm)*	*Total (ppm)*	*Induction period (min)*	*Protection factor*
α-TL 210 γ-TL 104		α-TL 210 γ-TL 104	165	—
α-TL 210 γ-TL 104	α-TL 200	α-TL 410 γ-TL 104	180	1·09
α-TL 210 γ-TL 104	γ-TL 200	α-TL 210 γ-TL 304	210	1·27
α-TL 210 γ-TL 104	α-TL 200 γ-TL 200	α-TL 410 γ-TL 304	205	1·24

Source: Pongracz *et al.*[99]

frying oils. Pongracz[70] compared various antioxidants, including α- and γ-tocopherol, by frying potato chips in a commercial frying fat.

Tocopherols are not volatile at the frying temperature of 180°C and are therefore suitable for this purpose. The test included the subsequent cooking of 35 batches of potato chips. The cooking oil as well as the products were analysed; and the following results were found:

—Cooking oil: only the tocopherol, which was present at the highest concentration (natural or added) was substantially reduced (e.g. when α-tocopherol was added, predominantly α-tocopherol was used up; when γ-tocopherol was added γ-tocopherol was used up).
—γ-tocopherol, as compared with α-tocopherol, provided good results.
—The figures in Table 18 give an overview concerning the quality of the products.

The use of ascorbyl palmitate in deep frying oils seems to be less attractive since brownish heat decomposition products are produced in the hot oil.

Summarizing, it can be said that natural antioxidants are capable of stabilizing all fats which do not contain too large amounts of tocopherols. Ascorbyl palmitate, which has an excellent synergistic activity, is used up first by scavenging oxygen (Table 15 and Fig. 13).

TABLE 18

STABILIZATION OF A COMMERCIAL DEEP FRYING FAT AND OF POTATO CHIPS

Antioxidant (ppm)	*Frying fat after 35 batches*		*Potato chips from batches 30–35*					
			Freshly after frying			*After 8 weeks at RT*		
	Peroxide value	*IP (min)*	*Peroxide value*	*IP (min)*	*Organoleptic quality*	*Peroxide value*	*IP (min)*	*Organoleptic quality*[a]
Fresh fat	1·8	430	—	—	—	—	—	—
Control	6·6	320	5·8	305	Good	8·6	270	4[b]
1 000 α-TL	5·8	340	5·8	345	Good	10·6	245	4
1 000 γ-TL	4·2	700	3·0	665	Good	6·2	520	6
500 OG	6·7	940	5·4	665	Good	8·2	540	4
500 TBHQ	5·8	660	4·9	460	Good	6·6	400	3[c]

[a] Number of persons who found the chips good (tested by 8 persons).
[b] Four persons found the control 'oily' or 'rancid'.
[c] Off-flavour was detected.
Methods: 3 litres of fat in a household frying pan; 180°C; 35 'batches' of potato chips produced.
IP: induction period of the oil measured in the Rancimat test, 120°C, in minutes.

Below the tocopherols the 'strongest' one sets the measure for the antioxidant activity provided a 'weaker' one is not present in substantially greater quantity.

Cholesterol

The oxidation of food components may not only reduce the palatability but also produce toxic or unwholesome compounds. Well known examples are heavily oxidized fats and overheated deep frying oils: another one is oxidized cholesterol.

Arteriosclerosis and myocardial infarction are the most frequent causes of death in Western countries. Oxidation products of cholesterol (Fig. 15) are supposed to be a possible primary cause besides others.[71–73] Some oxidation products have cytotoxic, atherogenic, mutagenic and carcinogenic effects, and inhibit cholesterol synthesis and membrane functions. Oxidized cholesterols have been detected in many foods of animal origin, and the oxidation of cholesterol has been studied in depth.[74]

The food industry can contribute to minimizing the problem of oxidation of dietary cholesterol by measures such as increased use of antioxidants, processing and packing under nitrogen.

Experiments[75] have demonstrated that the incorporation of 500 ppm

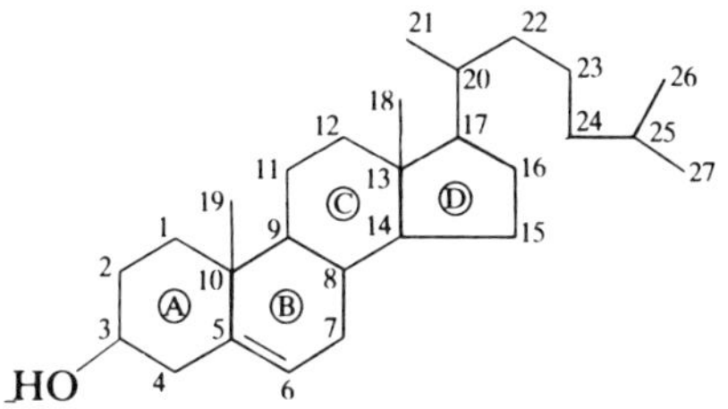

Cholesterol

Most preferred points of oxidation:

C-7 } at the B-ring
C-5 and C-6 }
C-20 and C-25

to some extent C-3 and C-4 at the A-ring

Some of the most suspect toxic oxidation products:

25-Hydroxycholesterol
$3\beta,5\alpha,6\beta$-Cholestantriol
7α-Hydroxycholesterol
7β-Hydroxycholesterol
7-Ketocholesterol
Cholestan-3,5-dien-7-one
5,6-Epoxycholesterol
20α-Hydroxycholesterol

FIG. 15. Some oxidation products of dietary cholesterol.

of ascorbyl palmitate and 100 ppm of dl-α-tocopherol in beef tallow decreased the formation of oxidized cholesterol derivatives during heat treatment.

The usual parameters also influence the autoxidation of cholesterol: catalysts, initiation by heat (especially over 50°C) and light and the presence of oxygen (major surfaces of the products exposed to air, e.g. as in powders).

Cholesterol and its oxidation products may find their way into foods via various pathways:[71,72]

—Egg, egg yolk → egg and egg yolk powders.
—Milk → butter oil, milk powder, evaporated milk, cheese and cheese powder.

—Animal fats→frying and cooking fats, margarine, shortenings, meat products.
—Meat and offal (liver, kidney)→sausages, pies, liver powder.

Through the above channels, oxidized cholesterols may enter a great variety of processed food such as bakery products, chocolate, pasta, soups, baby food, pudding powders, etc.

The formation of oxidized cholesterol and its transfer into many foods should be inhibited by various means. One of importance is the use of natural antioxidants to retard autoxidation, and the use of synergists and chelating agents to inactivate trace metals (especially Cu^{2+}, Fe^{2+}). Processing and packaging under nitrogen is also recommended for very sensitive food products.

Vitamins and Carotenoids

Vitamins

The vitamins most susceptible to oxidation are vitamins A, D (Figs. 16, 17), B_1 and C. To retain nutritional value, foods have to be protected by careful processing. The fat soluble vitamins A and D, and provitamin A (β-carotene) especially can be protected with antioxidants.

Table 19 shows the effect of various antioxidants on the stability of vitamin A palmitate in methyllinoleate at a concentration of 200 000 IU/g. Storage was in closed, ungassed testing tubes in the dark at 25°C.[99]

Table 19 shows that γ-tocopherol is superior to α-tocopherol, and that with the relatively high amount of 2% γ-tocopherol a stability comparable to the one obtained with BHA can be obtained.

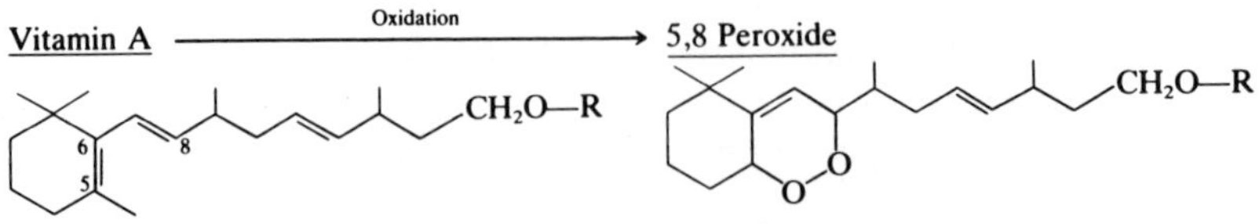

Vitamin A —Oxidation→ 5,8 Peroxide

$CH_2O—R$ 6 5 8 $CH_2O—R$ O O

R = H Retinol
R = $COCH_3$ A-Acetate
R = $CO(CH_2)_{14}CH_3$ A-Palmitate

Under more acid conditions:
5,6 Epoxide

FIG. 16. Oxidation of vitamin A.

Vitamin D_3 (Cholecalciferol) Provitamin D_3 (7-Dehydrocholesterol)

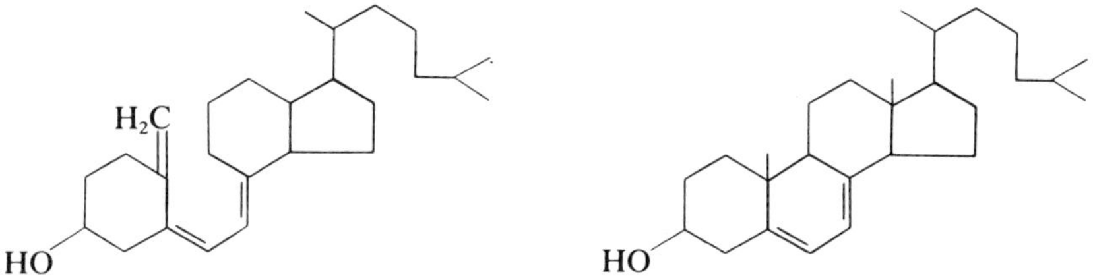

H_2C

HO HO

Vitamin D_3 $\rightleftharpoons$ Pre-Vitamin D_3 $\rightleftharpoons$ Provitamin D_3

↓ Oxidation (heat, 120°C)

various isomers

FIG. 17. Oxidation of vitamin D.

In emulsions with low pH values the oxidation of vitamin A is accelerated (Table 7). In such systems the addition of antioxidants is essential. Good results are obtained by combining tocopherols, ascorbic acid and chelating agents.[9]

Practical experience with natural vitamin A in cod liver oil confirmed the findings from the model: α-tocopherol is a less potent antioxidant for vitamin A compared with γ-tocopherol. The activity of the latter substance in cod liver oil is shown in Table 20.[70]

In contra-distinction to vitamin A, vitamin D is much less

TABLE 19

RETENTION (%) OF VITAMIN A-PALMITATE IN METHYLLINOLEATE (200 000 IU/g), 25°C)

Antioxidant	*Storage during (weeks)*					
	2	*4*	*5*	*6*	*7*	*8*
0	42	11	8	0		
1% α-tocopherol	92	45	31	20	0	
2% α-tocopherol	94	80	66	47	29	15
1% γ-tocopherol	91	73	70	59	46	37
2% γ-tocopherol	93	78	72	71	59	52
1% BHA	90	75	73	69	57	51
1% BHT	88	71	62	57	43	34
1% BHA + 1% BHT	93	78	72	71	59	52

Source: Pongracz *et al.*[99]

TABLE 20
STABILIZATION OF NATURAL VITAMIN A IN COD LIVER OIL

Antioxidant	*Quantity added (ppm)*	*Retention after storage at 50°C and times shown in an open petri-dish*	
		24 h	*48 h*
Control	0	72	1
γ-tocopherol	100	73	3
	200	78	26
	500	80	37
	1 000	83	51

Initial content: 950 IU vitamin A per g.
Initial peroxide value of the cod liver oil: 10·1.
Natural tocopherol content: 247 ppm α-tocopherol.
Source: Pongracz.[70]

sensitive to oxidation, and higher temperatures are required to initiate the deterioration (see Fig. 19). Table 21[99] shows the antioxidant activity of α-tocopherol in an oily vitamin D_3 solution. As with cholesterol, high processing temperatures irreversibly affect provitamin D_3 to give products such as 7-dehydroxy cholesterol.

Carotenoids

The colour of a wide variety of foods is due to their carotenoid content. The oxidation of carotenoids results in the fading of discoloration of such food products. The reaction is mainly induced by light.

Natural antioxidants are specially suitable for stabilizing carotenoid

TABLE 21
STABILIZATION OF VITAMIN D_3 (2% SOLUTION IN MEDIUM CHAIN TRIGLYCERIDES). RETENTION IN %

Storage at 45°C in ungassed bottles (months)	*Control, no antioxidant*	*Addition of 0·25% α-tocopherol*
3	95	99
6	84	93
12	54	86

coloured systems, as can be demonstrated in a simple model:

> White vaseline is warmed up to 60°C and coloured with an oily 30% β-carotene suspension (150 mg pure β-carotene per kg of vaseline), filled into white glass jars and exposed to daylight. The decomposition of β-carotene visibly starts on the surface of the vaseline where oxygen is available, leading to a complete loss of the orange colour. After 3 months the following β-carotene losses can be measured (total loss in the whole glass jar):
>
> Control sample without antioxidant: 33%
> With 500 ppm α-tocopherol: 7%
> With 500 ppm γ-tocopherol: 0%
> With 2000 ppm Ronoxan A®*): 0%

* Commercially available synergistic mixture of 25% ascorbyl palmitate, 5% dl-α-tocopherol and 70% lecithin.

After 6 and 12 months the γ-tocopherol sample will show losses of 3% and 4% respectively. Ascorbyl palmitate does not show any antioxidant activity since it is soluble in vaseline at 60°C. However, when dissolved in the lecithin of the synergistic mixture, an excellent effect is obtained.[99]

Meat and Meat Products

The colour of fresh and processed meat depends on the state of the myoglobin molecule (Fig. 18), which may be influenced with the aid of ascorbates.

The porphyrin ring of the molecule contains bivalent iron which may be linked with gases such as oxygen, nitric oxide or carbon monoxide. These links are reversible only as long as the iron is bivalent.

When exposed to oxidizing agents the iron is transformed into its trivalent state, which imparts a brownish colour to the meat. The various meat pigments can be stabilized by denaturation of the protein part of the molecule. In practice this involves cooking, salting and acidification. Figure 19 gives an overview on the possible changes in fresh meat as well as in the meat curing process.

In the following discussion the term 'ascorbate' is used for ascorbic acid as well as for sodium ascorbate. Most commonly, sodium ascorbate is applied in the meat industry. In contrast to ascorbic

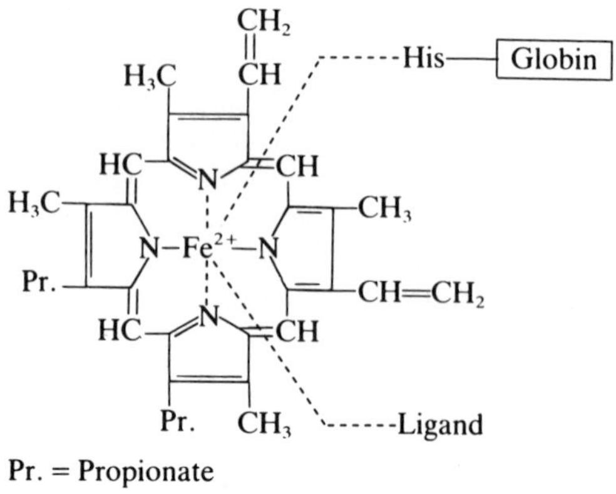

Pr. = Propionate

	Central atom	*Ligand*
Myoglobin	Fe^{2+}	$-(H_2O)$
Oxymyoglobin	Fe^{2+}	O_2
Metmyoglobin	Fe^{3+}	OH^-
Nitrosylmyoglobin	Fe^{2+}	NO
Nitrosylmyochrome	similar to nitrosylmyoglobin, globin denatured	
Myochrome	similar to metmyoglobin, globin denatured	

FIG. 18. The myoglobin molecule.

acid, sodium ascorbate does not react vigorously with nitrite in aqueous solutions releasing toxic nitrous gases.

Fresh Meat

When exposed to oxygen the purplish red myoglobin on the surface of fresh meat turns into the bright red oxymyoglobin. Below the oxymyoglobin layer little oxygen is available. Here the relatively stable brownish metmyoglobin is formed. This equilibrium between the three forms can be influenced by ascorbic acid (steps 2 and 3 in Fig. 19).

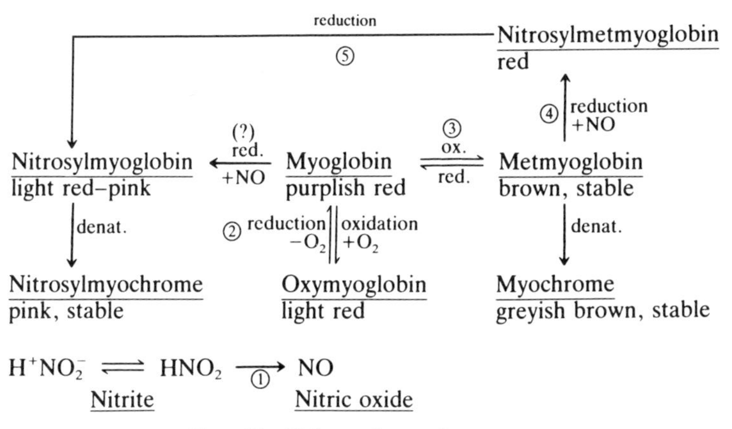

FIG. 19. Colour change in meat.

The practical use of ascorbate in fresh meat depends on national legislation:

The addition of ascorbic acid to fresh ground meat delays the formation of metmyoglobin. In practice, about 200 mg/kg is sufficient to extend the pleasant appearance of hamburgers or uncured sausages by 1 or 2 days. The surface treatment of fresh meat with ascorbic acid spray, or by dipping of whole pieces, also inhibits discoloration of the surface. In many countries this is considered as a deception of customers and therefore prohibited. It is important to ensure that such treatments do not delay microbial meat spoilage.

Ascorbic acid treatment of fresh meat not only retards the oxidation of the meat pigments but also delays fat oxidation. Optimum results in uncured sausages have been obtained with synergistic mixtures containing ascorbates or ascorbyl palmitate, tocopherol and a synergist.

In British pork sausages a condition is known as 'white spot'. It can occur after 2 or 3 days storage, especially in humid and cool conditions. Affected sausages develop round greyish white areas immediately under the skin. These rapidly increase in size and make the products look unappetizing. They rapidly become unsaleable. The problem seems to be caused by fat oxidation since high peroxide values in the affected areas have been detected.

Table 22 demonstrates the retardation of white spot formation in commercial recipe sausages.[76] Note the synergism between the antioxidants.

TABLE 22

RETARDATION OF WHITE SPOT FORMATION IN PORK SAUSAGES[76]

Antioxidant	*Antioxidant (mg per kg of fat)*	*Days storage at 5°C*						
		1	*2*	*3*	*4*	*5*	*6*	*7*
Control		0	00	000	000	0 000	0 000	0 000
Sodium ascorbate	500		0	00	000	000	0 000	0 000
Sodium ascorbate	1 000			0	00	000	000	0 000
Ascorbyl palm. +	25							
α-Tocopherol +	8				0	0	00	000
Citric acid	4							
Ascorbyl palm. +	50							
α-Tocopherol +	16					0	0	00
Citric acid	8							
Ascorbyl palm. +	125							
α-Tocopherol +	40							0
Citric acid	40							

Key: 0, few white spots. 00, intermediate number of white spots. 000, abundant white spots, partial coalescence. 0 000, excessive coalescence of white spots

Cured Meat

Meat curing involves the combination of nitric oxide gas with myoglobin to obtain finally the stable cured cooked meat pigment nitrosylmyochrome. The sequence is shown in Fig. 19:

(1) Nitrous acid is reduced to nitric oxide. The following equations have been suggested:[23,77]

$$3\,HNO_2 \rightarrow 2\,NO + HNO_3 + H_2O$$

In the presence of ascorbate the yield of nitric oxide is improved and the reaction accelerated:

$$\text{Ascorbate} + HNO_2 \rightarrow NO + H_2O + \text{dehydroascorbate}$$

(3) In the presence of nitrous acid Fe^{2+} is readily oxidized to Fe^{3+}.

(4,5) The two reducing reactions appear to be accelerated by ascorbates.

(6) The cured meat pigments are protected from oxidation by residual ascorbate. Its breakdown seems to be initiated by dissociation of nitric oxide from the pigment molecule, favoured by light. Furthermore, the destruction of the pigment is favoured by fat peroxides.

The curing process is improved by ascorbic acid in several ways:

—it accelerates the formation of nitric oxide;
—it increases the yield of nitric oxide, thus about 30% less nitrite is used, and there is less residual nitrite;
—the formation of nitrosyl myoglobin is improved and accelerated;
—the residual ascorbate protects the cured meat pigments from oxidation.

Advantages of ascorbate have been demonstrated by Borenstein[21] in a practical experiment comparing left and right side hams of pigs with various levels of addition of nitrite, ascorbate and erythorbate. The residual nitrite has been assayed, and the colour stability of the final products determined.

Meat Flavour

In cured meat products with a long shelf-life rancidity may become a problem. An example is salami, which is very susceptible, especially after being cut and packed. In a practical experiment[99] with salami 'Tipo Milano', the influence of added γ-tocopherol was investigated. The antioxidant was dispersed in the wine, followed by the usual manufacturing and fermentation processes.

The whole sausages were stored at 5°C with the bottoms cut off. To check the progress of oxidation a 1-cm slice was cut off and rejected. The peroxide value was then determined in the next 1-cm slice (Table 23).

The same sausages have been sliced and packed commercially in transparent vacuum bags. The added γ-tocopherol did not improve the

TABLE 23
PEROXIDE VALUES IN THE FAT PHASE OF SALAMI STABILIZED WITH γ-TOCOPHEROL (meq peroxides/kg)

Storage at 5°C (days)	*Control samples*		*150 ppm γ-Tocopherol without starter culture*	*450 ppm γ-Tocopherol without starter culture*
	Manufactured without starter culture	*Manufactured with starter culture*		
5	1·0	0·6	2·0	1·0
30	8·5	5·6	3·6	1·0

colour stability, but clearly retarded the fat oxidation and inhibited the formation of rancid off-flavours.

Pork products are susceptible to fat oxidation, which causes off-flavours, discoloration, and reduced keeping quality. Frozen pork especially is much less durable than beef. Many publications show that increased feeding of pigs with tocopherols can reduce this phenomenon. The influence of feed composition on lard quality is well known. To obtain good quality lard for raw sausage manufacture, pigs fed with cereals such as oats or barley are preferred.[8]

Birds whose feed contains fish lipids should receive supplements of tocopherols to increase the resistance of the meat to rancidity and off-flavour formation. According to Marusich[78] the development of rancid off-taste is reduced in proportion to the amount of tocopherol deposited in the tissues. To achieve good effects with vitamin E the following supplements are recommended:

—Chickens: 30–50 IU/kg fed continuously or 150–250 IU/kg for about 1 week prior to slaughter.
—Turkeys; 50–100 IU/kg fed continuously or 100–200 IU/kg for 1 month prior to slaughter.
—Pigs: 50 IU/kg fed continuously or 100–200 IU/head each day for 1 month prior to slaughter.

Nitrosamines
Beside the technological advantages previously indicated, based on the antioxidant activity of ascorbates, it should be mentioned that both ascorbates and α-tocopherol prevent the formation of nitrosamines in meat products.[79,80] In the USA the addition of 550 ppm ascorbate/erythorbate to bacon is mandatory for this purpose.

Fish Products
Fish and fish products suffer spoilage by various oxidation processes. They contain more highly unsaturated lipids than meat.

The oxidation of fish lipids proceeds by a free radical mechanism. Furthermore rancidity may be promoted by enzymatic oxidation, e.g. in frozen fish. In other products, e.g. non-frozen fish, enzymatic discoloration due to the oxidation of phenolic compounds takes place. Another source of discoloration is haeme pigment oxidation.[81]

Thus the full range of antioxidants and their combinations is used in the protection of fish products:

—Fat soluble antioxidants such as tocopherols may be used in fish and fish preserves.[82] Their synergism with ascorbates has been evaluated.[83]

—Ascorbates are widely used alone or in combination with citric acid or nicotinamide to protect seafood against various organoleptic and odour changes. Both act as oxygen reducing agents and chelators.[84] Nicotinamide reacts with the myoglobin and increases the formation of the stable pink haemochrome pigments by preventing the objectional discoloration of the myoglobin.[85]

The most effective treatment must be determined for each product and each species, having regard to its lipid composition, its enzyme systems and the processing conditions.

Milk Powder

The addition of ascorbates and sodium citrate to milk, prior to evaporation and spray drying, provides an antioxidant effect for the lipids and fat soluble vitamins A and D in the end product. This is of special interest in vitamin-fortified milk powders to keep the vitamin A overages lower. The shelf-life of such products is prolonged and organoleptic properties and nutritional value improved.

The method of addition depends on the type of product and manufacturing process. Approximately 50 ppm ascorbic acid or sodium ascorbate (to reduce the danger of curdling) and possibly 50 ppm sodium citrate are added to the milk prior to processing. Ascorbyl palmitate (approx. 200 mg per kg of end product) can be dissolved in coconut oil in which subsequently the oily vitamins A and D are diluted and then added to the concentrate before homogenization and spray drying.

Fruits and Vegetables

Enzymatic Browning of Fresh Fruits and Vegetables

When the tissues of fruits or vegetables are damaged during processing (peeling, size reduction, blanching, etc.), natural phenolic compounds are enzymatically oxidized to give yellowish quinone compounds.[86,87] The first step is reversible by the action of antioxidants. The following sequence of polymerization reactions gives rise to intense brownish products such as melanoidins. The flavour of the fruit or vegetable may also be altered by oxidation.

R–(o-dihydroxyphenyl: OH, OH) $\xrightarrow[\text{1/2 } H_2O_2 \text{, 2 } H_2O]{\text{1/2 } O_2^{a} \text{ PPO}^{b} \text{ } H_2O}$ R–(o-quinone: =O, =O) ⟶ brown or purple compounds

Peroxidase

polymerization with other *phenolic compounds*

oxidation

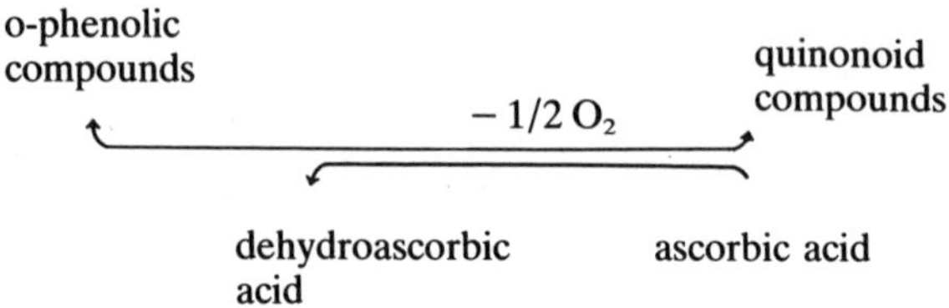

reduction

FIG. 20. Enzymatic oxidation of polyphenols and its inhibition. [a] Elimination with ascorbic acid. [b] Inhibition with ascorbic acid. R = Flavonoids, flavonols, anthocyanins, etc.

Enzymatic browning may be inhibited by various means (Fig. 20):

(1) elimination of undesirable reagents (e.g. oxygen);
(2) scavenging and reduction of intermediates before the irreversible condensation;
(3) inhibition of the involved enzymes.

Ascorbic acid may act in all three forms of inhibition: it reduces quinonoid compounds, it eliminates oxygen and it inhibits polyphenoloxidase (PPO), probably through its complexing properties on copper, which is contained in the PPO. However, with regard to inhibition of PPO some investigations are controversial, possibly as a result of the specific experimental conditions. Usually all three mechanisms contribute to the inhibition of enzymatic oxidation of fruits and vegetables.[86–89]

In practice, fruits and vegetables are protected during processing by the following means:

(1) Lowering of the oxygen concentration by processing/storing in water containing oxygen scavengers such as ascorbic acid, SO_2, etc.

(2) Application of reducing agents (ascorbic acid, SO_2).
(3) Inhibition of enzymes by lowering the pH value below 4 with edible acids; complexing of copper groups in the enzymes; addition of sodium chloride to dipping baths, since also the chloride anion inhibits PPO; heat treatment to inactivate PPO.

Usually such measures are combined to obtain optimum results. The levels of addition and the methods of application depend very much on the raw materials, processing steps and the end products.

Preserves
Oxidative deterioration may also affect fruits, vegetables or mushrooms after heat inactivation of the enzymes during storage in cans or glass jars. It may be inhibited by reduction of the oxygen in the head space of the container, or in the syrup or brine. Here again, ascorbic acid and its salts improve colour and flavour stability in combination with citric acid and sodium chloride.

Nuts
An interesting use for natural antioxidants has been found in the protection of whole or broken nuts (walnuts, almonds, hazelnuts). Rancidity, i.e. the oxidation of the fat on the surface, is delayed by surface treatment with ascorbyl palmitate and tocopherol dissolved in ethanol. The antioxidant solution is applied either by a spray technique or by dipping the nuts. Good results have been achieved by using solutions of 1% ascorbyl palmitate and 0·2% α-tocopherol in 96% ethanol.[99]

Potato Powders
Potato flakes are very susceptible to deterioration during manufacture and storage. Some of the problems are connected with 'Maillard' reactions and can be inhibited with sulphur dioxide or sodium metabisulphite, which evaporates to a great extent during drying. In spite of the low lipid content of potatoes (about 0·1%) the autoxidation of the mostly unsaturated fatty acids bring about serious off-flavour problems (rancidity or hay-like flavour) which are usually encountered with antioxidants. Furthermore metal ions accelerating autoxidation have to be inactivated. Beside the lipids phenolic compounds also have to be protected (Fig. 20).

Antioxidants for use in potato powders, flakes or granules are preferably evaluated by organoleptic testing. Analysis of the volatile compounds present in the head space over the product also gives reliable results.[90,91] The beneficial effect of carnosic acid, derived from rosemary extracts, have been described.[55,90]

Ascorbyl palmitate, ascorbic acid, tocopherol and citric acid have been particularly interesting as substitutes for synthetic antioxidants. Practical experience has shown that best results are achieved by using a combination of above-mentioned products. For potato flakes Bourgeois[92] suggested various combinations. The determination of optimum composition has to take into account processing properties (too high ascorbyl palmitate contents may influence the behaviour of the product in the dryer), and storage quality. The combination of about 250 ppm ascorbyl palmitate and 250 ppm ascorbic acid plus citric acid protected the lipid as well as the aqueous phase of the products.

Beverages

Soft Drinks

The deterioration of the flavour and colour of beverages during manufacture and storage is induced by dissolved and headspace oxygen.[29,93] It is obvious that the amounts of oxygen and trace metal (copper, iron) in the filled beverage container has to be kept as low as possible by adequate processing. As an antioxidant, ascorbic acid may exert a beneficial influence. By eliminating 1 mg of oxygen (= 3·3 ml of air) 11 mg of ascorbic acid are oxidized to give dehydroascorbic acid.

In canned carbonated beverages (pH 2·5–4·0) dissolved oxygen promotes metal corrosion, or even perforation of the container, resulting in flavour changes, discoloration and other objectionable effects, which again may be inhibited or delayed by ascorbic acid.

Bottled carbonated beverages coloured with carotenoids and protected with ascorbic acid are very resistant to fading when exposed to sunlight. In contrast, some azo dyes (e.g. tartrazine, sunset yellow) usually are not stable in such beverages when exposed to light, since this kind of colourant is not resistant to reducing agents under such conditions.

The necessary amount of ascorbic acid to be added to a beverage can be calculated from the amount of oxygen in the head space of the container, the oxygen dissolved and the oxygen penetrating through the packaging material during storage. If nearly all the ascorbic acid in

TABLE 24
COLOUR STABILITY OF β-CAROTENE AND APOCAROTENAL IN AN ORANGE DRINK

Antioxidant added to orange oil (1000 ppm)	*Colourant (20 ppm)*	
	β-Carotene	*Apocarotenal*
Control (no antioxidant)	Faded	Faded
α-Tocopherol	More stable	Slightly faded
γ-Tocopherol	Slightly faded	More stable
BHA	Slightly faded	More stable

Drink with 20 ppm carotenoids added bottled in clear glass bottles, carbonated, stored in a 'light box', 30°C. Determination of colour strength with photometer.

a beverage container is used up by oxygen, browning reactions may occur. Therefore it is recommended that the calculated necessary amount of ascorbic acid should be increased by approximately 10%.

Table 24 shows the influence on the colour stability of an orange drink in glass bottles[99], of the antioxidant used to stabilize the orange oil. In this example α- and γ-tocopherol behaved differently: β-carotene was better protected by α-tocopherol, whereas apocarotenal was better protected by γ-tocopherol.

Beer

The keeping quality of beer is strongly influenced by its oxygen content. Turbidity may occur owing to the oxidation of polyphenols, polymerization and association with proteins. The process is catalysed by metal ions. With regard to organoleptical properties, low oxygen concentrations are desirable to avoid oxidation of flavour components with low threshold values. Gray & Stone[94] obtained a US patent for the treatment of beer with the antioxidant ascorbic acid. The main advantage recognized was the considerably increased colloid stability and decreased gushing tendency.

Today's technological developments have produced bottled beer having 0·5–1 mg O_2 per litre, but practical oxygen contents are often considerably higher.[95] The amount of ascorbic acid to be added to the beer is determined by its oxygen content and the oxygen uptake during the bottling process. According to the theoretical value 11 mg ascorbic acid are added to eliminate 1 mg of dissolved oxygen. In addition approx. 10 mg residual ascorbic acid should be detectable in the bottled beer. In practice this means a dosage of 2–3 g of ascorbic acid per 100 litres of beer. This is preferably realized by dosing a 5% solution of ascorbic acid between Kieselguhr and layer filter.[96]

Flavours

Many flavours include compounds which are very sensitive to oxidation. Oxidation products may further initiate the decomposition of other components of foods. For example, the orange oil in orange drinks has to be protected adequately to avoid the decomposition of the carotenoids which give the yellow-orange hue to the product.[97]

Orange Oil

D-limonene is quantitatively the most important compound in orange oil. It may be present in quantities up to 95%. The flavour quality however is determined by some minor components.

For the evaluation of antioxidants in orange oil D-limonene is a useful model since the autoxidation of the other compounds can be disregarded. The terpenoid components are unstable and readily oxidized imparting a 'terpeney' off-flavour to the orange oil.

The determination of peroxides is a suitable criterion for the degree of oxidation of orange oil only at the beginning of the autoxidation. In the more advanced stages chromatographic methods and sensory evaluation give more reliable results.

The influence of various antioxidants on D-limonene has been evaluated (Table 25) and it was found that γ-tocopherol exerted an antioxidant activity similar to that of BHA or BHT. α-Tocopherol even showed under certain conditions a pro-oxidant effect. Correspondingly, the behaviour of mixed tocopherols was related to their contents of α-tocopherol in comparison with γ-tocopherol and δ-tocopherol.[99]

The stabilization of D-limonene with various fractions from rosemary extracts has been investigated by Hartmann *et al.*[98] A more practical example compares the efficacy of various antioxidants added to the orange oil used for the manufacture of orange drinks coloured with β-carotene and apocarotenal:[99] the colour stability has been tested in a 'light box'. The results are given in Table 24.

TABLE 25
PEROXIDE VALUES OF D-LIMONENE

Antioxidant	*Quantity added (ppm)*	*After 6 weeks (35°C)*	*After 12 weeks (35°C)*
Control	0	8·7	20·6
dl-α-Tocopherol	500	14·0	17·2
dl-γ-Tocopherol	500	4·4	7·8
Mixed tocopherols	1 000	8·2	10·3
BHA	500	6·2	10·1

Initial peroxide value of D-limonene: 1·3.
Mixed tocopherols: 50% total tocopherols, of which about 15% α, 55% γ, 30% δ.
Storage of samples in ampoule bottles.

Peppermint Oil

Peppermint oil is used in a range of food products, especially in confectionery. The compound mainly susceptible to autoxidation is menthol. The best results in peppermint oils are usually obtained with synergistic mixtures including tocopherols and ascorbyl palmitate. Ascorbates alone cannot be recommended as the sole antioxidant. Table 26 gives peroxide values of peppermint oil under various storage conditions. In practice the results can be confirmed with organoleptic

TABLE 26
OXIDATION OF PEPPERMINT OIL—PEROXIDE VALUES (meq peroxide/kg)

Antioxidant	*Quantity (mg/kg)*	*Storage conditions*			
		Transparent bottle, daylight, room temp., 4 days	*Dark, 45°C, 7 days*	*Dark, RT, 45 days*	*Dark, 5°C, 30 days*
Control	—	3·6	12·2	8·3	3·0
BHA	1 000	2·5	3·5	6·0	3·0
α-Tocopherol	1 000	3·2	5·0	8·0	4·5
Ascorbyl palm.	1 000	6·6	6·3	12·5	6·5
Mixture ascorbyl palmitate + α-Tocopherol + Lecithin	500 100 1 400	1·2	2·5	5·5	2·2

tests: best results have been found with the synergistic mixture in chocolate with a peppermint filling.[99]

Confectionery Products

Antioxidants are used in various confectionery products, viz.:

—Protection of the fat-phase (fats, emulsifiers, carotenoids, etc.) in fat containing products such as toffees and caramels to avoid rancidity and fading; the antioxidants tocopherol and ascorbyl palmitate or synergistic mixtures including additional lecithin or citric acid are added to the fat phase of the product.

—Hard candies or products without substantial amounts of lipids may include ascorbic acid and citric acid to protect the carotenoids used for colouration purposes.

—Chewing gums may include tocopherols to avoid bitterness resulting from the oxidation of abietic acid.

Chewing gum is composed of a water insoluble gum base (15–30%) and edible water soluble components such as sugar, acids, etc. Two components have to be protected against oxidation: the flavour/aroma components and the ester gums which give elasticity and plasticity to the gum base. Ester gums are glyceryl esters of rosin acids, and 90% are the glyceryl esters of abietic acid. Colophony (rosin) is the basic raw material for the manufacture of ester gums.

Figure 21 shows the possible oxidation pathways for abietic acid and its derivatives in chewing gum. In general the decomposition products are bitter. The bitterness may be hidden by flavours, sweeteners, etc., but it can be detected organoleptically after the disappearance of the aroma and sweet taste.

From a commercial point of view the use of antioxidants is the most economic way in which to avoid the formation of bitterness in chewing gums.

Table 27 shows the efficacy of antioxidants on colophony and ester gum.[99] Organoleptical tests confirmed that α-tocopherol not only kept the peroxide value low, but also substantially reduced the sensorial deterioration of ester gum.

Two types of chewing gums, manufactured either with α-tocopherol or BHT as the only antioxidants (1 g/kg gum base), have been tested by a panel. The chewing gums were stored under pure oxygen at room temperature. The figures given in Table 28[99] indicate the number of

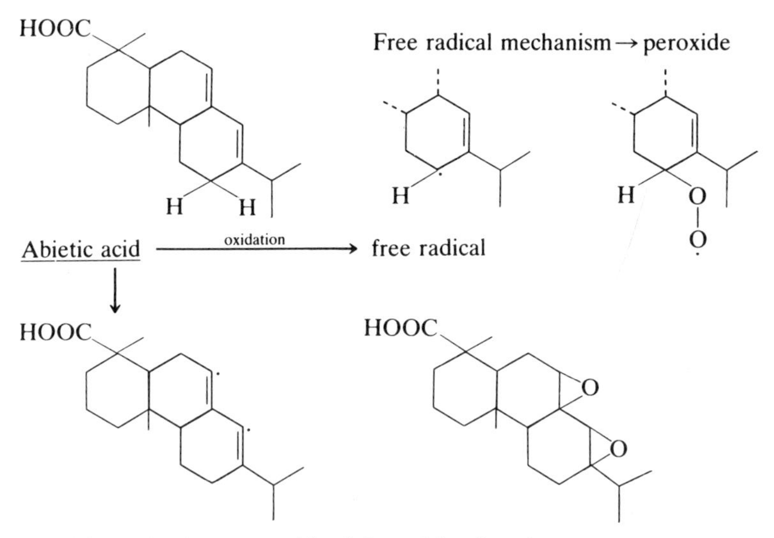

FIG. 21. Oxidation of abietic acid.

TABLE 27

STABILISATION OF COLOPHONY AND ESTER GUM (PEROXIDE VALUE)

Antioxidant	*ppm*	*Storage at RT (weeks)*							
		Colophony				*Ester gum*			
		2·5	*5*	*6*	*7*	*2*	*4*	*7*	*10*
Control	—	15	45	79	120	12	19	27	44
α-Tocopherol	1 000	9	18	28	40	9	13	20	27
	2 000	8	20	23	32	—	—	—	—
γ-Tocopherol	1 000	12	28	43	57	10	15	22	34
	2 000	5	19	31	35	—	—	—	—
BHA	1 000	7	187	41	100	10	17	25	35
BHT	1 000	6	14	31	40	9	13	21	32

Incorporation of antioxidants in colophony: melting at 150°C, addition, cooling, breaking of the sheets, sieving, storing of the fraction between 0·16 and 1·0 mm in open beakers. Incorporation of antioxidants in ester gum: melting at 200°C, cooling to 160°C, addition.

TABLE 28

	Antioxidant (1000 ppm on gum base)			*Total number of testers*
	BHT	*α-Tocopherol*	*Control (no antioxidant)*	
After 2 weeks				
Bubble gum	3	6	14	16
Sticks	4	6	10	15
After 4 weeks				
Bubble gum	0	3	12	12
Sticks	6	5	12	13
After 6 weeks				
Bubble gum	Among 11 testers 5 correctly identified the identical pair in a triangle test with samples stabilized with BHT, and one with the α-tocopherol samples.			

negative comments such as 'bitter', 'strange', 'bad', etc. Under the testing conditions given in Table 28 α-tocopherol clearly improved the keeping quality of the two kinds of chewing gum. After 6 weeks the bubble gum samples with BHT could not be separated significantly from those with α-tocopherol in a triangle test.

Ascorbic acid and its derivatives are less suitable as antioxidants in chewing gums for the following reasons:

—Ascorbic acid may provoke browning reactions. In the manufacture of vitamin C enriched chewing gums the formulation must be carefully adapted. It is only compatible with gum bases which include talc as filler.

—Sodium and calcium ascorbates are unstable in chewing gums.

—Ascorbyl palmitate may also act as an emulsifier, and the gum base is squashed during mastication.

REFERENCES

1. Berger, K. G., Catalysis and inhibition of oxidation processes. *Chem. Ind.*, **3** (1975) 194–9.
2. Hall, G. & Andersson, J., Volatile fat oxidation products. *Lebensmittel-Wissenschaft und -Technologie* **16** (1983) 354–61.

3. Rawls, H. R. & van Santen, P. J., A possible role for singlet oxygen in the initiation of fatty acid autoxidation. *JAOCS*, **47** (1970) 121–5.
4. Kurze, W., Antioxidantien. In *Ullmanns Encyclopaedie der technischen Chemie*, Vol. 8. Verlag Chemie, Weinheim, 1974, pp. 19–45.
5. *Recommended Dietary Allowances*, 9th edn. National Academy of Sciences, Washington, 1980.
6. Tappel, A. L., Vitamin E and free radical peroxidation of lipids. *Ann. NY Acad. Sci.*, **203** (1972) 12–28.
7. Ishikawa, Y. & Yuki, E., Reaction products from various tocopherols with trimethylamine oxide and their antioxidative activities. *Agric. Biol. Chem.*, **39** (1975) 851–7.
8. Schriener, Verhinderung von Ranzigkeit bei Rohwurst. *Fleischwirtschaft*, **47** (1967) 55–6.
9. Fatterpekar, M. S. & Ramasarma, G. B., Stability of vitamin A in aqueous dispersions. *Ind. J. Pharm.*, **24** (1962) 159–62.
10. Isler, O., Progress in the field of fat-soluble vitamins and carotenoids. *Experientia*, **33** (1977) 555–73.
11. Liebing, H. & Karwiese, R., Möglichkeit der Gewinnung von Sterol-Tocopherol-Konzentraten bei der Desodorierung von Pflanzenölen. *Seifen, Öle, Fette, Wachse*, **110** (1984) 573–5.
12. Ernst, H. G., Vitamin E. In *Ullmanns Encyclopaedie der technischen Chemie*, Vol. 23. Verlag Chemie, Weinheim, 1983, pp. 643–9; Reiff, F., Vitamin C. In *Ullmanns Encyclopaedie der technischen Chemie*, Vol. 23. Verlag Chemie, Weinheim, 1983, pp. 685–92.
13. Mattikow, M. & Perlman, D., Treatment of fatty material. US Patent 2,704,764, 1955.
14. Brown, W. & Meng, K. H., Process for recovery of tocopherols and sterols. US Patent 3,108,120, 1963.
15. Brown, W., Process for separating tocopherols and sterols from deodorizer sludge and the like. US Patent 3,153,054 and 3,153,055, 1964.
16. Kasparek, S., Chemistry of tocopherols and tocotrienols. In *Vitamin E*, ed. L. J. Machlin. Marcel Dekker, New York/Basel, 1980, pp. 7–65.
17. Yourga, F. J., Esselen, W. B. & Fellers, C. R., Some antioxidant properties of d-isoascorbic acid and its sodium salt. *Food Res.*, **9** (1944) 188–96.
18. Esselen, W. B., Powers, J. J. & Woodward, R., D-isoascorbic acid as an antioxidant. *Ind Engng Chem.*, **37** (1945) 295–9.
19. Schulte, K. E. & Schillinger, A., Comparison of the kinetics of non-fermentative oxidation of d-isoascorbic and l-ascorbic acids. *Zeitschrift für Lebensmittel-Untersuchung und -Forschung*, **94** (1952) 77–87.
20. Kajita, T. & Senda, M., *Nippon Shokuhin Kogyo Gakkaishi*, **16** (1969) 259–65; *Chemical Abstracts* **73** (1970) 94404s.
21. Borenstein, B., Improving colour and flavour stability of cured meats. *Meat Magazine*, **29** (11) (1963) 30–1.
22. Borenstein, B., The comparative properties of ascorbic acid and erythorbic acid. *Food Technol.*, **19** (11) (1965) 115–7.
23. Ranken, M. D., The use of ascorbic acid in meat processing. In *Vitamin*

C, ed. J. N. Counsell, & D. Hornig. Applied Science Publishers, London, 1981, pp. 105–22.

24. Hornig, D., Weber, F. & Wiss, O., Influence of erythorbic acid on the vitamin C status in guinea pigs. *Experientia,* **30** (1974) 173–4.
25. Hornig, D. & Weiser, H., Interaction of erythorbic acid with ascorbic acid catabolism. *Int. J. Vit. Nutr. Res.,* **46** (1976) 40–7.
26. Moser, U., *Uptake of Ascorbic Acid by Leucocytes.* New York Academy of Sciences, New York, 1987.
27. Arakawa, N., Suzuki, E., Kurata, T., Otsuka, M. & Inagaki, C., Effect of erythorbic acid administration on ascorbic acid content in guinea pig tissues. *J. Nutr. Sci. Vitaminol.,* **32** (1986) 171–81.
28. Thewlis, B. H., Fate of ascorbic acid in the Chorleywood bread process. *J. Sci. Food Agric.,* **22** (1971) 16–9.
29. Kefford, J. F., McKenzie, H. A. & Thompson, P. C. O., Effects of oxygen on quality and ascorbic acid retention in canned and frozen orange juices. *J. Sci. Food Agric.,* **10** (1959) 51–63; Robertson, G. L. & Samaniego, C. M. L., Effect of initial dissolved oxygen on the degradation of ascorbic acid and the browning of lemon juice. *J. Food Sci.,* **51** (1986) 184–92.
30. Huelin, F. E., Coggiola, I. M., Sidhu, G. S. & Kennett, B. H., The anaerobic decomposition of ascorbic acid in the pH range of foods and in more acid solutions. *J. Sci. Food Agric.,* **22** (1971) 540–2.
31. Seib, P. A., Oxidation, monosubstitution and industrial synthesis of ascorbic acid. *Int. J. Vit. Nutr. Res. Suppl.,* **27** (1985) 259–306.
32. Michal, G., *Biochemical Pathways.* Boehringer, Mannheim, 1972.
33. Loewus, F. A., Wagner, G. & Yang, J. C., Biosynthesis and metabolism of ascorbic acid in plants. *Ann. NY Acad. Sci.,* **258** (1975) 7–23.
34. Chatterjee, I. B., Majumder, A. K., Nandi, B. K. & Subramanian, N., Synthesis and some major functions of vitamin C in animals. *Ann. NY Acad. Sci.,* **258** (1975) 24–47.
35. Wintermeyer, U., *Vitamin C.* Deutscher Apotherkerverlag. Stuttgart, 1981.
36. Reichstein, T. A., Grüssner, A. & Oppenhauer, R., Synthese der D- und L-Ascorbinsäure. *Helv. Chim. Acta,* **16** (1933) 1019–33.
37. Grüssner, A. & Schlegel, W., Vitamin C (Chemie und Synthese). In *Ascorbinsäure,* Bd 14, ed. K. Lang. Wiss. Veröffentl. der DGE. D. Steinkopff Verlag, Darmstadt, 1965. pp. 1–16; Brubacher, G. & Vuilleumier, J. P., Biologische Wirksamkeit und Stoffwechsel der Ascorbinsäure und verwandter Substanzen. In *Ascorbinsäure,* Bd 14, ed. K. Lang. Wiss. Veröffentl. der DGE. D. Steinkopff Verlag, Darmstadt, 1965, pp. 61–79.
38. Hudson, B. J. F. & Ghavami, M., Stabilising factors in soybean oil—natural components with antioxidant activity. *Lebensmittel-Wissenschaft und -Technologie,* **17** (1984) 82–5.
39. Hayes, R. E., Bookwalter, G. N. & Bagley, E. B., Antioxidant activity of soybean flour and derivatives. *J. Food Sci.,* **42** (1977) 1527–32.
40. Souci, S. W. & Mergenthaler, E., Weitere chemische Zusatzstoffe. In

Die Bestandteile der Lebensmittel, ed. J. Schormüller. Springer Verlag, Berlin, 1965, pp. 1159–78.
41. Cantarelli, C. & Montedoro, G., Extraction des antioxydants naturels des olives. In *Proceedings 14th Int. Symp. Saarbrücken, 1972.* ed. R. Ammon & J. Hollo, Darmstadt, D. Steinkopff Verlag, 1974, pp. 84–93.
42. Japanese Patents 7,22,250/9,067, 884/9,126,578 to Tanabe.
43. Seher, A. & Löschner, D., Natürliche Antioxidantien. V: Antioxidantien und Synergisten aus antarktischem Krill. *Fette-Seifen-Anstrichmittel,* **87** (1985) 454–7; VI: Aminosäurengemische als effiziente Synergisten. *Fette-Seifen-Anstrichmittel,* **88** (1986) 1–6.
44. Oliveto, E. P., Nordihydroguaiaretic acid. A naturally occurring antioxidant. *Chem. Ind.* (2nd Sept. 1972) 677–9.
45. Stan, H. J. & Huni, W., Flavonole-Mutagene in unserer täglichen Nahrung. *Deutsche Lebensmittel-Rundschau,* **80** (1984) 85–7.
46. Herrmann, K., Über die antioxidative Wirkung von Gewürzen. *Deutsche Lebensmittel-Rundschau,* **77** (1981) 134–9.
47. Chipault, J. R., Mizuno, G. R., Hawkins, J. M. & Lundberg, W. P., The antioxidant properties of natural spices. *Food Res.,* **17** (1952) 46–55.
48. Chipault, J. R., Mizuno, G. R. & Lundberg, W. O., The antioxidant properties of spices in foods. *Food Technol.,* **10** (1956) 209–11.
49. Gerhardt, U. & Blat, P., Dynamische Messmethode zur Ermittlung der Fettstabilität. Einfluss von Gerwürzen und Zusatzstoffen. *Fleischwirtschaft,* **64** (1984) 484–6.
50. Palitzsch, A., Schulze, H., Methl, F. & Baas, H., Untersuchungen über die Wirkung von Naturgewürzen. I. Naturgewürze und Gewürzextrakte. *Fleischwirtschaft,* **49** (1969) 1349–54.
51. Palitzsch, A., Schulze, H., Lotter, G. & Steichele, A., Untersuchungen über die Wirkung von Naturgewürzen. III. Gewürzextrakte und Extraktionsrückstände. *Fleischwirtschaft,* **54** (1974) 63–8.
52. Gerhardt, U. & Schröter, A., Antioxidative Wirkung von Gewürzen. *Gordian,* **83** (1983) 172–6.
53. Griffiths, B. & McDonald, B., The natural answer. *Food Flavourings, Ingredients & Processing,* **7** (5) (1985) 44–7.
54. Chang, S. S., Ostric-Matijaseric, B., Hsieh, O. A. L. & Huang, C., Natural antioxidants from rosemary and sage. *J. Food Sci.,* **42** (1977) 1102–6.
55. Bracco, U., Löliger, J. & Viret, J. L., Production and use of natural antioxidants. *JAOCS,* **58** (1981) 686–90.
56. Brieskorn, C. H. & Dömling, H. J., Carnosolsäure, der wichtigste antioxidativ wirksame Inhaltsstoff des Rosmarin- und Salbeiblattes. *Zeitschrift für Lebensmittel-Untersuchung und -Forschung.* **141** (1969/79) 10–6.
57. Gerhardt, U. & Schröter, A., Rosmarinsäure, ein natürlich vorkommendes Antioxidans in Gewürzen. *Fleischwirtschaft,* **63** (1983) 1628–30.
58. Griffith, T. & Johnson, J. A., Relation of the browning reactions to storage stability of sugar cookies. *Cereal Chem.,* **34** (1957) 159–69.

59. Kessler, H. G., *Food Engineering and Dairy Technology*. Verlag A. Kessler, Freising (FRG), 1981.
60. Evans, C. D., Moser, H. A., Cooney, P. M. & Hodge, J. E., Amino-hexose-reductones as antioxidants. *JAOCS,* **35** (1958) 84–8.
61. Rhee, C. & Kim, D. H., Antioxidant activity of acetone extracts obtained from a caramelisation-type reaction. *J. Food Sci.,* **40** (1975) 460–2.
62. Packer, J. E., Direct observation of a free radical interaction between vitamin E and vitamin C. *Nature,* **278** (1979) 737–8.
63. Lambelet, P., Saucy, F. & Löliger, J., Chemical evidence for interactions between vitamins E and C. *Experientia,* **41** (1985) 1384–8.
64. Lundberg, W. O., Dockstader, W. B. & Halvorson, H. O., The kinetics of the oxidation of several antioxidants in oxidizing fats. *JAOCS,* **24** (1947) 89–92; Lundberg, W. O., *Autoxidation and Antioxidants.* Interscience Publishers, New York, 1962.
65. Heimann, W. & von Pezold, H., Über die prooxygene Wirkung von Antioxygen. *Fette-Seifen-Anstrichmittel,* **59** (1957) 330–8.
66. Cillard, J., Cillard, P., Cormier, M. & Girre, L., α-tocopherol prooxidant effect in aqueous media: increased autoxidation rate of linoleic acid. *JAOCS,* **57** (1980) 252–5.
67. Cillard, J., Cillard, P. & Cormier, M., Effect of experimental factors on the prooxidant behaviour of α-tocopherol. *JAOCS,* **57** (1980) 255–61.
68. Kanner, J., Mendel, H. & Budowski, P., Prooxidant and antioxidant effects of ascorbic acid and metal salts in a β-carotene-linoleate model system. *J. Food Sci.,* **42** (1977) 60–4.
69. Grosch, W., Abbau von Linol- und Linolensäure hydroperoxiden in Gegenwart von Ascorbinsäure. Zeitschrift für Lebensmittel-Untersuchung und Forschung, **163** (1977) 4–7.
70. Pongracz, G., γ-Tocopherol als natürliches Antioxidans. *Fette-Seifen-Anstrichmittel,* **86** (1984) 455–60.
71. Sieber, R., Oxidiertes Nahrungscholesterin—eine Primärursache der Arteriosklerose? *Ernährung/Nutrition,* **10** (1986) 547–56.
72. Appelqvist, L. A. & Nourooz-Zadeh, J., The content of some products of cholesterol oxidation in Swedish food. In *Proceedings Lipidforum Göteborg,* April 22–23, 1985. Kompendietryckeriet, Kallered, 1986.
73. Eneroth, P., Studies on autoxidation of cholesterol. In *Proceedings Lipidforum Göteborg,* April 22–23, 1985. Kompendietryckeriet, Kallered, 1986.
74. Smith, L. L., *Cholesterol Autoxidation.* Plenum Press, New York, 1981.
75. Won Park, S. & Addis, P. D., Further investigation of oxidized cholesterol derivatives in heated fats. *J. Food Sci.,* **51** (1986) 1380–1.
76. Roche Products Limited, *Control of White Spot . . . Control of Colour . . .* Roche Products Ltd, Welwyn Garden City, Hertfordshire, July 1971.
77. Möhler, K., *Das Pökeln.* Rheinhessische Druckerwerkstätte. Alzey, 1980.
78. Marusich, W. L., Vitamin E as an in vivo lipid stabilizer and its effect on

flavour and storage properties of milk and meat. In *Vitamin E,* ed. L. J. Machlin. Marcel Dekker, New York/Basel, 1980, pp. 445–72.
79. Bauernfeind, J. C., Ascorbic acid technology in agricultural, pharmaceutical, food and industrial applications. In *Ascorbic Acid,* ed. P. A. Seib & B. M. Tolbert. Adv. Chem. Ser. No. 200. American Chemical Society, Washington, 1982, pp. 395–497.
80. Mergens, W. J., Efficacy of vitamin E to prevent nitrosamine formation. In *Vitamin E, Biochemical, Hematological and Clinical Aspects,* ed. B. Lubin & L. J. Machlin. New York Academy of Sciences, New York, 1982.
81. Naughton, J. J., Zeitlin, H. & Frodyma, M. M., Spectral reflectance studies of the heme pigments in tuna fish. *Agric. Food Chem.,* **6** (1958) 933–8.
82. Olcott, H. S., Oxidation of fish lipids. In *Fish in Nutrition,* ed. H. Eirik. Fishing News Ltd, London, 1962.
83. Asahara, M., Matsuzaki, Y. & Matsumori, S., Antioxidant effect of natural tocopherol mixture on salted and dried marine fish. *Nippon Shokuhin Kogyo Gakkaishi,* **22** (1975) 467–73.
84. Bauernfeind, J. C. & Pinkert, D. M., Food processing with ascorbic acid: *Adv. Food Res.,* **18** (1970) 219–315.
85. Koizuma, C. & Nonaka, J., Pink colour fixation of the salted product of Alaska pollack roe with nicotinic acid and nicotinamide. *Bull. Jap. Soc. Scient. Fisher.,* **40** (1974) 789–97; Spectroscopic observation on the orange discoloured meat of canned tuna fish. *Bull. Jap. Soc. Scient. Fisher.* **39** (1973) 237.
86. Herrmann, K., Über die Verfärbungen des Gemüses durch phenolische Inhaltsstoffe. *Ind. Obst. und Gemüseverwertung,* **61** (1976) 257–61.
87. Vamos-Viyazo, L., Polyphenol-oxidase and peroxidase in fruits and vegetables. *CRC Crit. Rev. Food Sci. & Nutr.,* **15** (1981) 49–127.
88. Dimpfl, D. & Somogyi, J. C., Enzymatische Bräunung und ihre Hemmung durch verschiedene Substanzen. *Mitt. Gebiete Lebens. Hyg.,* **66** (1975) 183–90.
89. Matheis, G., Enzymatic browning of foods. *Zeitschrift für Lebensmittel-Untersuchung und -Forschung,* **176** (1983) 454–62.
90. Löliger, J. & Jent, A., Analytical methods for quality control of dried potato flakes. *Am. Pot. J.,* **60** (1983) 511–25.
91. Buttery, R. G., Autoxidation of potato granules. *Agric. Food Chem.,* **9** (1961) 245–52.
92. Bourgeois, C., Utilisation du palmitate d'ascorbyle et de l'acide ascorbique antioxygènes dans les purées deshydratées. Presented at the Additives in Food Industries Symposium, Madrid, October 15–17, 1986.
93. Henshall, J. D., Ascorbic acid in fruit juices and beverages. In *Vitamin C,* ed. J. N. Counsell & D. Hornig. Applied Science Publishers, London, 1981, pp. 123–37.
94. Gray, Ph. P. & Stone, I., *J. Inst. Brew.,* **45** (1939) 443–52; *Wallerstein Lab. Comm.,* **24** (1961) (84) 179–92.

95. Ullmann, F., Massnahmen zur Verminderung der Sauerstoffaufnahme auf dem Abfüllweg. *Brauerei-Rundschau,* **92** (1981) 143–7.
96. Anderegg, P. & Hug, H., Ascorbinsäure als Antioxidans für Bier. *Brauerei-Rundschau,* **92** (1981) 241–3.
97. Crandall, P. G., Kesterson, J. W. & Dennis, S., Storage stability of carotenoids in orange oil. *J. Food Sci.,* **48** (1983) 924–7.
98. Hartmann, V. E., Racine, Ph., Garnero, J. & Tollard d'Audiffret, Y., Les extraits de romarin, antioxygènes naturels utilisables dans la protection des huiles essentielles. *Parfums, Cosmétiques, Arômes (Paris),* **36** (December 1980) 33–40.
99. Pongracz, G., Kracher, F. & Schuler, P., Unpublished results from Technical Services for the Food and Pharmaceutical Industries, Vitamin Division, Roche Basel (1978–1987).
100. Allen, J. C. & Hamilton, R. J. (Eds.) *Rancidity in Foods,* 2nd edn. Elsevier Applied Science, London/New York, 1989.

Chapter 5

NATURAL ANTIOXIDANTS NOT EXPLOITED COMMERCIALLY

DAN E. PRATT

Department of Foods and Nutrition, Purdue University, West Lafayette, Indiana 47907, USA

&

BERTRAM J. F. HUDSON

Department of Food Science and Technology, University of Reading, Whiteknights, Reading RG6 2AP, UK

INTRODUCTION

Reports in recent years both in the popular and scientific press have stressed the value and advantages of natural ingredients as food preservatives. There is an implied assumption of safety for compounds that occur naturally in foods and that have been consumed for many centuries. It is not the intent of the authors to debate the issue of superiority of either natural or synthetic food components as to the safety or functional properties. It is preferable, however, to use substances that do not pose problems of proof of safety. Caution should be employed in the use of natural compounds: except for the major commercial synthetic versions (tocopherols, ascorbic acid) they have not usually been subjected to scrutiny and scientific evaluation as have the artificial synthetic compounds (BHA, BHT). Their potential as mutagens, carcinogens, teratogens, or as other pathogens must be investigated.

Use of natural compounds to prevent oxidative deterioration of lipids and other organic chemical is not new. Perhaps the first comprehensive study and certainly one which first stimulated interest in the area of antioxidation was by Morreu and Dufraise (Blanck[1]). Some 70 years ago, these researchers investigated over 500 natural and synthetic compounds for antioxidant activity. Most of the studies immediately following were centred on ingredients to inhibit oxidation

TABLE 1
SOME SOURCES OF NATURAL ANTIOXIDANTS

Algae	Plant extracts
Cereals	Protein hydrolysates
Cocoa products	Resins
Citrus	Various peppers
Herbs and spices	Onion and garlic
Legumes	Olives
Oil-seeds	

in rubber, gasoline, plastic, and other non-food materials. However, it was these early studies that initiated our interest in the continuing search for chemicals to regulate oxidation.

Musher[2] demonstrated that rancidity in lard could be inhibited by soy flour, oat flour and sesame flour. His greatest antioxygenic activity was from suspending a bag of defatted soy flour in melted lard.

Natural antioxidants in foods may be from (a) endogenous compounds in one or more components of the food; (b) substances formed from reactions during processing; and (c) food additives isolated from natural sources. Most natural antioxidants are from plants. Of the nearly 300 000 species of angiosperms in the plant kingdom, less than 500 have been gathered or cultivated as human foods. Some common sources of natural antioxidant are presented in Table 1. Most of these contain compounds that possess antioxidant activity. They are primarily polyphenolics that may occur in all parts of the plant—wood, bark, stems, leaves, fruit, roots, flowers, pollen and seeds (Table 2). The antioxidant activities in these plants range from extremely slight to very great.

TABLE 2
RELATIVE CONCENTRATION OF FLAVONOIDS AND RELATED COMPOUNDS IN PLANT TISSUE

Tissue	*Relative concentrations*
Fruit	Cinnamic acids > catechins ≃ leucoanthocyanins (flavan-3,4-diols) > flavonols
Leaf	Flavonols ≃ cinnamic acids > catechins ≃ leucoanthocyanins
Wood	Catechins ≃ leucoanthocyanins > flavanols > cinnamic acids
Bark	As wood but greater concentrations

Natural antioxidants may function in one or more of the following ways: (a) as reducing agents, (b) as free radical scavengers, (c) as complexers of pro-oxidant metals, and (d) as quenchers of the formation of singlet oxygen. These compounds are most commonly phenolic or polyphenolic from plant sources. The most common natural antioxidants are flavonoids (flavanols, isoflavones, flavones, catechins, flavanones), cinnamic acid derivatives, coumarins, tocopherols, and polyfunctional organic acids.

FLAVONOIDS

This term embraces a wide variety of compounds which result in plants from photosynthesis. For example, in the case of one of the best known of them, quercetin, the sequence is as follows:

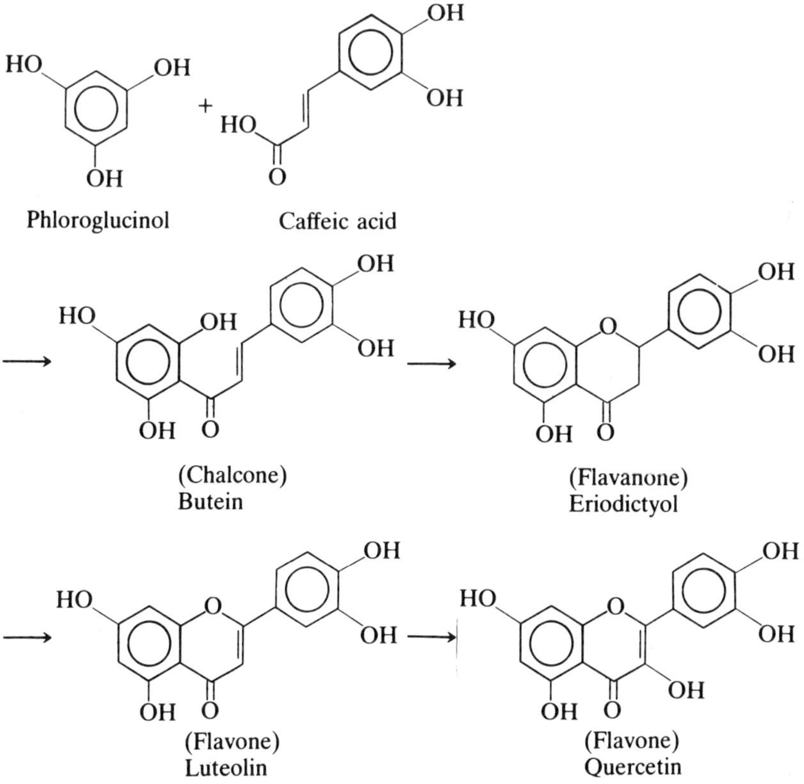

Phloroglucinol + Caffeic acid

→ (Chalcone) Butein → (Flavanone) Eriodictyol

→ (Flavone) Luteolin → (Flavone) Quercetin

The flavonols, as well as both their precursors and the corresponding dihydro-flavonols, are universally present and widespread in plant material, especially in leaves and fruit, whether as glycosides or aglycones. Typically, brussels sprouts can contain 50 mg/kg of quercetin, lettuce 200 mg/kg and apricots 50 mg/kg, though onions, containing as much as 10 g/kg, are one of the richest sources. In addition to such sources, the ingestion of cereal grains, tubers, tea and coffee provides significant supplementation of flavonoids and related compounds in a normal mixed, healthy diet. The occurrence of flavonoids in common edible fruits, leaves and other parts of food plants is well documented by Herrmann.[3]

Antioxidant Activity Assessment

In considering the antioxidant properties of the whole flavonoid group of compounds it must be kept in mind that a wide range of methods of assessment, even in in-vitro studies, has been used, no single one of which is completely free from criticism if comparisons are desired. In general, means have had to be devised for artificially accelerating the oxidation or auto-oxidation processes so that quantitative data can be obtained reasonably quickly. For example, among the most popular methods of evaluation is the use of unrealistically high temperatures. This can be misleading, since oxidation mechanisms can change as temperatures are raised.[4,5]

In this chapter, a comprehensive account of the many studies of the antioxidant properties of food flavonoids and related compounds will not be attempted. Rather, a general survey will be made, including judgements on the comparative efficacies of the various types of compounds falling under the general heading, and the mechanisms involved. Structures of the more significant groups of flavonoids are shown in Fig. 1.

Many of the flavonoids and related compounds have marked antioxidant characteristics in lipid–aqueous and lipid–food systems (Tables 3, 4, Fig. 1). As may be seen, certain flavones, flavonols, flavonones, flavanones, and cinnamic acid derivatives have considerable antioxidant activity. The very low solubility of these compounds in lipids is often considered a disadvantage and has been reported as a serious disadvantage if an aqueous phase is also present.[6] However, flavonoids suspended in the aqueous phase of a lipid–aqueous system offer appreciable protection to lipid oxidation.[7–9] Also, Lea & Swoboda,[10] over 30 years ago, found that flavonols were effective antioxidants when suspended in lipid systems.

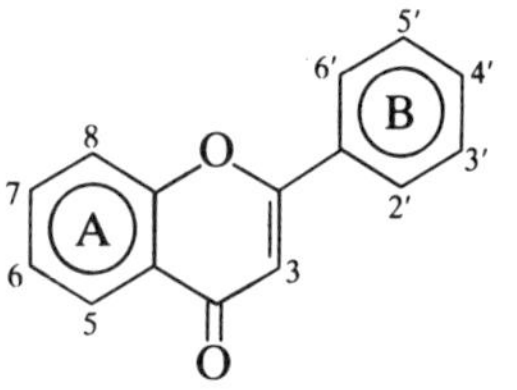

Flavones

Quercetin 3,5,7,3′,4′-penta OH
Fisetin 3,7,3′,4′-tetra OH
Luteolin 5,7,3,4-tetra OH
Quercitrin 5,7,3,4-tetra OH
3-O rhamnoside

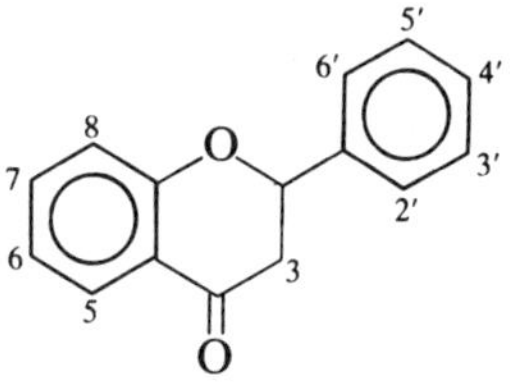

Flavanones

Taxifolin 3,5,7,3′,4′-penta OH
Fustin 3,7,3′,4′-tetra OH
Eriodictyol 5,7,3,4-tetra OH

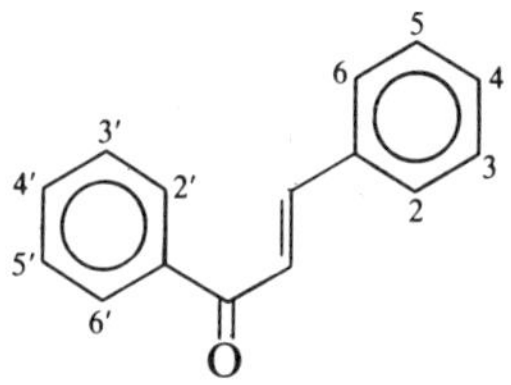

Chalcones

Butein 2′,4′,3,4-tetra OH
Okanin 2′,3′,4′,3,4-penta OH

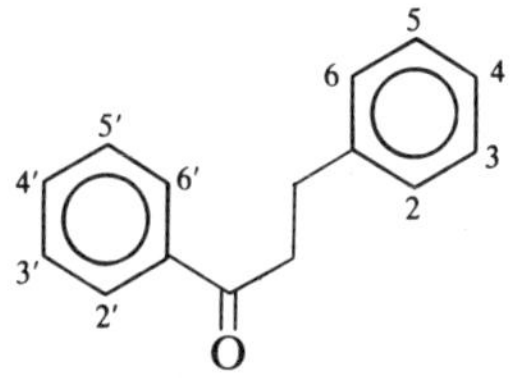

Dihydrochalcones

Phloretin 2′,4′,6′,4-tetra OH

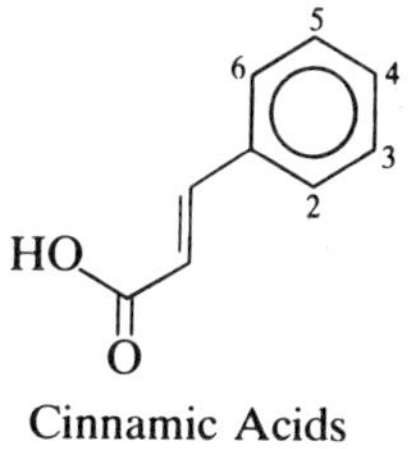

Cinnamic Acids

Caffeic acid 3,4-di OH
Ferulic acid 4-OH, 3-OMe

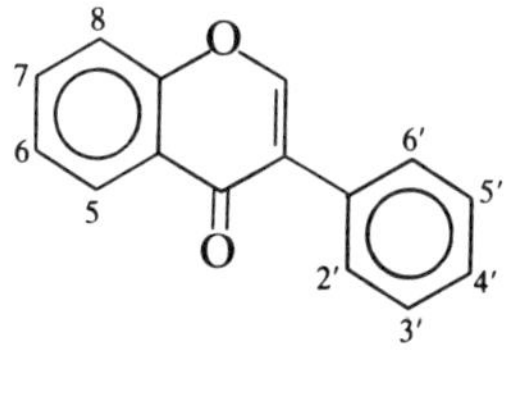

Iso-flavones

Daidzein 7,4′-di OH
Genistein 5,7,4′-tri OH

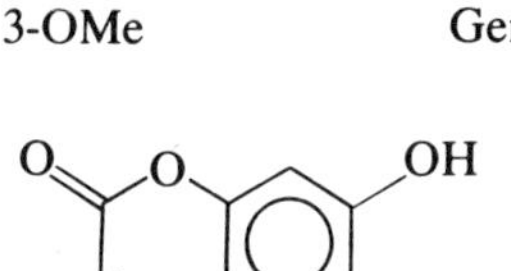

Aesculetin

FIG. 1. Structures of flavonoid and related compounds.

Antioxidant Mechanism

The antioxidant action of flavonols is bi-modal. Flavanols are known to form complexes with metals. Chelation occurs at the 3-hydroxy,4-keto grouping and/or at the 5-hydroxy,4-keto group, when the A ring is hydroxylated in the 5-position.[11] An *o*-quinol grouping on the B-ring can also demonstrate metal-complexing activity.[7,12] However, the major value of flavonoids and cinnamic acids is in their primary antioxidant activity (i.e., as free radical acceptors and as chain breakers).

The main evidence that these compounds work mainly as primary antioxidants is their ability to work equally well in metal catalysed and uncatalysed systems. They are also efficient antioxidants in systems catalysed by relatively large molecules, such as haem and other porphyrin compounds. They are also effective against lipoxygenase catalysed reactions. These compounds cannot be envisaged as forming complexes with flavonols. In addition, hesperetin (5,7,3′-trihydroxy-4′ methoxyflavone), which possesses an active metal complexing site, has demonstrated negligible antioxidant activity.

Structure–Activity Relationships

The position and the degree of hydroxylation is of primary importance in determining the antioxidant activity of flavonoids proper (see Tables 3 and 4). There is general agreement that orthodihydroxylation of the B ring contributes markedly to the antioxidant activity of flavonoids.[13] The para-quinol structure of the B ring has been shown to impart even greater activity than the orthoquinol structure; while the meta configuration has no effect on antioxidant activity.[9] However, para and meta hydroxylation of the B ring apparently do not occur commonly in nature.

All flavonoids with the 3′,4′-dihydroxy configuration possess antioxidant activity. Two (robinetin and myricetin) have an additional hydroxyl group at the 5′ position, which increases the antioxidant activities over those of the corresponding flavones without the 5′-hydroxyl group, fisetin and quercetin. Two flavanones (naringenin and hesperetin) having only a single hydroxyl group on the B ring possess only slight antioxidant activity. Hydroxylation of the B ring is a major consideration for antioxidant activity.

Meta 5,7-hydroxylation of the A ring apparently has little, if any, effect on antioxidant activity. This is evidenced by the findings that quercetin and fisetin have relatively the same activity and myricetin possesses the same activity as robinetin. Heimann and his

TABLE 3
ANTIOXIDANT ACTIVITY OF FLAVONES

Compound	*Time to reach a peroxide value of 50*[a] *(h)*	*Induction period (h) by Rancimat*[b]
Control		
Stripped corn oil	105	
Lard		1·4
Aglycones		
Quercetin (3,5,7,3′,4′-pentahydroxy)	475	7·1
Fisetin (3,7,3′,4′-tetrahydroxy)	450	8·5
Luteolin (5,7,3′,4′-tetrahydroxy)		4·3
Myricetin (3,5,7,3′,4′,5′-hexahydroxy)	552	
Robinetin (3,7,3′,4′,5′-pentahydroxy)	750	
Rhamnetin (3,5,3′,4′-tetrahydroxy 7-methoxy)	375	
Glycosides		
Quercitrin (Quercetin 3-rhamnoside)	475	1·9
Rutin (Quercetin 3-rhamnoglucoside)	195	

[a] 5×10^{-4} M in stripped corn oil.
[b] 2.3×10^{-4} M in lard.

associates[12,13] reported that meta 5,7-hydroxylation lowered antioxidant activity. On the other hand, Mehta & Seshadri[6] found quercetin to be a more effective antioxidant than 3,3′,4′-trihydroxyflavone. Data from our laboratory support the finding of Mehta & Seshadri.

The importance of other sites of hydroxylation were studied by Lea & Swoboda;[10] Mehta & Seshadri;[6] Simpson & Uri;[14] and Uri.[7] The two former groups found quercetagetin (3,4,5,7,3,4′-hexahydroxyflavone) and gossypetin (3,4,7,8,3′,4′-hexahydroxyflavone) to be very effective antioxidants. Uri[7] found that the ortho-dihydroxy grouping on one ring and the paradihydroxy grouping on the other (i.e. 3,5,8,3′,4′- and 3,7,8,2′,5′-pentahydroxyflavones) produced very potent antioxidants. These four polyhydroxyflavones

TABLE 4

ANTIOXIDANT ACTIVITY OF FLAVANONES

Compound	*(Time to reach a peroxide value of 50[a] (h)*	*Induction period (h) by Rancimat[b]*
Control		
Stripped corn oil	105	
Lard		1·4
Aglycones		
Taxifolin (dihydroquercetin) (3,5,7,3′,4′-pentahydroxy)	470	8·2
Fustin (3,7,3′,4′-tetrahydroxy)		6·7
Eriodictyol (5,7,3′,4′-tetrahydroxy)		6·7
Naringenin (3,5,3′-trihydroxy)	198	
Hesperetin (5,7,3′-trihydroxy 4′-methoxy)	125	
Glycosides		
Hesperidin (Hesperitin 7-rhamnoglucoside)	125	
Neohesperidin (Hesperitin 7-glucoside)	135	

[a] 5×10^{-4} M in stripped corn oil.
[b] $2{\cdot}3 \times 10^{-4}$ M in lard.

are the most potent flavonoids, as antioxidants, yet reported in non-aqueous systems. Simpson & Uri[14] found 7-*n*-butoxy-3,2′,5′-trihydroxyflavone to be the most effective antioxidant of 30 flavones studied in aqueous emulsions of methyl linoleate.

The 3-glycosides possess approximately the same antioxidant activity in some tests, but less in others, as the corresponding aglycone when the glycosyl substitution is with monosaccharide. In the case of rutin, in which the substitution is with a disaccharide, antioxidant activity is reduced. The antioxidant capacity of a commercial preparation of rutin is considerably lower than that of the corresponding aglycone, quercetin. Kelley & Watts[15] studied the antioxidant effect of several flavonoids and found rutin somewhat inferior to quercetin and quercitrin but the differences were not as great as we have found. Chromatographic purification and the use of several commercially available samples (to eliminate the effect of possible contamination) did not alter the finding. Kelley & Watts,[15] using a carotene–lard system also found that quercitrin had approximately the same protective effect as quercetin. Glycosides and aglycones are likely, however, to exert comparable effects *in vivo*.

Crawford *et al.*[8] found that methylation of the 3-hydroxyl group of quercetin only slightly lowered antioxidant activity. However, considerable importance has been attached to the free 3-hydroxyl by others.[12,13] Mehta & Seshadri[6] postulated that the 3-hydroxyl and the 2,3 double bond allowed the molecule to undergo isomeric changes to diketo forms which would possess a highly reactive CH group in position 2.

Dihydroquercetin was found to have the same antioxidant activity as quercetin, indicating either that the 2,3 double bond is not of major importance to antioxidant activity or that conversion of dihydroquercetin to quercetin took place while the compound was in contact with the oxidising fat. Mehta & Seshadri[6] suggested that conversion might account for the antioxidant activity of dihydroquercetin. However, chromatographic tests demonstrated that dihydroquercetin is not converted to quercetin by the hydrolysis procedure, nor could quercetin be chromatographically detected in the carotene–lard system in which dihydroquercetin was used as antioxidant. Dihydroquercetin was still present after 12 h in the system. More recent findings suggest that neither the 3-hydroxyl group nor the 2,3 double bond are essential for high activity, taxifolin and luteolin both being highly active.[11]

Perhaps the best potential source of flavonoids for food antioxidants is from wood as a by-product of lumber and pulping operations. Whole bark of the Douglas fir contains about 5% of taxifolin or dihydroquercetin (3,4,7,3′,4-pentahydroxyflavonone). The cork fraction, readily separated from the bark, contains up to 22% dihydroquercetin.[8] Kirth[16] reported that approximately 150 million pounds of dihydroquercetin are potentially available annually in Oregon and Washington alone. Quercetin (3,4,7,3′,4′-pentahydroxyflavone) has been produced commercially as an antioxidant from wood sources.[16] Quercetin is present in much lower amounts in wood and bark than is dihydroquercetin, but it can be obtained in quantity from it by oxidation.

FLAVONOID-RELATED COMPOUNDS

High levels of antioxidant activity are not confined to the flavones, flavanones and their glycosides. Consideration of the biosynthetic

TABLE 5
ANTIOXIDANT ACTIVITY OF SOME FLAVONOID-RELATED COMPOUNDS

Compound	*Time to reach a peroxide value of 50[a] (h)*	*Induction period (h) by Rancimat*	
Control			
Stripped corn oil	110	*A*	*B*
Lard		1·3[b]	0·35[c]
Iso-flavones			
Daidzein		1·4	
(7,4′-dihydroxy)			
Genistein		2·6	
(5,7,4′-trihydroxy)			
Chalcones			
Butein		94	
(2′,4′,3,4-tetrahydroxy)			
Okanin		97	
(2′,3′,4′,3,4-pentahydroxy)			
Phenolic acids			
Protocatechuic acid			4·8
(3,4-dihydroxy-benzoic acid)			
Gallic acid			28·6
(3,4,5-Trihydroxy-benzoic acid)			
Coumaric acid	120		0·8
(*p*-hydroxy-cinnamic acid)			
Ferulic acid	145		2·0
(4-hydroxy-3-methoxy-cinnamic acid)			
Caffeic acid	495		23·3
(3,4-dihydroxy-cinnamic acid			
Dihydrocaffeic acid			31·4
(3,4-dihydroxyphenyl-propionic acid)			
Chlorogenic acid	505		
(caffeoyl quinic ester)			
Quinic acid	105		
Phenolic ester—Propyl gallate	435		21·8
Miscellaneous			
D-Catechin	410		
Hesperidin methyl chalcone	135		
Aesculetin (6,7-dihydroxycoumarin)			15·5

[a] 5×10^{-4} M in stripped corn oil.
[b] 0·05% of test compound in lard, batch A.
[c] 0·05% of test compound in lard, batch B.

pathways leading to these compounds, and to variants of them, suggests that their precursors, which are well known to survive along with the end-products, in natural food commodities, should also be evaluated. These include phenolic acids and chalcones. The variants include isoflavones and others. The observed antioxidant activities of many of these compounds are recorded in Table 5.

Phenolic Acids

A major portion of the antioxidant activity of oil-seeds and oil-seed flours and concentrates is attributable to flavonoids and hydroxylated cinnamic acids. The combined influence of these isolated phenolic compounds accounts for nearly all antioxidant activity of soybeans, soy flours and concentrates, cottonseed, and cottonseed flour. They account for an appreciable amount of the activity of peanuts and peanut flour.

Phenolic acids, including chlorogenic, isochlorogenic, caffeic, ferulic, *p*-coumaric, syringic, vanillic, and p-hydroxybenzoic acids are present in soybeans,[17,18] (Tables 5, 6, Fig. 2) cottonseeds[19] and peanuts. The cinnamic acid derivatives, chlorogenic, isochlorogenic,

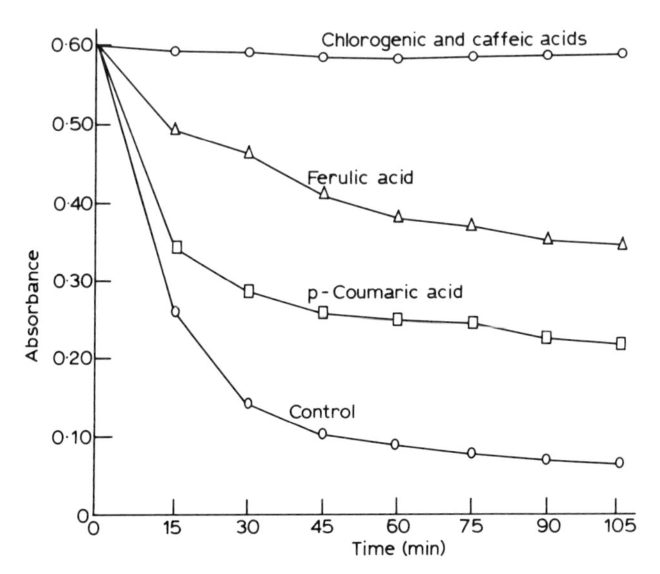

FIG. 2. Antioxidant activity of cinnamic acids isolated from soybeans. (The method is explained in the text.)

TABLE 6
ANTIOXIDANT ACTIVITY OF AQUEOUS EXTRACTS OF SOYBEANS AND SOYBEAN PRODUCTS

Extract	*Concentration (g of soybean/100 ml extract)*		
	0	*10*	*20*
Soybean (fresh)			
peroxide no	>1000	62	41
A.I.[a]	1·0	8·7	11·2
Soybeans (dried)			
peroxide no.	>1000	55	37
A.I.[a]	1·0	10·3	14/5
Soy concentrate			
peroxide no.	>1000	57	49
A.I.[a]	1·0	7·2	10·1

[a] Coupled oxidation of carotene and linoleic acid. Antioxidant index = rate of bleaching of control/rate of bleaching of β-carotene in test solution. Bleaching rate measured at 270 nm.

and caffeic acids were found in significant concentrations (Table 7) in the three oil-seeds (Table 5). These hydroxylated cinnamic acid derivatives possess appreciable antioxidant activity in lipid-aqueous systems.[9,20] Another possible oil-seed plant, the desert plant chia, has been shown to be an excellent source of antioxidant (Figs 3 and 4).

Leafy material, generally, is well known as a rich source of both

TABLE 7
PHENOLIC ACIDS WITH ANTIOXIDANT ACTIVITY OF SOY HYDROLYSED VEGETABLE PROTEIN[a]

Acid	*Concentration (mole/kg)*
Caffeic	$3{\cdot}6 \times 10^{-3}$
Ferulic	$1{\cdot}5 \times 10^{-4}$
p-coumaric	Trace
Syringic	$1{\cdot}8 \times 10^{-4}$
Vanillic	$1{\cdot}2 \times 10^{-5}$
Gentisic	Trace
p-hydroxybenzoic	Trace

[a] Pratt *et al.*

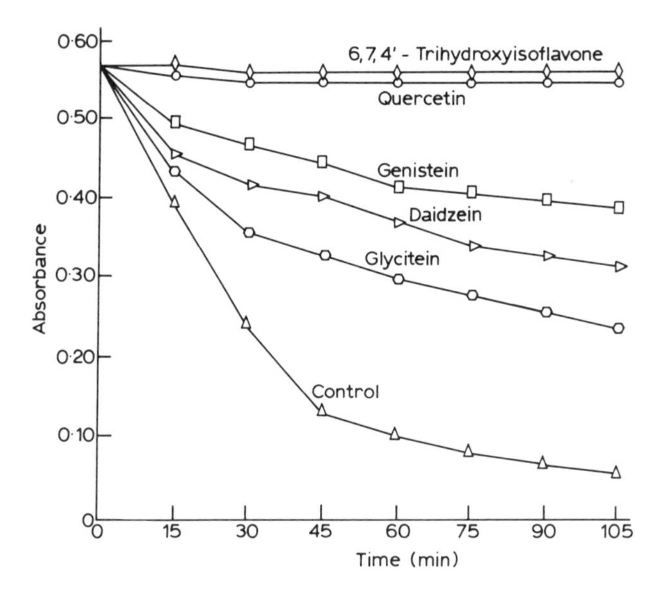

FIG. 3. Antioxidant activity of selected flavonoids (genistein, daidzein and glycitein are isolated from soybeans; 6,7,4′-trihydroxyisoflavone is from tempeh; quercetin is a commercial sample).

flavones and phenolic acids. Though some of the commonest of the phenolic acids, such as coumaric and ferulic acids, have little activity, when more phenolic hydroxyls are present in their molecules, as in caffeic and chlorogenic acids, activity becomes quite marked.[21,22]

Gallic acid and its esters are of course well known as potent antioxidants, showing that the cinnamic acid skeleton, as distinct from that of benzoic acid, is not essential for activity. Dihydrocaffeic acid is even more active than caffeic acid, but all of these compounds fall short in this respect of the most active flavones and flavanones.[23]

Chalcones

These are of special interest as the natural precursors of the flavones and flavanones themselves. Under acid conditions they readily cyclise to these end-products. They have been found to be exceptionally potent antioxidants (Table 5), butein, for example, showing about twice the activity of the corresponding flavanone, butin.[24] Dihydrochalcones are even more active than the corresponding chalcones.[25]

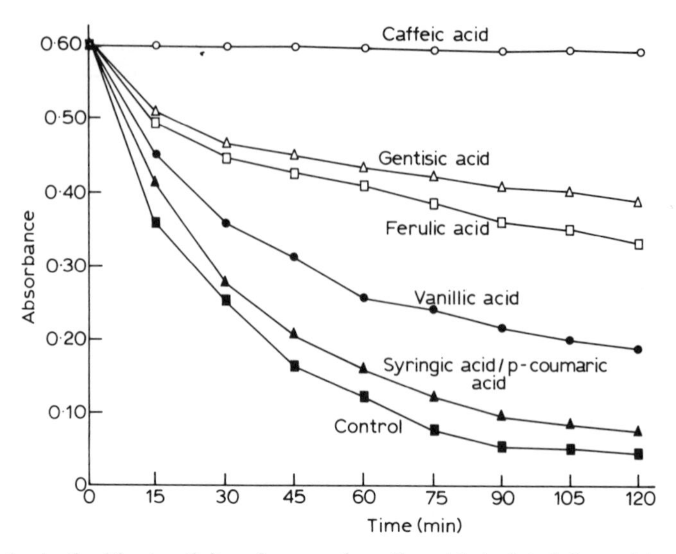

FIG. 4. Antioxidant activity of some phenolic acids isolated from chia seeds.

Isoflavones

Although flavonoids and hydroxylated cinnamic acids are present in soy isolates, their combined antioxidant power by no means accounts for all of the antioxidant activity.

Several phenolic compounds possessing antioxidant activity have been identified and isolated from soybeans.[9,17,18,26] The flavonoids of soybeans are unique in that all identified and isolated flavonoids are isoflavones. The isoflavones occur in soybeans primarily as 7-0-monoglucosides of three isoflavones (Fig. 3). The glycosides are present in concentrations of approximately 100 times those of the corresponding aglycones. The 7-0-monoglucosides of 5,7,5′-trihydroxyisoflavone (genistein) and 7,4′-dihydroxyisoflavone (daidzein) account for nearly 90% of the flavonoids. The genistein glucoside ($3{\cdot}5 \times 10^{-3}$ m/kg) was present in 0·35 times the concentration of the daidzein glucoside ($1{\cdot}0 \times 10^{-3}$ m/kg). Only one other isoflavone glycoside has been found in fresh or dried soybeans, i.e. 7,4′-dihydroxy, 6-methoxyisoflavone-7-0-monoglucoside (Fig. 3). This compound was characterised and identified by Naim *et al.*[26] and the aglycone named glycitein. Another isoflavone, 6,7,4′-isoflavone, has been shown to be present in several fermented soybean products[28,29] and in extremely 'browned' samples of soybean flakes. Several investigators have attempted to identify this compound in fresh and/or dried soybeans without success. TLC, HPLC and GLC analyses, in our laboratory failed to demonstrate the presence of 6,7,4′-isoflavone in any unfermented soybean product.

In comparison with the flavones, flavanones and chalcones, however, the isoflavones show a relatively low order of antioxidant activity. This is not surprising, since in no known natural products of this class does the important 3,4-dihydroxyphenyl structure occur.[30]

There have been considerably fewer reports on flavonoids of cottonseed and peanuts than there have been on soybeans. The work in our laboratory has not been as comprehensive on these oil-seeds as they have been on soybeans. Four flavanol aglycones and one flavanonol aglycone has been identified in cottonseed. The flavonols are quercetin, kaempferol, gossypetin and heracetin. The flavanonol is dihydroquercetin.[18] The flavonol glycosides that have been identified are rutin (quercetin 3-rhamnoglucosides) quercetrin (quercetin 3-rhamnoside) and iso-quercetrin (quercetin 3-glucoside). The flavonoids of peanuts are apparently in very low concentrations. The only flavonoid that we have found in Spanish peanuts is dihydroquercetin.

AMINO-ACIDS AND PEPTIDES

Hydrolysed proteins from soybeans (Fig. 5), yeasts, certain leaves, fish and other sources have been tested in foods and model systems and found to possess marked antioxidant activity. The properties of protein hydrolysates and the quantities required to provide a substantial antioxidant effect rule against their use for this purpose in many systems. On the other hand, the studies with hydrolysates suggest that unexpected keeping quality may result in a food from processing in which substantial hydrolysis of proteins may occur. Protein hydrolysates and extracts from them have been used to stabilise various oils.

Autolysed yeast protein and hydrolysed soybean protein were effective antioxidants in freeze-dried model systems consisting of stripped corn oil on carboxymethyl cellulose and protein. Large quantities (10%, 25%) of the hydrolysates were required to inhibit oxidation. The protein hydrolysates were synergistic with BHA, BHT, and tocopherols in the system.[31]

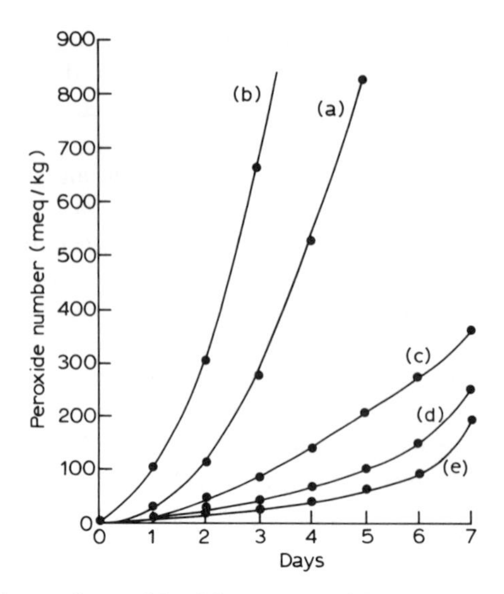

FIG. 5. Peroxide numbers of lard in contact with test solutions. (a) Control; (b) fresh soybeans (cold-water extract); (c) fresh soybeans (hot-water extract); (d) defatted soy flour; (e) soy protein concentrate.

Amino acids have been frequently tested as antioxidants and Marcuse[32] found that histidine and tryptophan had good activity in linoleic acid, methyl linoleate and methyl linolenate. Merzametov & Gadzhieva[33] established that a combination of tryptophan at 0·20% and lysine at 0·15 and 0·20% were most effective in butterfat. Revankar[34] found proline at 0·02% to be equivalent to BHA in sardine oil and five times more effective at 0·1%. Olcott & Lin[35] suggested that the nitroxide of proline is responsible for its antioxidant activity. The possibility that nitroxide radicals may be in that form is consistent with theories involving radicals as terminating agents.

Peptides and amino acids possess both antioxidant and pro-oxidant activity. Marcuse[32] found that most amino acids had a significant antioxidative potential. Cysteine was a notable exception and was found to be normally a pro-oxidant. Even cysteine under certain conditions possessed antioxidant activity.[32] Specific amino acids may serve as antioxidants under some conditions and pro-oxidants under others. At very low concentrations most amino acids have marked antioxidant acitivity, but with an increasing concentration many become pro-oxidative.[32] Also pH plays a role in the oxidative influence of amino acids. At low pHs most amino acids are pro-oxidants while higher pHs favour antioxidant activity. Methionine is perhaps the most important amino acid antioxidant. Of course methionine, as well as cysteine, is deficient in soybeans; however, it is conceivable that methionine as a nutritional supplement may contribute to antioxidant activity in products containing soybeans.

Amino acids function both as primary antioxidants and as synergists with phenolic antioxidants.[32] Bishov & Henich[31,36] showed low molecular weight peptides (<700) were superior as primary antioxidants to those of higher molecular weight. Marcuse[32] reported that chelation of pro-oxidant trace metals could explain the synergistic action of amino acids. On the other hand Bishov & Henich[31] reported that metal chelation was not the sole synergistic mechanism and the mode of action still remained obscure.

SPICES AND HERBS

Spices and herbs have been used for many centuries to enhance flavour and extend the keeping times of various foods. However, it was not until the work of Chipault *et al.*[37] that spices were compared as antioxidants in various fat sources. These authors demonstrated that 32 spices could behave as antioxidants to prime steam lard at 98°C. Rosemary and sage were the most effective antioxidants of the spices tested. Alcoholic and ethereal extracts possessed activity but the activity was considerably lower than for the whole spice.

Chipault *et al.*[37] showed the effects of spices in several food systems. In oil-in-water emulsions cloves appeared to be the most impressive as they did in ground pork. Allspice, cloves, sage, oregano, rosemary and thyme possess antioxidant properties in all fats in which they were tested.

Hiraharce & Tokai[42] compared various spices at 0·0003% with 0·02% BHA in olive, soybean, sesame and linseed oil and found several to be more effective than BHA. Clove performed well in all oils but was especially effective in olive oil.

The antioxidant activities reported in Figs 2, 3 and 4 were measured using 20 mg of linoleic acid, 200 mg of Tween 40 and 1 ml of 0·02% β-carotene in chloroform. The chloroform was removed by evaporation on a water bath at 50°C, using a rotary evaporator. 50 ml of oxygenated water was added, and 5 ml aliquots of this emulsion were placed in spectrometer tubes with 2 ml of the antioxidant solution under test. For the control, 2 ml of deionised, distilled water, or ethanol, as appropriate, were added to the emulsion. Readings at 470 nm were taken immediately. The tubes were stoppered and placed in a water bath at 50°C. Absorbance readings were taken at regular intervals until the control was bleached. The absorbance readings were plotted against reaction time.[9]

More recently, comprehensive studies have sought to identify and evaluate the compounds primarily responsible for antioxidant activity in spices and herbs, especially in rosemary, probably one of the most active of them. A useful account of this has been given by Löliger[38] and methods have been developed for the isolation of active ingredients and concentrates. Among the main ones are carnosic acid, a complex and highly sterically hindered dihydroxy phenolic acid, and rosmarinic acid, clearly a derivative of caffeic acid.

SYNERGISM OF FLAVONOIDS AND RELATED COMPOUNDS WITH OTHER FOOD COMPONENTS

In food commodities, as in most natural products, especially plant products, two or more components often interact to produce a protective antioxidant effect which is greater, sometimes dramatically greater, than one would expect if the total effect was merely additive. For instances, in the case of leaves, tocopherols (primary antioxidants), phospholipids (proton donors), ascorbic acid (oxygen scavenger) and flavonoids (primary antioxidants and metal chelators) cooperate to give a high degree of protection.[39] Similar considerations apply to seeds (see under 'Isoflavones' above). As a generalisation it can be confidently said that synergism will always occur between components exerting their protective effects by different mechanisms.

The degree of synergism observed between phospholipids, especially phosphatidyl ethanolamine, which is always present in unrefined vegetable products, has been documented for flavones,[40] isoflavones[30] and phenolic acids.[23] The mechanism has been elucidated in general terms.[41]

WHY ARE THESE ANTIOXIDANTS NOT EXPLOITED COMMERCIALLY?

Evidence has been presented in this chapter that there are, in various natural products, many very effective antioxidants. Most of the natural products mentioned have been used as foods or as food supplements for hundreds of years, without fear of toxicity or harmful side effects. Why therefore are they not used more freely in procesed foods instead of the popular antioxidants favoured by processers which do not occur in nature? The answers fall into three categories.

First, availability—natural products must be either (a) easily extracted in a pure form from the raw material in which they occur, or (b) synthesised readily in a form identical with that in which they occur naturally. As has been noted, taxifolin and quercetin on the one hand and carnosic acid on the other, offer possibilities, but much more development work will be required before full realisation.

Second, functional properties—adequate solubility, especially oil solubility, stability to heat and light, freedom from colour and bland aroma and flavour are necessary for general food use. The flavones, flavanones and their glycosides and phenolic acids have extremely low solubilities. Their chelation properties, important for antioxidant activity are marked in the most active compounds. Thus, 3,4-dihydroxyphenyl in the B ring leads to serious discoloration in products which contain trace heavy metals or are packaged in metal containers. Gallic acid is an obvious example. Equally it is very difficult to remove aromas and flavours from herb and spice extracts.

Third, toxicity—though there have been many reports suggesting that some flavonoids can have chemotherapeutic benefits, others have been less favourable (see Chapter 6). Any major use of such compounds in food products would certainly need to be preceded by both acute and chronic toxicity studies. However, it must always be remembered that nearly all of us ingest a wide variety of such compounds in food regularly and as a matter of course.

REFERENCES

1. Blanck, F. C., *Handbook of Food and Agriculture*. Reinhold, New York, 1955.
2. Musher, S., *Food Ind.*, **7** (1935) 167.
3. Herrmann, K., *J. Food Technol.*, **11** (1976) 433.
4. Dziedzic, S. Z. & Hudson, B. J. F., *J. Am. Oil Chem. Soc.*, **61** (1984) 1042.
5. Hudson, B. J. F., Evaluation of oxidative rancidity techniques. In *Rancidity in Foods*, ed. J. C. Allen & R. J. Hamilton. Elsevier Applied Science, London and New York, 1989, p. 53.

6. Mehta, A. C. & Seshadri, T. R., *J. Sci. Ind. Res.*, **18B** (1959) 24.
7. Uri, N., Mechanism of antioxidation. In *Autoxidation and Antioxidants*, Chapter 4, ed. W. O. Lundbert. Interscience Publishers, New York, 1961.
8. Crawford, D. L., Sinnhuber, R. O. & Aft, H., *J. Food Sci.*, **26** (1962) 139.
9. Pratt, D. E., In *Phenolic, Sulfur and Nitrogen Compounds in Food Flavor,* ed. G. Charalambous & I. Katz. ACS Symposium Series No. 26, American Chemical Society, Washington, D.C., 1976, Chap. 1.
10. Lea, C. H. & Swoboda, P. A. T., *Chem. Ind.* (1956) 1426.
11. Hudson, B. J. F. & Lewis, J. I., *Food Chem.*, **10** (1983) 47.
12. Heimann, W., Heimann, A., Gremminger, M. & Holland, H., *Fette u. Seifen,* **55** (1953) 394.
13. Heimann, W. & Reiff, F., *Fette u. Seifen,* **55** (1953) 451.
14. Simpson, T. H. & Uri, N., *Chem. Ind.*, (1956) 956–7.
15. Kelley, G. G. & Watts, B. M., *Food Res.*, **22** (1957) 308.
16. Kirth, E. F., *Ind. Engng Chem.*, **45** (1953) 2096.
17. Hammerschmidt, P. A. & Pratt, D. E., *J. Food Sci.*, **43** (1978) 556.
18. Pratt, D. E. & Birac, P. M., *J. Food Sci.*, **44** (1979) 1720.
19. Pratt, D. E., In *Flavor Chemistry of Fats and Oils,* ed. D. B. Min & T. M. Smouse. American Oil Chemical Society, Champaign, IL, 1985.
20. Pratt, D. E., In *J. Food Sci.*, **30** (1965) 737.
21. Hudson, B. J. F. & Mahgoub, S. E. O., *J. Sci. Food Agric.*, **31** (1980) 646.
22. Thumann, I. & Herrmann, K., *Deutsche Lebensm.-Rundschau,* **76** (1980) 344.
23. Dziedzic, S. Z. & Hudson, B. J. F., *Food Chem.*, **14** (1984) 45.
24. Dziedzic, S. Z. & Hudson, B. J. F., *Food Chem.*, **12** (1983) 205.
25. Dziedzic, S. Z., Hudson, B. J. F. & Barnes, G., *J. Agric. Food Chem.*, **33** (1985) 244.
26. Naim, M., Gestetner, B., Zilkah, S., Birk, Y. & Bondi, A., *J. Agric. Food Chem.*, **22** (1974) 806.
27. Pratt, D. E., *J. Food Sci.*, **37** (1972) 322.
28. Gyorgy, P., Murata, K. & Ikehata, H., *Nature,* **203** (1964) 870.
29. Ikehata, H., Wakaizumi, M. & Murata, K., *Agric. Biol. Chem.*, **32** (1968) 740.
30. Dziedzic, S. Z. & Hudson, B. J. F., *J. Food Chem.*, **11** (1983) 161.
31. Bishov, S. J. & Henich, A. S., *Food Technol.*, **40** (1975) 345.
32. Marcuse, R. J., *J. Am. Oil Chem. Soc.*, **39** (1962) 97.
33. Merzametov, M. M. & Gadzhuva, L. I., *J. Am. Chem. Soc.*, **81** (1976) 4979.
34. Revankar, G. D., *J. Food Sci. Technol., Mysore,* **64** (1974) 10.
35. Olcott, H. S. & Lin, J. S., *Proc. IV Int. Congr. Food Sci. Technol.*, Vol. 1, p. 482.
36. Bishov, S. J. & Henich, A. S., *Food Technol.*, **37** (1972) 873.
37. Chipault, J. R., Mizun, G. K., Hawkins, J. M. & Lundberg, W. O., *Food Res.*, **17** (1952) 46.
38. Löliger, J., Natural antioxidants. In *Rancidity in Foods,* ed. J. C. Allen & R. J. Hamilton. Elsevier Applied Science, London and New York, 1989, p. 105.

39. Hudson, B. J. F. & Mahgoub, S. E. O., *J. Sci. Food Agric.*, **32** (1981) 208.
40. Hudson, B. J. F. & Lewis, J. I., *Food Chem.*, **10** (1983) 111.
41. Dziedzic, S. Z., Robinson, J. L. & Hudson, B. J. F., *J. Agric. Food Chem.*, **34** (1986) 1027.
42. Hiraharce, T. & Tokai, H., *Japan. J. Nutri.*, **32** (1974) 1.

BIBLIOGRAPHY

Pratt, D. E., diPietro, C., Porter, W. L. & Giffee, J. W., *J. Food Sci.*, **47** (1982) 24.

Daniels, D. G. H., King, H. G. C. & Martin, H. F., *J. Sci. Food Agric.*, **14** (1963) 385.

Daniels, D. G. H. & Martin, H. F., *Chem. Ind.* (1964) 2058.

Daniels, D. G. H. & Martin, H. F., *J. Sci. Food Agric.*, **18** (1967) 589.

Daniels, D. G. H. & Martin, H. F., *J. Sci. Food Agric.*, **19** (1968) 710.

Hergert, H. L. & Kurth, E. F., *Tappi*, **35** (1952) 59.

Phillip, F., An investigation of the antioxidants in a textured vegetable protein product from soy flour. PhD thesis, Purdue University, W. Lafayette, IN, 1974.

Pratt, D. E., In *Antioxidation and Antioxidants*, ed. M. G. Simic. Plenum Publishing Corp., New York, 1980.

Pratt, D. E. & Watts, B. M., *J. Food Sci.*, **29** (1964) 27.

Ramsey, M. B. & Watts, B. M., *Food Technol.*, **17** (1963) 1056.

Sangor, M. R. & Pratt, D. E., *J. Am. Diet. Assoc.*, **64** (1974) 268.

Chapter 6

BIOLOGICAL EFFECTS OF FOOD ANTIOXIDANTS

P. BERMOND
Binningerstrasse 12, 4123-*Allschwil, Switzerland*

INTRODUCTION

The human organism, as well as that of animals, is oxygen-dependent. This implies that oxygen, essential for life, works through a succession of mechanisms which indeed have their limits and their side effects. The survival of the species means that biochemical protection systems have developed in parallel with those promoting oxygen utilisation.

Respiratory reactions occurring in the biological cell include:

—Fixation of molecular oxygen, for instance $C + O_2 \rightarrow CO_2$; this reaction is a true oxidation.

—Loss of hydrogen, which will combine with oxygen to produce water, e.g. $CH_3—CH_3 \rightarrow CH_2═CH_2 + 2\,H$; such a reaction is also called dehydrogenation;

—Further oxidations may also occur without direct participation of oxygen or hydrogen. This is the case for instance when, under the influence of an oxidant, an atom of Fe takes an additional valence, its salt being changed from ferrous to ferric: $2\,Fe^{2+} + Cl_2 \rightarrow 2\,Fe^{3+} + 2\,Cl^-$.

More generally it may be considered that oxidation involves a loss of one or more electrons, i.e. of negative electric charges.

However, as we shall see further, more complex oxidation reactions occur, with the production of toxic radicals which, in the absence of antioxidant mechanisms, would soon destroy the vital elements of the cell.[1] The accumulation of hydroperoxides, for instance, requires the intervention of catalase, tocopherol, selenium or reduced glutathione and its conversion enzymes; their deficiency would promptly lead to an attack on membranes.[2,3]

This chapter will discuss the biological oxidative environment, the different types of food antioxidants and their action in cellular protection, the risk situations, possible prophylaxis and the future fields of research.

THE BIOLOGICAL OXIDATIVE ENVIRONMENT

The multiplicity of in-vivo cellular reactions allowing aerobic life to develop, implies extremely active binding as well as exchanges of oxygen molecules with other radicals. Whereas some of these metabolic sequences require radicals interacting with reactive oxygen—the synthesis of prostaglandins, the metabolism of molecules with quinone structures or the activity of macrophages for instance—these radicals may also be at the origin of chain reactions eliciting deleterious effects at the level of cell particles. The survival of tissues requires in turn that other molecules either terminate oxidative chain reactions or scavenge the excess of the generated free radicals.

A number of phenomena, whether physical or chemical, are able to initiate peroxidation procedures which, in the presence of a susceptible medium, proceed continuously until a blocking defence intervention occurs, viz.:

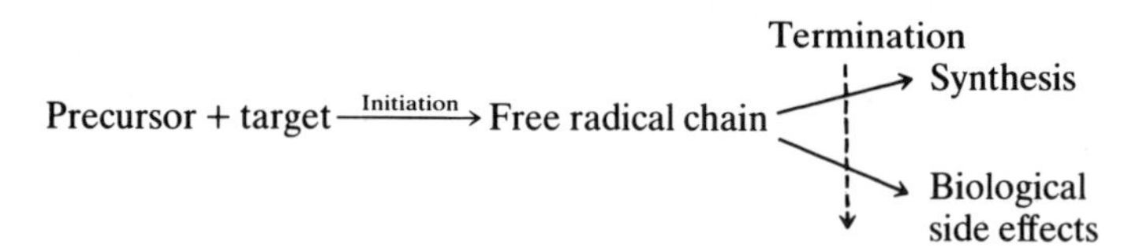

Precursors of reactive oxygen radicals or of lipid peroxidation may be of chemical, electrochemical, photolytic or ionising radiative origin. The most frequently mentioned are UV, X-rays, heavy metals, ozone, ethanol, nitrous oxide, halogen compounds, CCl_4, mutagens, chemotherapeutic drugs with quinone structures (e.g. adriamycin, bleomycin, anthracyclin . . .), bipyridyl phenols (Paraquat), xenobiotics, alloxan, tobacco and air pollutants, among others.[(4–6)] Priviliged target substances, other than oxygen itself, are polyunsaturated fatty acids (PUFA), phospholipids, nucleotides, DNA, some extracellular macromolecules of heteroproteins such as hyaluronic acid, membrane constituents, free cholesterol and lung tissue as exposed to high pressures of oxygen.[6]

PUFA, phospholipids and free cholesterol are fundamental and permanent constituents of cellular membranes (nuclear, mitochondrial, reticular or lysosomal); these membranes are bilayers in which protein macromolecules such as receptors, specific carriers and enzymes are inserted. Membranes also occur as monolayers, such as in lipoprotein membranes in which apolipoproteins are embedded. Because of their structure, membranes are a priviliged site for deleterious peroxidation reactions.[4]

Although not yet fully elucidated, knowledge of the mechanism of action exerted at the level of peroxidation phenomena has grown during recent years. Free radicals are short-life items which may act as oxidative or reducing agents and result in the transfer of one electron. They can derive among others from:

—enzymatic generation of R-OO° by cyclo-oxygenase or lipoxygenase, and of O_2 by NADPH oxidase (phagocytes), cytochrome P450, xanthine-oxidase;
—generation of excited oxygen (singlet $^1O_2$, hydroxyl OH°, superoxide anion O_2^-) by photochemical activation of O_2 or by the effect of ionising radiation on H_2O.[4]

A complete cycle from initiation to quenching can thus be schematised as follows:

$$O_2 \xrightarrow{(+e^-)} O_2^- \xrightarrow{(+e^-)+2H^{2+}} H_2O_2 \xrightarrow{(+e^-)+H^+} H_2O + OH\cdot \xrightarrow{(+e^-)+H^+} H_2O$$

The three unstable intermediates which occur mainly at the level of the mitochondrial respiratory system[6] constitute initiators for oxidative chain reactions to the detriment of other PUFA molecules with such effects as fragmentation, side-chain branching, alteration of protein-SH radicals, or generation of deleterious molecules: such as 4-hydroxynonenal, hydrocarbons, oxysterols, malonyldialdehyde (MDA) which, in turn, may react with apolipoproteins preventing identification by their receptors, or changing the properties of the lipid films of membranes.[4,7]

Such transformations may then produce cellular lesions at the level of tissues either exposed to a high oxygen circulating concentration, such as lungs or vessel walls, or subject to major storage of pro-oxidative intermediates (e.g. liver, kidneys, erythrocytes) or more generally composed of target molecules (cell membranes).

Later sections will deal with the clinical pathology linked to organ or systemic lesions resulting from oxidative or peroxidative mechanisms as well as with the food antioxidant possibilities available for their prophylaxis or their treatment. The question now arises, on a practical basis, whether laboratory methods allow a correct appraisal of the degree of oxidative damage in the organism. Attempts have been made to assay the concentrations of the different oxidation radicals.

Of all the methods studied so far in experimental research, those concerning the malonaldehyde molecule (MDA) are the most commonly used.[8] They are based either on the direct assay of MDA by HPLC[9] or on colorimetric measurement of thiobarbituric acid,[8,10] an end product of MDA. The plasma concentration of MDA according to our own experience, as well as to that of most authors, is clearly increased in cases of severe oxidation situations and it comes back to normal values ($\simeq 250 \pm 125$ nmol/g Hb) whenever sufficient doses of antioxidant substances are administered.

Among research approaches aimed at the measurement of a different index, some investigate directly the concentration of free radicals, other indirect methods measure concentrations of antioxidant enzymes (catalase, glutathione peroxidase, superoxide dismutase) in serum as well as in RBC,[9,11,12] or vitamins (tocopherol, ascorbic acid, total carotenoids).[13–15] These methods, however, may sometimes yield contradictory data, or depend critically on recent food intake.

A recently developed technique for the measurement of lipid peroxidation *in vivo* consists of the analysis of expired air for volatile hydrocarbon products of lipid hydroperoxide decomposition (ethane, pentane).[16,17] Further methods aiming at the evaluation of oxidative radicals are the appearance of chemiluminescence,[18,19] photobleaching,[20] detection of $^1O_2$ at 1268 nm in solution[21] and measurement of H_2O_2.[22]

BIOLOGICAL ANTIOXIDANT SYSTEMS

In the presence of the very complex hyperoxidation reactions occurring in biological media, a correct metabolic balance requires an antioxidant mechanism. The dual nature of the pro-oxidant attack calls for a dual metabolic defence, namely by the quenching of lipid superoxides on the one hand, and enzymatic termination or modification of excited pro-oxidative radicals (O_2, OH, H_2O_2) on the other. The word 'system' used in the title means that an association of so called antioxidants of exogenous origin (food antioxidants) and of endogenous enzymes thus forming synergistic multilevel defence systems is generally, if not always, necessary for the accomplishment of the antioxidative process.

The most prominent antioxidants, naturally present in food are ascorbic acid (vitamin C), tocopherols (including vitamin E), flavonoids, selenium, carotenoids and glutathione.

Some synthetic antioxidants such as butylated-hydroxytoluene (BHT), butylated-hydroxyanisol (BHA), erythorbic acid, may in circumstances be added to certain food preparations. Here we shall concentrate on natural food antioxidants.

Although the antioxidant mechanism of action is extremely complex, it is important to note that some of the substances mentioned are water soluble whereas some are fat soluble, these characteristics giving each of them a relative medium specificity; relative only, since synergism has been observed between both groups of substances.

Ascorbic Acid

Ascorbic acid, a water-soluble vitamin, has been known for a long time to play a dominant part in diverse biological functions, among which are hydroxylation reactions necessary for collagen formation and carnitine synthesis, as well as the facilitation of iron absorption[23] and occasionally chelating.[24] One of its main mechanisms of action however involves its participation in reduction–oxidation reactions by loss or reintegration or an atom of hydrogen as well as in the balance between its reduced and oxidised form, i.e. between ascorbic acid and dehydroascorbic acid. This transformation has been shown to undergo a two-step reversible oxidation process, each step corresponding to the emission (or integration) of an electron, the intermediate being the ascorbyl radical. (For structures see Chapter 4.)

The ascorbic free radical, which is moderately reactive and may have a protective role in scavenging other free radicals, is reduced by a microsomal NADH-dependent enzyme.[25]

The detailed metabolism has been extensively reviewed by Counsell and Hornig in 1982[26] and the reduction–oxidation process by Bendich *et al.* in 1986.[5] In this dual interchange reaction, it is interesting to consider the essential antioxidant contribution that ascorbic acid provides to the organism; the fact that high levels of ascorbic acid in circulating cells (neutrophils, macrophages) protect the organism against bacterial invasion or against inflammation actually has been proposed as a support mechanism, namely on the occasion of the respiratory burst.[27] In animal and human studies, ascorbic acid has actually been shown to stimulate neutrophil activity, including mobility towards chemotactic factors.[28,29]

Vitamin C has further been shown to protect human α-1-proteinase inhibition from direct deactivation by cigarette smoke,[30] smokers needing to consume approximately twice the vitamin C RDA compared with non-smokers to maintain similar serum concentrations.[31] The role of ascorbic acid in the defence against free radicals in the lung has also been claimed[32] as well as its release of ozone induced bronchoconstriction in non-smoking adults.[33] Similarly, in-vitro studies[11] as well as in-vivo experiments[34] have contributed to show that ascorbic acid may protect the retina[35] and the lens tissue against photoperoxidation and help to delay cataract formation. Since in humans the vitamin C content is significantly lower in lenses taken from elderly individuals with senile cataracts than in lenses without cataracts, it is proposed that ascorbic acid may prevent age-related changes in lens tissues.[36] Administration of 400 mg/day of ascorbic acid for one year led to a 13% reduction in serum lipid peroxides in an elderly population.[37]

The frequently reported preventive action of ascorbic acid against the mutagenic and carcinogenic action of N-nitroso compounds in the stomach is a further example of the role of this vitamin as antioxidant. If in general, ascorbate on its own has no effect on the preformed N-nitroso compounds, it is nevertheless able to prevent nitrosation of precursors.[38,39] According to Mirvish *et al.*[40] the ascorbate ion is more active than ascorbic acid and the efficacy of both of them is pH related. The same author finds that ascorbic acid can be effective against nitrosation as a scavenger of nitrite and as an antioxidant acting as a free radical quenching agent, at degrees varying according to the compound considered, minimal for N-methylaniline (45–60%) and total for anthracyclin antibiotics at pH range 2·8–3·8.

Vitamin C can conversely undergo autoxidative destruction, for instance in the presence of some transition metal ions, leading subsequently to the promotion and acceleration of the autoxidation of any polyunsaturated material.[38,41] A unique case of autoxidative reaction, manifested as acute haemolysis, was reported in an individual suffering from genetic glucose-6-phosphate dehydrogenase (G6PD) deficiency, following intravenous injection of massive amounts of ascorbic acid, i.e. 80 g/2 days,[42] such circumstances indeed being exceptional.

Enhancement of lipid peroxidation, hydrogen peroxide generation, DNA damage or mutagenesis by ascorbate in vitro are mostly controlled under normal conditions in vivo, presumably by the protective action of enzymes such as superoxide dismutase, catalase and glutathione peroxidase, and by vitamin E.[25] The unpaired electron of dehydroascorbic acid actually, due to its delocalised nature makes it relatively unreactive or mostly reactive with other free radicals, thereby effectively terminating the free radical chain.[5]

The three mechanisms of the antioxidant contribution of ascorbic acid have been observed:

—as quencher of singlet oxygen, as formed when ionising radiations or UV incident on skin containing photosensitising molecules (e.g. porphyrin in different forms of porphyrias)[43]

—as scavenger of peroxyl free radicals occurring in cases of lipoperoxidation, a more surprising intervention for a water-soluble molecule. It is assumed however that vitamin C intercepts free radicals generated in the aqueous phase, thus preventing their further attack on lipid membranes[44]

—vitamin C helps the reduction of α-tocopheroxyl to α-tocopherol, thus showing an indirect contribution in the prevention of fat autoxidation: a series of in-vitro experiments have confirmed this inter-vitaminic cooperation[5,45] which, according to Kläui & Pongracz[15] may be tabulated as:

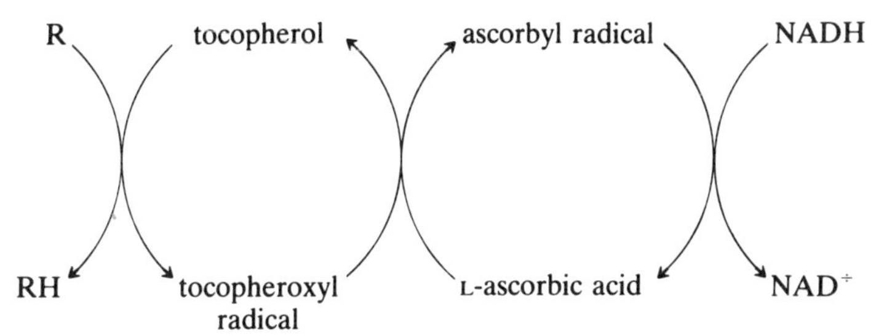

In this field, vitamin C has also been observed to quench the electron spin resonance spectrum of the α-tocopheroxyl radical generated in monolayers of autoxidised methyl linoleate on silicagel[46] and to significantly extend the induction periods of autoxidising substrate in the presence of vitamin E.[47]

Autoxidation of the phosphatidylcholine liposomes, started with a lipid-soluble initiator, was practically unaffected by the addition of sodium ascorbate alone to the aqueous phase, whereas α-tocopherol alone inhibited the oxidation. Nevertheless the induction period was increased by the addition of ascorbate.

Two studies in guinea pigs have shown that animals fed high levels of ascorbic acid had significantly higher tocopherol concentrations in their lungs, thus showing a sparing effect of the first vitamin for the second one.[5] In one human study, the cumulative effect of vitamin C and E in reducing serum lipid peroxides was higher than that of either ascorbic acid or tocopherol supplement given alone.[37] It can thus be concluded that regeneration of vitamin E by ascorbate helps to spare the former in its antioxidant role. Reciprocally, it could be shown that the balance between pro-oxidant and antioxidant activity of vitamin C was in some way dependent on the membrane tocopherol concentration.[48]

Entering the vitamin E antioxidant process, ascorbic acid thus belongs indirectly to a complex antioxidant system including also glutathione (see vitamin E)[49] a speculation that has been supported experimentally by the fact that Mg ascorbate administration helps to protect reduced glutathioine (GSH) against disintegration by peroxides.[50]

The recommended daily allowance (RDA) for vitamin C in human adults ranges from 30 (UK) to 100 mg (USSR), thus varying largely from country to country. The body pool has been shown to be saturated with a daily intake of 100 mg (140 mg in smokers).[51]

Flavonoids

Flavonoid molecules are polyphenols. A large number have been identified, though with varying metabolic properties. Like tocopherols, flavonoids have had to overcome a period of scepticism as regards their biological utility. Considered for some time as vitamins (vitamin P), then rejected as such, they have further been accepted in some countries as playing an auxiliary part in the redox function of vitamin C, mainly at the level of the vessel wall. Their role however remained controversial until recently, but now they have been recognised as potent antioxidants though, as we shall see, with polarities varying according to the form, the dose, the enzyme or the deoxidation system considered and, as ever, the animal model used. As for ascorbic acid, the mode of action of flavonoids is of the redox type:

$$\text{HO},\text{HO-substituted ring} \underset{+2H}{\overset{-2H}{\rightleftharpoons}} \text{O=},\text{O=-substituted ring}$$

The fact that many of them have poor intestinal absorption and thus low bioavailability may also have been, in some cases, at the origin of the controversy on their metabolic integration.

As expressed by Deby, some flavonoids are predominantly radical traps, others anti-lipoperoxidants.[52] Using as a model erythrocyte membrane damage consisting of lipid peroxidation, and increase in passive K^+ permeability after blocking superoxide dismutase, Maridonneau-Parini *et al.* were able to classify the various flavonoids into four groups:

—those decreasing the oxygen free radical stimulated K^+ permeability, e.g. kaempferol, naringenin, apigenin, naringin;
—those increasing the deleterious effect of oxygen-free radicals, e.g. myricetin, delphinidin, quercetin;
—those characterised by opposite effects, depending on the concentration, e.g. phloretin, cyanin, catechin, morin;
—those remaining inactive, e.g. rutin, phloridzin.[53]

Actually, as can be judged from the literature, the reality is not always so unequivocal. Using H_2O_2-induced histamine release from human basophils, Ogasawara *et al.* find that quercetin and apigenin (5–50 μM) inhibit histamine release in a concentration-dependent manner; these two flavonoids as well as taxifolin inhibit the generation of H_2O_2.[54] In other conditions, i.e. the inhibition of the cytochrome P450 mediated reaction in rat liver microsomes, Sousa & Marbetta using concentrations of 10–250 nM, reach the conclusion that quercetin is a potent uncoupler of P450 reactions, elevating the rates of H_2O_2 formation almost twofold.[55]

Testing the effects of naturally occurring and of synthetic flavonoids on the metabolism of the carcinogens benzo(a)pyrene and aflatoxin B_1, Buening *et al.* found that the addition of apigenin, chrysin, fisetin, flavonone, galangin, hesperitin, kaempferol, morin, myricetin, naringenin or quercetin to human liver microsomes inhibits the hydroxylation of benzo(a)pyrene. Flavone, nobiletin, tangeretin or 7,8-benzoflavone cause manifold stimulation in the hydroxylation and activation of aflatoxin to mutagenic substances. Whereas quercetin, morin and kaempferol inhibit cytochrome C (P450) reductase, flavone and 7,8-benzoflavone have no effect.[56]

Flavonoids have also been shown to exert antimutagenic activity. Myricetin, for instance, acts against the mutagenicity elicited by the bay-region diol epoxides of benzo(a)pyrene.[57] Testing the mutagenic potential of carcinogenic N-nitrosamines by means of Ames preincubation assay using liver supernatant S9 fractions from different animals and human origin, Yamazaki *et al.* find that metyrapone but not 7,8-benzoflavone decreases the mutagenic activity of five nitrosamine compounds (out of six tested) by 29–71%.[58]

By assaying the malonyldialdehyde formed in a model using human platelets submitted to peroxidation, Koch & Löffler observed an inhibitory effect of flavonoids.[59] Rosin found that the pH may influence considerably the metabolic behaviour of certain phenolic derivatives (among others, those extracted from the betel nut) thus leading to the generation of H_2O_2 at alkaline but not at acid levels.[60]

Aqueous alcoholic extracts of *Ginkgo biloba* (quercetin, kaempferol) in contact with polymorphonuclear leukocytes (polymorphs) + luminol cause a binary reaction, luminescence being decreased in in-vitro procedures, whereas in-vivo administration of these constituents, paradoxically increases polymorph luminescence.[61] Studying the flavonoids contained in grapes and wine, Masquelier stresses a double prerequisite for the metabolic efficiency of flavonoids: their intestinal absorbability and their compability with the plasma pH, two conditions which not all of them are able to fulfill.[62]

In the field of haemolysis of human erythrocytes by haematoporphyrin, and photosensitisation, Sorata *et al.* find that quercetin and rutin in submillimolar concentrations suppress photohaemolysis and inhibit lipid peroxidation. The authors conclude that flavonols can function as antioxidants in biological systems by terminating radical chain reactions and removing singlet molecular O_2.[63] Mun *et al.* find that hesperitin is an inhibitor of the oxidative metabolism of exogenous substances catalysed by the microsomal P450 dependent mono-oxygenases, namely benzo(a)pyrene hydroxylase and aniline hydroxylase in Dawley rats.[64] When testing 14 flavonoids in beef heart mitochondria, Hodnick *et al.*, find the following decreasing order of potency: chalcone > flavone > flavonol > dihydroflavonol > anthocyanidin. Two flavonols with a pyrogallol configuration (myricetin and quercetagetin) were found to autoxidise, resulting in the generation of H_2O_2.[65] The decrease of reduced glutathione (GSH), a system which contributes strongly to reducing reactions, is accompanied by a marked increase in pulmonary lipid peroxidation. Operating in such conditions with rats, Videla *et al.* concluded that dextro-cyanidanol-3 abolishes completely such peroxidation.[66] Rathi *et al.*, assume that certain flavonoids like quercetin and hesperidin may exhibit their antioxidant action by restoring the tissue content of glutathione and the activity of glutathione-peroxidase, by retarding the depletion of selenium,[67] whereas Nakadate *et al.* ascribe the inhibitory effects of flavonoids and other antioxidants on tumour promotion to their inhibition of the lipoxygenase activity.[68]

Readily oxidisable fat, such as lard, may be effectively stabilised by chalcones in the concentration range of 0·025–0·1%, more so than by the corresponding flavonones, as shown by Dziedzic and Hudson using induction periods at 100°C. Moreover they found a pronounced antioxidant synergism when phosphatidyl-ethanolamine is added.[69–71] Younes & Siegers working on the antioxidant role of the rat glutathione conclude that the only structure of flavonoids supporting their antioxidant function is the 3-4′-dihydroxy-grouping.[72]

The considerable number of publications appearing recently on the antioxidant potential of flavonoids,[72–77] as well as those cited above, tend to confirm the role of these substances as redox agents, as shown many years ago,[1] rather than as unequivocal and exclusive antioxidants. This ambivalence may account for the observed antioxidant effects only under definite conditions of concentration, temperature, pH and chemical association. Moreover, the type of enzymatic reaction, the experimental species and mode chosen, and of course the chemical type of flavonoid considered, as seen above, determine antioxidant potency. When the appropriate conditions are not adhered to, flavonoids may be either inactive or even pro-oxidant, as is ascorbic acid, which shows some limited parallels.

This glance over the recent literature leads to at least three conclusions:

—the speculation that flavonoids are generally inactive must be definitely rejected
—extrapolations from studies carried out on specific species or particular biological models, in restricted physico-chemical conditions and with specific flavonoids should be particularly carefully considered

—although accompanied by somewhat confusing and occasionally controversial results, the unquestionable evidence of very promising biological and clinical effects from flavonoids deserves further research, especially in the fields of capto-donors and of dichotomous molecules able to selectively scavenge the appropriate free radicals.[4,61,78]

Vitamin E

From Speculative Extrapolations to Sound Research

Since the results of animal experiments showing that a vitamin E-deficient diet resulted in necro atrophy of the ovarian duct in rats, and the speculative as well as unfounded extrapolation of this observation to the gestational function of human females, there was doubt for some time concerning the role of tocopherols, even as vitamins in humans.

Research, however, in the field of peroxidative reactions and the mechanisms able to control their effects, proved vitamin E (see below) to be a major radical trap.[79] This property will be described, in relation to the whole antioxidant system in which vitamin E is involved.

Vitamin E exists in four natural forms: α, β, γ and δ-tocopherol. The basic structure consists of a hydroxylated ring and of an isoprenoid side chain, both implicated in the antioxidant function of vitamin E.

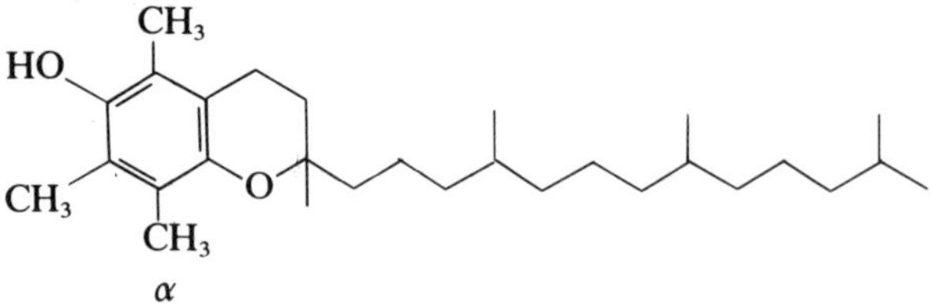

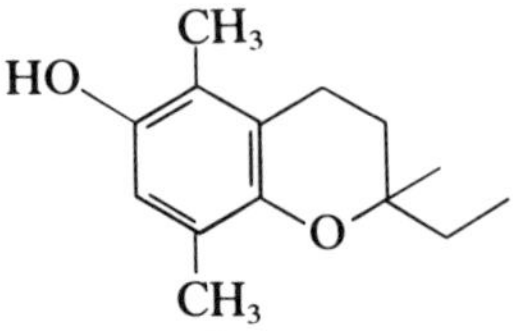

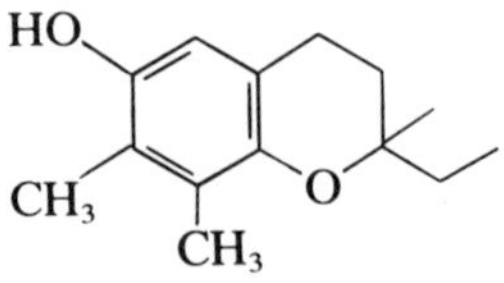

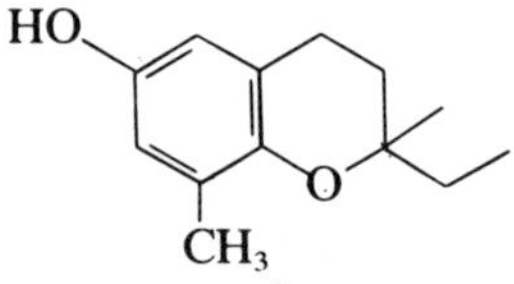

Vitamin E is a fat-soluble vitamin existing in several chemical forms, the most common one being α-tocopherol.

Vitamin E is bound to the cellular and subcellular membranes by specific physico-chemical interactions between its phytyl side chain and the fatty acyl chain of the membrane polyunsaturated phospholipids, particularly those derived from arachidonic acid. This interaction may account for at least one of the mechanisms by which vitamin E contributes to the stabilisation of phospholipids.[80]

Assay Methods

The appraisal of the status of tocopherol, whether in humans or in animals, is obviously one of the first approaches to understanding the role of vitamin E in the organism. The assay of tocopherol in plasma may be carried out by fluorimetric analysis[14,81] or by an HPLC method.[13] Normal plasma values in humans are in the range 13·1 + 3·2 mg/litre (♂) and 12·1 + 2·5 mg/litre (♀).[13] Functional investigations regarding the erythrocyte tocopherol content refer to the determination of the minimal level of H_2O_2 concentration causing haemolysis.[82]

It has been observed that a certain parallelism exists between the plasma levels of vitamin E and of lipids, regardless of the tocopherol status. Most experts therefore agree on the need to consider plasma lipids with standardised vitamin E rather than the mere level of the vitamin itself.[83–88] Tocopherol may also be assayed in erythrocytes[89,90] and in animal tissue.[91–93]

More direct methods have been made available to measure the degree of oxidation in the experimental medium, and the ability of vitamin E, or of the potentiating contributors of the system, to antagonise peroxidation radicals.

Estimation of the level of thiobarbituric acid-reactive substances, like malonyldialdehyde, is one of the most common methods.[8,10] Further indirect methods used for other substances and allowing an estimation of the antioxidant potential such as chemiluminescence,[18,19,94] measure of the changes occurring in light absorption[64] or the composition of the alkanes in the exhaled air[16,17,95] are mentioned elsewhere in this chapter. Others have been developed individually by authors investigating specific animal experiments.

Results of Metabolic Studies

A number of in-vivo experiments have been carried out in animals made deficient in vitamin E. Although most of them advocate an antioxidative role for vitamin E in the different biological peroxidative situations, discrepancies in the conclusions, likely to be linked to the experimental conditions, have shown a need to review the literature of recent years on this topic.

Deprivation of vitamin E has been found to be accompanied by a discoloration of fat with accumulation of peroxides in mink and pigs.[96] Mice fed a vitamin E deficient diet showed a 65·7% depletion in cardiac muscle after 4 weeks and the susceptibility of muscle homogenates to in-vitro lipid peroxidation increased by 101·8% during the corresponding period.[97] The peroxidative damage to rat brain DNA initiated by methyl ethyl ketone peroxide were compared in groups of animals fed casein-based diet containing 10% of tocopherol-stripped corn oil supplemented with either 0; 3; 5; or 10 IU of dl α-tocopheryl acetate/kg. Both template activity and bound tryptophan and cross links favoured the maximal protection in the group receiving 30 IU of tocopheryl acetate/kg feed, a concentration then considered an adequate one.[98]

A common technique mainly promoted by Japanese authors, triggering peroxidative damage, consists of reperfusing tissues made partially ischaemic by clamping, an operation which results in a rapid rise of lipid peroxides in tissues and in serum (and in the occurrence of neurological signs when brain is taken as experimental tissue). Intravenous injection of 20 mg/kg tocopherol 30 min prior to the experiment significantly suppressed the rise in lipid peroxides both in brain and serum, improved the severely expressed neurological signs and promoted resynthesis of ATP in rats.[99] Whereas clamping liver arteries in rats, then reperfusing the organ after 90 min did not permit survival, pre-injection of 10 mg/kg α-tocopherol for 3 days increased the survival rate to 45·5%, accelerated the resynthesis of ATP, suppressed the elevation in lipid peroxides and prevented the decrease in glutathione.[100]

Reperfusion was found to increase malonyldialdehyde in rat brain, the effect being largest in tocopherol deficient groups, intermediate in controls and smallest in a vitamin E supplemented group of animals; conjugated dienes changed in parallel with thiobarbituric acid reactants (MDA); α-tocopherol decreased after aerobic incubation and polyunsaturated fatty acids fell significantly in the vitamin E deficient group after reperfusion. Thus, cerebral reoxygenation after ischaemia propagates peroxidative reactions within esterified polyunsaturated fatty acids. The decrease of vitamin E in the non-supplemented groups suggests free radical mediation and increase of tocopherol consumption.[101,102]

Lipid peroxidation may also be estimated by measuring volatile hydrocarbons (ethane and pentane). Such peroxidation initiated by methyl ethyl ketone peroxide was found to produce more pentane exhalation in vitamin E deficient rats than in those pretreated with 10 IU/kg dl α-tocopherylacetate, the increased peroxidation thus eliciting a vitamin E requirement increase.[103]

Measuring MDA, the most common test for lipid peroxidation, and lipid epoxide production, in vitamin E deficient rat lung tissue exposed to $3 \pm 0{\cdot}1$ ppm NO_2 for 7 days, an inverse relation was found between tissue vitamin E content and the degree of peroxidation.[104] Anaemia occurring in rabbits fed a cholesterol rich diet can be prevented by administering vitamin E concomitantly.[105]

Further studies were carried out with labeled markers, which help to elucidate the metabolisms involved. As determined by in-vivo experiments using ^{14}C L-leucine and glycine, vitamin E deficiency in young rabbits caused a higher turnover rate of liver proteins and of plasma albumin as well as globulin fractions.[106] Specific binding sites on human erythrocytes were demonstrated for tritium-labelled α-tocopherol.[107]

Many more recent studies confirm the antioxidant role of vitamin E as a scavenger of free radicals at the level of peroxidation of lipid chains, a mechanism which is no longer contested, whether in terms of in-vivo (animal and human) or in-vitro investigations or as concerns the type of tissue investigated.[108–117]

The Tocopherol–Selenium–Glutathione Enzyme System

An additional aspect of recent research has been to focus on mechanisms of action, and on the study of compound antioxidant systems in which vitamin E is involved.

Antiperoxidation has been studied extensively during recent years. The intervention of the different intermediates involved has been summed up as follows.

Superoxide dismutase controls the concentration of the superoxide anion (O_2^-); the mitochondrial form of the enzyme contains Mn and the cytosolic form Cu and Zn; these trace elements are thus associated with antiperoxidative intervention in the different cellular organelles. Several enzymes in turn control the accumulation of hydrogen peroxide catalase which, at the level of peroxisomes, catalyses its disproportionation to water and oxygen, and glutathione-peroxidase which, at the level of mitochondria and the cytosol, catalyses the reduction of H_2O_2 to water by reducing equivalents derived from glutathione.

One of the forms of glutathione-reductase involved is selenium dependent, its activity being linked to the presence of four atoms of selenium at the active site per mole of enzyme. Membrane macromolecules such as polynucleotides and unsaturated phospholipids are subject to the attacks by oxygen radicals which cause severe disruption to the membrane architecture. Tocopherol prevents such damage by donation of its phenolic hydrogen atom to peroxyl radicals formed from unsaturated fatty acyl residues, thus effecting termination of the free radical chain. Peroxisome membranes and enzymes seem to be the most sensitive to the variations of the vitamin E status.[118,119] The antioxidant function in which vitamin E is involved thus appears to be integrated in an extremely complex mechanism requiring the presence of trace elements, among others selenium, glutathione, a tripeptide, and different enzymes, only the main ones being mentioned above.

Deficiency of one of these contributing elements entails an increased consumption of the others, a substitution confirmed by more than one study, at least within certain limits.[3,101,102,120] Recent literature in this field in no way contradicts the summary cited above and further clarifies the mechanism of action of vitamin E. Slight variations arise according to the organs or to the cellular elements investigated, whether in humans or in animals.

Studies of erythrocytes of neonates and adults incubated with increasing concentrations of H_2O_2 in the presence of a catalase inhibitor lead to the conclusion that in such experimental conditions, oxidation of glutathione precedes that of vitamin E and tocopherol is the last antioxidant to be consumed before the autocatalytic generation of MDA.[121] Concomitant selenium and tocopherol deficiency in the rat shortens the time of onset of vitamin E deficiency symptoms, including the decline in body weight gain, centrobulbar, hepatic and renal necrosis, as well as skeletal muscle degeneration. Selenium deficiency was found to be followed by a subsequent decrease in the Se-dependent glutathione peroxidase.[122] Glutathione delays α-tocopherol oxidation and subsequent lipid peroxidation in rat liver microsomes.[123] Conversely, the parenteral administration of high doses of vitamin E (30 mg/kg/day) to rabbits significantly decreases the ratio of oxidised to reduced glutathione (from 2% to 1·3%) in erythrocytes, suggesting a reduced utilisation of this tripeptide for free radical detoxifying activity.[105,124]

Monitoring catalase, glutathione peroxidase, vitamin E and malonyldialdehyde in the laying hen proved the antioxidant system to be able to eliminate free radical formation induced by hormonal effects.[125] Comparing four groups of trout given either normal diet or a diet deficient in vitamin E, Se or both, showed a significant interaction between the two nutrients with respect to packed cell volume and to malonyldialdehyde formation in the in-vitro NADPH-dependent microsomal lipid peroxidation system. Plasma pyruvate-kinase increased significantly in the group deficient in both nutrients, a fact thought to be due to a leakage of the enzyme from muscle, and thus indicative of incipient muscle damage.[12] Administration of $HgCl_2$ at a dose of 5 mg/kg/day for 15 days to male albino rats brought about a marked depression of the scavenging enzymes glutathione peroxidase and glutathione S-transferase in the kidney, with an adaptive rise of catalase and of superoxide dismutase. Vitamin E supplement was effective in bringing back glutathione levels to normal.[126] Alpha-tocopherol (5 mg/kg/bw) or Selenium (1 mg/kg bw Na_2SeO_3) remarkably depressed the formation of lipid peroxide in tumour-bearing rats. Whereas doses of 0·5 mg/kg bw of the first nutrient or 0·01 mg/kg bw of the second had no action on lipid peroxide formation, the same doses of these two drugs administered together had a synergistic effect.[127]

Human cultured skin fibroblasts infected by *Mycoplasma pneumoniae* and with added buthionine sulphoximine, show a drastic lowering of reduced glutathione due to an elevation of H_2O_2 in the medium and 140% enhancement of malonyldialdehyde (MDA) concentration. Cells enriched with 0·25 or 2·25 μg of vitamin E per mg of protein prior to infection revealed a lesser elevation of MDA (55 and 20% respectively). These data show that the increase in intracellular levels of H_2O_2 requires a higher contribution of the GSH redox cycle and that vitamin E supplementation helps in reducing the concomitant lipid peroxidation.[128] More generally, in many circumstances of vitamin E deficiency, an adaptive response of the enzymic antioxidant defence system is evident in certain tissues.[129] Further investigations confirmed that vitamin E deficiency, both in humans and animals, was accompanied by an increase of prostaglandin synthesis, of MDA formation and of platelet aggregability, whereas reintroduction of the vitamin reversed these alterations.[130] Conversely, arterial generation of prostacyclin was shown to be decreased in vitamin E deficiency and restored after administration of the vitamin.[131,132]

Referring to the well known intervention of tocopherol in the limitation of lipid chain peroxidation, mainly at the level of membrane lipids, several authors have tried to determine whether the polyunsaturated/saturated fatty acid ratio P/S might play a part in vitamin E requirement. Several groups of rats were fed diets containing different P/S ratios, and within each group, supplemented with different vitamin E doses (0–100 IU/kg). They showed increases in the amounts of pentane expired, of the liver MDA levels and of the haemolysis in the unsupplemented animals but normalisation of such values in the high vitamin E groups, whatever the P/S ratio. A supplement of 40 IU/kg was considered adequate. The fatty acid composition of phospholipids, especially the content of arachidonic acid, seems to be more influential in determining the effect of tocopherol in improving membrane fluidity.[133,134] Vitamin E completely inhibits arachidonic acid induced human platelet aggregation at high concentration, thus supporting the beneficial role of this vitamin when its use is advocated for restraining peroxidation phenomena of atherosclerosis.[135]

Further authors investigating in rats the comparative effects of a strong synthetic antioxidant (ethyl methyl hydroxypyridine HCl) and vitamin E on ubiquinone-dependent enzymes present in liver mitochondria came to the conclusion that in oxidative phosphorylation vitamin E may play a part both as an antioxidant and as a regulator of the metabolism of ubiquinone.[136]

Major variations have been observed in the oxidative mechanism and the antioxidant defence according to the tissue studied. For instance, most of the peroxidation in lung microsomes appears to proceed non-enzymatically whereas peroxidation in liver is largely enzymatic; actually the margin for protection from vitamin E in lung is less than in liver.[137] Further studies have shown that it is possible to distinguish a membrane effect from the phytol side chain of α-tocopherol and an antioxidative effect from its chromane structure.[95]

As far as the chemical form of vitamin E is concerned, several experiments sought to compare the antioxidant potencies of α-, γ-, and δ-tocopherol and α-tocopheryl acetate. Due to the diversity of methods, techniques (in-vitro, in-vivo) and animals used, it may be hazardous to draw any conclusion. The majority of authors find, though with exceptions,[138] γ-tocopherol has a higher antioxidant effect than α-tocopherol[139–142] which in turn is more potent than α-tocopheryl acetate.[111] Many more papers have been published which nowadays leave no doubt about an antioxidant effect of vitamin E.

Clarification of its precise mechanism of action and of its effects, however, remains fragmentary, due to the diversity of the reported situations, depending on the variety of tissues involved, the multitude of toxic substances triggering peroxidation phenomena, the concentration levels required, and the different oxidative radicals or peroxidative chains, among others. The rich variety of antioxidant systems to which vitamin E contributes seems to be able to adapt for most of the aggressive situations, at least as far as a sufficient supply in each of the contributive elements of the system is available. The complexity of the metabolisms involved does not yet make it possible to decide whether a compound antioxidant system in which vitamin E is implicated acts as a mere antioxidant barrier or whether it is able to control and balance the indispensable oxidative reactions. Enough evidence has been accumulated showing that, when deficient, vitamin E antioxidant fails to oppose, and thus to control, oxidative tissue damage. Whether, conversely, an excess of antioxidant more than fulfills its role and blocks completely the necessary intervention of peroxide radicals is not proved. Much work remains to be done in this field. Recommended daily allowances for human adults range from 7 to 16 mg α-tocopherol;[143] due to the low foetal to maternal diffusion of tocopherol, premature infants are at risk of oxidative accidents.

Carotenoids

Two Totally Different Functions

Carotenoids (see page 212 for structures) are natural fat soluble substances of which more than 500 different molecules have been identified so far.

Firstly, the identification of β-carotene and of a few other carotenes (α, γ, δ, . . .) as precursors of vitamin A. The transformation of β-carotene to vitamin A occurs mainly by cleavage of the molecule at the central double bond by the action of the enzyme carotene deoxygenase, which has been isolated from rat and from human intestinal mucosa and liver.[144]

However, due to the poor food extraction and conversion potential of the digestive tract, an average of 6 μg β-carotene, the main source

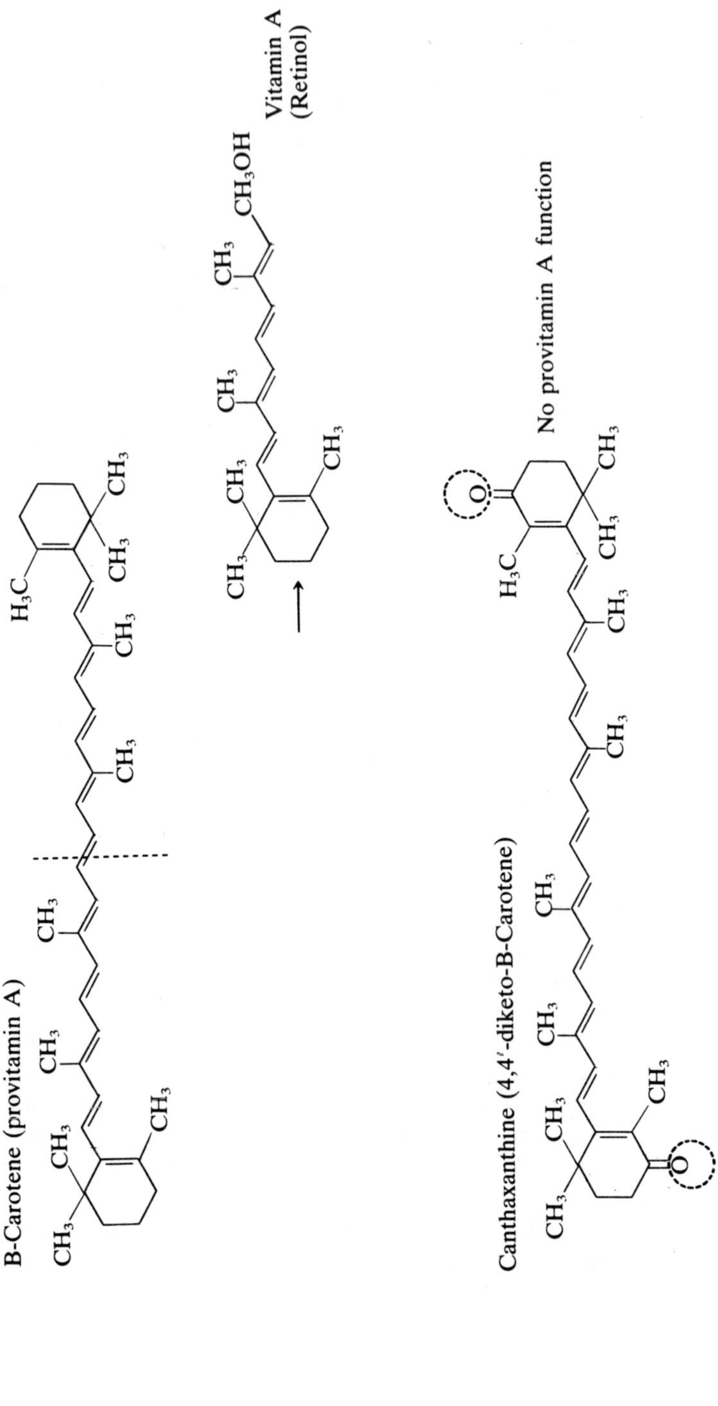

B-Carotene (provitamin A)

Vitamin A (Retinol)

CH_3OH

Canthaxanthine (4,4′-diketo-B-Carotene)

No provitamin A function

of non-preformed vitamin A[145] is required to obtain 1 μg retinol, daily ingestion ranging from 1500 to 4000 μg. The conversion from β-carotene to retinol varies inversely with the amount ingested[146] and has been found to be dependent on vitamin A requirement. No excess of vitamin A nor direct toxicity linked to β-carotene, apart from carotinodermia, has been reported after ingestion of large amounts of this carotenoid in its pure synthetic form[147–152] although large amounts of carotene-containing foods have in some rare cases, elicited toxic symptoms likely to be linked to accompanying substances.[147,153,197]

Secondly, during the 1960s, β-carotene, as well as other carotenoids which have no provitamin functions (xanthophylls), were recognised to exhibit potent protective properties against the deleterious effect of radiation on light-sensitized cells.[154–156] The mediation of a quenching effect of such carotenoids was suggested as a possible mechanism for these properties.[157–159]

Carotenoids, not Retinoids

Only the antioxidant function of carotenoids, which has raised much interest in biology and in medicine, will be discussed in these pages. But before proceeding it must be emphasised that antioxidant properties have been shown to extend to non-provitamin carotenoids,[160] thus indicating that the antioxidant function of carotenoids is totally independent of vitamin A activity. The terminology 'retinoid' sometimes used in the literature for carotenoids is therefore confusing and should be avoided. Further aspects of this biological system, such as the absence of mutagenic effects of carotenoids, or even their ability to reduce or prevent the mutagenicity of other compounds,[161,162] which retinoids are unable to achieve, largely contribute to contrast carotenoids with retinoids.[163]

The Antioxidant Function

Carotenoids are known to deactivate free radicals and excited oxygen, both of which are implicated in a multitude of degenerative diseases and in inborn errors of metabolism.[164] Such mechanisms and implications have been clarified by in-vitro as well as in-vivo animal studies.[165–167]

Much of the investigations regarding the role of carotenoids have been carried out in the course of studies in the field of protection against diseases in which uncontrolled oxidations may be mechanisms (tumours, cardiovascular diseases) or the consequence of an error of metabolism (erythropoietic protoporphyria) or even the fact of a metabolic impairment through intoxications or radiations.[155,168,169]

As long ago as 1956, Sistrom *et al.*, working with a wild type of the purple photosynthetic bacterium *Rhodopseudomonas spheroides*, and with a mutant lacking coloured carotenoid pigments, observed, in the mutant type, a photodynamic effect which did not occur in the wild type. They proposed that the carotenoid pigments present in the latter type might have acted as protective agents.[156] These observations were confirmed later by Mathews and Sistrom in the non-photosynthetic *Sarcina lutea*, containing carotenoids, and its mutant, lacking pigments.[170] Several authors have depicted these reactions as being mediated by oxygen radicals under the influence of a sensitiser and showed that β-carotene was able to quench the reactive form of singlet oxygen $^1O_2$.[158,171]

Studies carried out later demonstrated that carotenoids are also able to inhibit pro-oxidative mechanisms in circumstances where light radiations are not involved.[160,172] Studying the toxicity of aflatoxin B_1 a pro-oxidative poison in chickens, Dalvi & Ademoyero found that β-carotene (as well as charcoal, GSH, selenium, cysteine and fisetin) considerably reduces its toxicity.[173] The addition of autoxidised linseed oil in rat liver homogenate shows a clear increase of chemiluminescence emission with lines indicative of $^1O_2$ production. Further addition of oxygen quenchers suppresses lines corresponding to the simultaneous transition of O_2, β-carotene proving to be the most effective quencher.[174] Studying the lethal photodynamic action of endogenous photosensitisers, some authors presumed that the $^1O_2$ generated by the sensitiser was acting as the lethal agent and they found carotenoids to act as protective inhibitors of such reactions.[154,155,175,176] The fact that the protection *in vivo* depends on the number of conjugated double bonds is considered as supporting the hypothesis that these pigments function by quenching $^1O_2$.[175] Further studies have shown that carotenoids may also react directly or in joint action with superoxide dismutase, against O_3^- (Ref. 300) and against H_2O_2 (Ref. 301). Larvae of the domestic fly, taken as an in-vivo model for protection against photosensitisation, were shown to receive some protection from β-carotene but not from BHT, ascorbate or diazabicyclo-octane.[177]

Non-Provitamin Carotenoids

Many papers in recent years on carotenoids refer to β-carotene. Although fewer in number, the publications dealing with non-provitamin carotenoids must however be considered as having major importance, since the implication of these other carotenoids in the protection against the deleterious effects of pro-oxidation reactions allows one to relate this essential protection mechanism to a function totally independent of the vitamin A activity of the carotenes.

Foote *et al.* were the first to assume that the ability of carotenoids to quench singlet oxygen is related to the number of double-bonds.[158] Extracting the different kinds of carotenoids contained in the wild-type of *Sarcina lutea,* Mathews-Roth confirmed that carotenoids with 9, 10 and 11 conjugated double bonds (CDB) are better quenchers than those with 8 and less CDB.[170] Among them, P-438 with 9 CDB, P-422 with 8 CDB, *Sarcina phytofluene* (5 CDB) and *Sarcina phytoene* (3 CDB) were compared with β-carotene (11 CDB), isozeaxanthin (11 CDB) and lutein (10 CDB). All showed quenching ability, but those carotenoids with 9 CDB or more, had a three-fold activity level when compared with those having 8 CDB or less.[175]

Further in-vitro work by the same author showed that administration of the carotenoid pigments β-carotene, canthaxanthin and phytoene can quench the photochemical reactions occurring in skin on exposure to UV-B.[178] Using a quantitative method, she measured the photochemical reactions induced in epidermis prepared from hairless mice made porphyric and demonstrated protection from β-carotene ($p < 0{\cdot}014$), canthaxanthin ($p < 0{\cdot}014$) and phytoene ($p < 0{\cdot}029$).[178] Of all the three carotenoids cited only β-carotene has provitamin A activity in humans. Foote has calculated the constant for quenching of $^1O_2$ by β-carotene as approaching a diffusion-controlled limit of $1{\cdot}3 \times 10^{10}\,m^{-1}s^{-1}$ in benzene solution[159] and Krinsky considers this value to be similar for many carotenoid pigments which contain at least nine conjugated double bonds.[179]

Further Mechanisms of Action

Some evidence has been provided, not only that different types of carotenoids develop antioxidant functions in quenching oxygen radicals (type 2 reaction) but also that both provitamin and non-provitamin carotenoids are able to limit the consequences of lipid peroxidation in tissues (type 1 reaction). Burton and Ingold regard the fact that β-carotene is particularly effective at partial pressures of oxygen as low as those under physiological conditions in most tissues as an indication that it probably complements the chain-breaking properties of vitamin E in the lipid phase.[180]

Using liposomes with trapped glucose and generating radical species by the decomposition of K_3CrO_8 under UV-C irradiation, Krinsky and Deneke showed a strong reduction of the reaction by both canthaxanthin and β-carotene. These authors, generating lipid peroxidation by incubating egg phosphatidyl choline lipsomes with $FeCl_2$ measured MDA in the medium and found that β-carotene and canthaxanthin allowed a marked delay in the appearance of MDA when compared with a preparation free of carotenoids.[160] MDA, a product of peroxidation reactions, was found to be inversely related to the β-carotene concentration[181] and also to the canthaxanthin concentration.[182] Such an inverse correlation is an indication that these carotenoids counteract the oxidation reactions occurring *in vitro*[181] or *in vivo*, as tested in patients suffering from β-thalassaemia.[182]

Reduced-glutathione (GSH), a potent antioxidant, is lowered in brain 24 h after a single injection of 40 mg/kg of N-methyl-4-phenyl-1,2,3,6-tetrahydropyridine. Large doses of vitamin E or of β-carotene pretreatment prevent GSH destruction, thus indicating a possible participation of β-carotene and of other carotenoids in the tocopherol–glutathione–selenium peroxidase antioxidant system, and thus in protection against lipid peroxidation.[183,184] The sparing effect provided by carotenoids in favour of other antioxidants in the course of reactions involving generation of free radicals is a further indication of the participation of carotenoids in complex antioxidant systems.

β-Carotene is destroyed by oxidation, as shown in an in-vitro model using phagocytic cells in the presence of muscle food and halogen ions or H_2O_2. The reaction is evaluated by bleaching of the carotenoid as an index. If halogen ions and H_2O_2 are omitted, no discoloration occurs; it can thus be speculated that β-carotene offers itself for oxidation in the place of lipid radicals and other antioxidants.[185] This hypothesis is corroborated by the fact that pretreatment with isoascorbic acid, reduced glutathione, α-tocopherol or β-carotene suppressed pentane and ethane production in vitamin C deficient guinea-pigs injected with CCl_4.

Highlights Summarised

The studies cited above, and a number of others, provide evidence that both provitamin and non-provitamin carotenoids are potent quenchers of oxygen radicals. They further support the hypothesis that carotenoids can also quench the sensitiser triplet state, such as chlorophyll[186] and that they also take part in other complex antioxidant systems involving glutathione, superoxide dismutase, vitamin E and selenium, among others, the function of which is to oppose lipid peroxidation. Carotenoids might thus provide a logical and helpful prevention or even a complementary therapy against diseases in which epidemiological surveys have shown that these diseases may be accompanied by relatively low levels of serum carotenoids, such as for instance tumours,[187,188] light dermatoses[189] or congenital haemoglobinopathies.[182,190,191] These indications will be developed in the section on diseases with pro-oxidative mechanisms.

The assay method most commonly used now for the measurement of plasma concentration of carotenoids is high performance liquid chromatography, which provides good separations of the different carotenoids.[192]

The plasma carotene concentration of healthy adults averages 80–300 μg/dl (1·5–5·6 μmol/litre) and is subject to seasonal variations with the highest values in summer. According to our own experience, low values are not uncommon, this point being confirmed by other authors.[193] Cord β-carotene concentration is directly correlated to maternal values ($p < 0{\cdot}01$). Bottlefed infants were found to have much lower values at day 3 (13·3 ± 5·3 μg/dl) than breast-fed infants (30·0 ± 11·1 μg/dl), the former thus lacking one of the antioxidants as compared to the latter.[157,194] Certain diseases may have an influence on the plasma level, some of them, like thyroid deficiency or diabetes, being associated with higher values, those accompanied by a decrease of fat absorption or by an increase of lipid peroxidation being associated with lower values.

A normal and varied food supply provides 1000–4000 μg of carotenoids daily. As mentioned above, no vitamin A excess was ever observed in humans after administration of carotene supplements unless vitamin A is ingested as well.[147–152,195] Carotenoids have been shown to have very low toxicity.[196] The rare cases of intoxication in humans were reported on the occasion of extreme diets (very rich carrot and very low protein intakes)[153] in which accompanying substances, such as nitrates, may have played a part.[197] No such observations have been reported after supplementation with synthetic carotenoids.[147] Several papers mention the occurrence of crystal deposits in the macula after long intakes of high doses of canthaxanthin[198,199] although no major permanent visual alteration was noted.[200,201] The histological aspects of one case have been published, confirming the canthaxanthin nature of the crystals.[202]

DISEASES INVOLVING INCREASED OXIDATION

It is not unlikely that practically all human diseases involve oxidation at the subcellular level, whether as causal or as accompanying phenomena. The subsequent tissue damage can thus be at the origin of part or of the whole of the pathological symptoms. The sometimes lengthy oxidative exposure of the organism before pathological symptoms occur, should offer the possibility for applying timely prophylaxis or to compensate for hyperoxidative influences (drugs, intoxication, chronic pathological mechanisms) as long as the situation is diagnosed and antioxidant substances are available. Such substances exist in food items along with other components with pro-oxidative properties. A healthy organism supplied with a balanced and varied diet should thus ingest sufficient antioxidant in quantity and in variety or type.

It is well known however that groups of populations are at risk from living under predominantly pro-oxidative conditions, whether as a consequence of unbalanced diet habits, or of consumption of excess of fat of animal origin, smoked food or alcohol, lack of vegetables, or environmental contamination (mine workers, traffic policemen, smokers, albinos). Others at risk include patients submitted to chronic pro-oxidative or to radiative therapy, and individuals suffering from congenital or acquired diseases with peroxidation mechanisms and subsequent increased consumption of antioxidants (see risk situations, p. 230).

This section will deal with the main pathological situations in which the use of food antioxidants may be justified either for prophylaxis or as accompanying therapy. The literature is extremely rich in papers strongly suggestive of the higher risk of degenerative diseases among populations with low antioxidant blood concentrations, mainly regarding vitamin E, vitamin C, glutathione and selenium; such publications have appeared more recently for carotenoids, but are based on equally suggestive evidence.[6,187]

The conditions most frequently mentioned in the literature for their relation to oxidative mechanisms are ageing, tumours, haematological, cardiovascular, gastroenterological and neuromuscular pathologies, metabolic disorders, inflammation, rheumatism, genetic errors, prematurity, iatrogenicity and of course manifestations directly linked to oxygen involvement (oxygen therapy, decompression sickness).

Hyperoxygenation

In the treatment of hyaline membrane disease, pneumothorax, apnoea and more generally in premature infants with respiratory problems, hyperbaric oxygenation accompanied by poor reserves of antioxidant substances can be the cause of various conditions during the first days of life, namely pulmonary dysplasia, retrolental fibroplasia, manifestations of intraventricular haemorrhagia, and, in adults, the decompression syndrome.

The benefits of the early administration of vitamin E to premature infants in improving the incidence and gravity of retrolental fibroplasia and subsequent blindness remains controversial. However, randomised double-blind studies have shown that the incidence and prognosis of the illness can be improved when intramuscular injections of dl-tocopheryl acetate (>25 mg/kg) supplement oral administration of the vitamin.[203] In addition to its antioxidant properties, vitamin E has been reported to have an inhibitory effect on neovascularisation, thus contributing to long-term improvement.[204]

In cases of cerebral ischaemia as well as of intraventricular haemorrhagia, the advantage of early intramuscular injections of vitamin E is supported by randomised double-blind studies.[205,206] An alternative of a 10 mg injection followed by a 5 mg injection every 48–72 h is also effective.[207]

The fact that the reserves of all antioxidants are low in low-weight neonates should not be disregarded, since bottle-fed as well as breast fed infants during the first 6 days of life have very poor reserves of β-carotene, known to be a strong antioxidant of a complementary type.[194]

The decompression disease has been shown to entail excessive production of arachidonic acid metabolites, prostaglandin endoperoxides PG G2 and thromboxane A which favour platelet aggregation. Vitamin E is therefore proposed in high dose as a rationale for prevention and treatment.[208]

Metabolic Disorders (Genetic or Acquired)

Photohaemolysis occurring in erythropoietic protoporphyria (and other porphyrias) has been related to direct attack on double bonds by $^1O_2$.[61] Spectacular improvements in children suffering from these diseases came from the administration of carotenoids, which proved to enhance considerably their tolerance to sunlight, as well as to improve other light dermatoses.[149–152,189,209,210]

The β-form of thalassaemia, and several other haemoglobinopathies, have an increased generation of oxygen radicals as a result of the action of incomplete haemoglobin chains and of the transfusion related hypersiderosis.[211] A clear improvement of the condition of these patients has been observed under combined therapy including vitamin E and canthaxanthin.[190,191] The aggravation by sun radiation of degenerative skin diseases, such as xeroderma pigmentosum, may also be alleviated by carotenoids (see Tumours). Sickle cell anaemia gave rise to enzyme assays and in-vitro studies which suggest a possible palliative effect from vitamin C and vitamin E.[212,213]

Enzyme modifications at the time of the favism haemolytic crisis suggest a possible role of active oxygen species, so that an improved antioxidant supply is indicated.[214] The utilisation of antioxidants (vitamin E by the IM route) for the prevention and therapy of the neurological abnormalities caused by vitamin E deficiency in chronic child cholestasis may also be beneficial.[215] Vitamin E and selenium administration have been shown to improve erythrocyte survival in children with G6PD deficiency.[216]

Investigation on inborn human defects of sulphur metabolism in children with glutathione synthetase deficiency, and in those with G6PD deficiency, in which the capacity to maintain glutathione in the reduced state is compromised, indicate that pharmacological doses of vitamin E can correct some functional consequences of the disease at the level of erythrocytes and of polymorphonuclear leukocytes.[217] Another inborn metabolic disorder, cystinosis, with cystine accumulation may benefit from the administration of cysteamine and of vitamin C which both deplete the cystine content of fibroblasts.[217] An open study carried out in 16 children suffering from haemolytic uraemic syndrome and administered vitamin E supplements, resulted in a much better survival of 15 of them compared with non-supplemented patients.[218]

Hyperlipoproteinaemia is indeed a condition favouring the formation of free radicals and leading to arterial damage and platelet aggregation. The administration of pharmacodynamic doses of vitamin E (300 mg/d) or of vitamin C[219] was found to lower the serum peroxide concentration, increase high density lipoprotein (HDL) and lower density lipoprotein (LDL);[132,219–221] reduction of platelet aggregation, easily demonstrated *in vitro*, required 600 mg/d of vitamin E *in vivo*.[132] Fewer publications refer to diabetes *per se*, some of them however report the intervention of free radicals in the mechanism of complications. Sustained treatment with flavonoids such as quercetin or anthocynanin decrease the occurrence of cataracts and of retinopathy.[222–224]

Stress

Stress increases lipid peroxidation and O_2 production.[225,226] At the level of the myocardium, ischaemia and hypoxic damage were found to be increased by stress and alleviated by preliminary administration of α-tocopherol to rats.[227] Similar results were observed *in vitro* comparing O_2 production of polymorphs from stressed neonates with those from controls.[226]

Exercise-training experiments in rats were followed by vitamin E depletion at the level of skeletal muscle mitochondria.[228] Further studies suggest that an adequate ascorbic acid intake could be of special significance for patients whose cardiac function is under stress.[229] Exercise experiments in humans reveal a sustained oxidation of glutathione during exercise and a compensatory reduction during recovery, thus suggesting an increased consumption of antioxidant.[229,230]

Stress situations are frequent in the current life of humans for instance in cases of psychopathological instability, surgery, strain, overwork and trauma. Vitamin C and vitamin E supplementation may be indicated especially when the stress is chronic or recurrent and when debilitated or deficient organisms are involved.

Toxic Substances (Environmental, Pharmaceutical, Alimentary)

Addictive Compounds

Among the multitude of toxic substances the organism has to cope with, those associated with smoking, alcohol consumption, air pollution, side effects of drugs, pesticides and food poisoning are the most frequently reported. Their effects have been proved to be mediated by the intervention of free radicals at various levels. The placental response to maternal smoking involves the induction of cytochrome P450, a mediator of pro-oxidant radicals.[231] The vitamin E concentration has been found to be increased in the lungs of young cigarette smokers and conversely decreased in the alveolar fluid.[232,233] Plasma and leukocyte vitamin C concentrations are decreased by 10–15% in smokers[234] and the turnover of this vitamin averages 90 mg/d in contrast to 60 mg/d in non-smokers,[51] thus proving increased ascorbate requirement in subjects exposed to smoking.[234,235]

The metabolism of alcohol generates acetaldehyde, which is a mutagen, a teratogen and a co-carcinogen; it also increases the production of oxygen radicals and accordingly lipid peroxidation.[236,237] The origin of pro-oxidant radicals involved in this metabolism are not only of microsomal but probably also of mitochondrial and cytosolic origin.[238] Alcoholic patients suffering from chronic hepatitis have significantly higher MDA and lower glutathione peroxidase values compared with controls.[9] Such an increase of oxidative reactions will overwhelm the antioxidant systems, which are at risk of long-term depletion, especially because alcoholism is frequently associated with smoking.

Environmental Contaminants

Air pollution has gradually increased, in parallel with energy requirements for transport, industry and individual homes. This is the main source of the SO_2, NO_2, CO, hydrocarbons and soot involved in human intoxication by air pollutants.

These various toxic substances are all involved in reactions generating oxygen radicals, such as lipid peroxidation, methaemoglobin and local irritation. Adaptation requires, as ever, an overconsumption of antioxidants to control such metabolic deviations.[239] Benzo(a)pyrene, one of these pollutants well known for its carcinogenic properties could be counteracted and tumour growth reversed by supplements of carotenoids in experiments on mice.[168,240] The relevance to human chemotherapy has accordingly been suggested.[241]

Pharmaceutical Products

Certain pharmaceuticals like antitumour agents, analgesics, antiphlogistic or antiuric medicines, heavy metals and antibiotics, exert their therapeutic activity through such mechanisms as generation of active oxygen radicals, peroxidation of unsaturated membrane lipids, stimulation of prostaglandin secretion or inactivation of prostacyclin, all transformations which are accompanied by the consumption of antioxidant substances and the risk of uncontrolled chain reactions.[116,242,243]

Most of the side effects triggered by substances with quinone or phenol structures as well as by chlorinated radicals are linked to these mechanisms.[4,243–245] Iron preparations, a necessary treatment in iron-deficiency anaemia, are often associated with ascorbic acid to improve intestinal absorption. However such an association is known to increase peroxidation reactions and is currently used for this purpose by investigators of animal or in-vitro experiments. It should therefore be kept in mind that the long-term therapeutic use of Fe preparations may gradually escape from the control of oxidative reactions and be an indication for the administration of antioxidants of another type such as tocopherol or carotenoids. Conversely, ascorbic acid has been shown to decrease the toxicity of hexavalent chromium compounds[246] and of aspirin.[243] Tocopherol lessens the cardiotoxicity of adriamycin in antitumour chemotherapy.[247]

Intrauterine contraceptive devices are thought to increase the endometrial H_2O_2 formation responsible for menorrhagia. Indian authors tried therefore, with total success, vitamin E prevention of this dysfunction.[248] Vitamin E and selenium have also been shown to be helpful in acute copper toxicity.[242] The field of the counteraction of drug side effects linked to free radical production may also be relevant, mainly in so far as these side effects can be alleviated by association with non-toxic antioxidant vitamins.[235]

Food Toxins

Food toxins such as nitrite, amine and amide radicals, when present in the acid gastric medium generate N-nitrosamines, highly carcinogenic molecules. The effect has been shown to be inhibited by high concentrations of ascorbate.[39,249] A ratio of ascorbate to nitrite as high as 16 to one may be required in some circumstances thus giving a further indication for a sufficient vitamin C supply.[39]

Carotenoids are capable of protecting plants from light induced oxidative damage, but herbicide residues have been found to decrease carotenoid concentrations in plants,[169] thus providing an explanation for the toxic mechanism of herbicides. The cytotoxic effects of paraquat, a phenol bipyridyl herbicide, have been shown to be decreased by vitamin E,[250] although results remain at present contradictory. This vitamin, when applied to the skin, was shown to reduce the local irritation which follows contact with synthetic pyrethroids used as insecticides.[251] Vitamin C, which is necessary for the synthesis of many enzymes, protects against liver, kidney, blood and other organ damage when given after exposure to a number of pesticides including Parathion, Malathion,[252] Toxaphene,[253] DTT, Lindane, Chlordane and Aldrin.[254] Protection of individuals from pesticide poisonings is an area in which the encouraging preliminary results provided by antioxidant vitamins deserve further research.

Radiation

Radiative influences may be classed as external interventions or deleterious side effects: they are experienced by the organism either as unavoidable environmental pollution, or as the unwanted effect of certain therapies. Whether electromagnetic (X-rays or γ-rays) or corpuscular (α-rays, β-rays or neutrons), radiation interacts with living tissues to generate free radicals, mainly H and OH, then HO_2 and H_2O_2 respectively.[255] Very positive results have been obtained by the use of carotenoids in light-induced dermatoses.[149,150–152,209,256]

Depending on the need to obtain a radio-sensitisation or a radio-protection effect, practitioners may modulate oxygen supply and pressure or increase antioxidant supplements. Trials are still running in the field of radiotherapy and in the developing area of puvatherapy.[110,257]

Inflammation, Arthritis

Free radicals have been associated with the inflammatory response seen in arthritic joints.[258] Prostaglandins, leukotrienes and other lipid peroxides such as those of arachidonic acid, linoleic acid and eicosa-penta-enoic acid, play an important part in the establishment of inflammatory reactions. Platelets, which are rich in unsaturated fatty acids, and polymorphs which, under the influence of the O_2 radical, adhere to the vessel wall, contribute strongly to the mechanism of vascular inflammation.[259] All these oxidative metabolic pathways account for the therapeutic effects of corticoids, which inhibit the 'arachidonic acid cascade', and of non-steroidal anti-inflammatory agents, which are cyclo-oxygenase inhibitors, in the treatment of inflammation and arthritis.

The fact that gold compounds, which are potent singlet oxygen quenchers, relieve several types of rheumatism, supported the hypothesis that antioxidant treatment might be a logical intervention to control these cumulative pro-oxidant pathological mechanisms. Several investigations supporting this speculation showed that carotenoids are even stronger oxygen quenchers than gold compounds,[260] thus supporting their potential role in the field of rheumatism. An evaluation of the clinical efficacy of carotenoid supplements in the prevention, the treatment or the recurrence of arthritis therefore warrants consideration.[261] The production of collagen, an important constituent of connective tissue and a target for rheumatic damage, is further known to be mediated and controlled by vitamin E and vitamin C.[262]

Finally, it should be kept in mind that heavy metals and several other anti-arthritic agents used for long periods entail a depletion of certain antioxidants, such as for instance the selenium dependent glutathione peroxidases, thus providing a further reason to supply arthropathic patients with enough antioxidant material.[242,245]

Infection and Immunity

As many antibacterial drugs exert their anti-infection activity through the generation of free radicals that are toxic for bacteria, the question arises whether antioxidants may not develop an antagonising effect capable of counteracting the bacteriostatic action of these pharmaceuticals. The answer might rely on the evidence that a number of the investigations cited in this chapter advocate primarily a modulation or a control of oxidative reactions by the so-called antioxidants rather than pure, unlimited antioxidative action, a fact which would question the terminology 'antioxidant' for substances which appear above all to be oxidation modulators. A more specific answer may result from experiments showing the protection afforded by antioxidants (vitamin E, vitamin C) against phagocytosis suppression induced, for instance, by hydroxyurea.[263]

Vitamin E exerts a protective effect on immune triggered endothelial damage partly by increasing the endogenous antioxidant potential, partly by modulating intrinsic endothelial prostaglandin production.[264] The intravenous injection of 1 g vitamin C in volunteers was found to be accompanied by a stimulation of lymphocyte reactivity to mitogens, while in-vitro experiments showed vitamin C dose-dependent protection of lymphocytes against the HRP/H_2O_2/halide system,[28] thus supporting the property of this vitamin to protect or stimulate immuno-defence reactions.

Ageing

Trisomy 21 is an interesting model for the study of ageing, not only because the senescence sequences are shortened in individuals suffering from this genetic disorder, but also because the structural gene of the cytosolic Cu–Zn–superoxide dismutase, an enzyme which might play a part in the ageing process, is also localised on band q 221 of chromosome 21. There is thus a possible link with the hypothesis current since the 1960s that ageing results from the action of free radicals at the level of nerve and endothelial cell membranes and from the gradual weakening of the antioxidant defence.[265] Oxygen radicals and lipid peroxidation are confirmed by other authors to be the major sources of DNA alterations,[266] this damage being considered the major cause of ageing.[267] Experiments in rats moreover show that DNA damage correlates with vitamin E depletion.[268]

Vitamin E, selenium, dimethyl sulphoxide (DMSO) and reduced glutathione were found to decrease the rate of formation of lipofuscin, a waste product of lipid peroxidation found in ageing cells,[269] although another experiment using mice did not confirm lipofuscin reduction by vitamin E.[270] Foetal rat brain cultures of neurons were prevented from degeneration, in comparison with controls, after 27 days by adding 5–15 μg vitamin E/ml of diet.[271,272] These experimental data provide a sound and encouraging basis for human applications.[120,273] However, even in trisomic patients, in-vivo trials would be of an unrealistic duration, thus confining the observations either to human cell cultures or to empirical, but safe and logical, administration of the antioxidant vitamins. Actually they strengthen the plea for more scientific studies as well as recommending a diversified diet, including ample fruits, cereals and vegetables, and restricting animal fats, but maintaining an adequate protein supply.

Cardiovascular Diseases

Dietary polyunsaturated fatty acids, which have long been recommended for lowering plasma cholesterol levels, and consequently decreasing the risk of atherosclerosis, are actually extremely subject to peroxidation reactions if a sufficient antioxidant defence, mainly in the form of tocopherol, is not available. Epidemiological studies suggest the ratio cholesterol/antioxidant may be considered an important risk factor for ischaemic heart disease.[6]

Publications have accumulated in recent years which support the decisive role of peroxides in the establishment and progress of vascular abnormalities and heart disease.[6] In the human, the degree of atherosclerosis is positively correlated with the accumulation of lipid peroxides and hydroperoxides.[274] Circulatory shock in humans induces an increase in MDA levels and in α-tocopherol utilisation.[117] The influence of vitamin C on the distortion of lipid metabolism has long been stressed.[219]

Vitamin E at a dose of 300 mg/d in 20 volunteers during 20 days significantly increased HDL and decreased LDL. However, it did not succeed in lowering platelet aggregation, as was observed *in vitro*.[275] In another in-vivo human study, platelet aggregation inhibition occurred only at minimal daily doses of 600 mg/d, whereas 300 mg/d lowered elevated peroxide concentrations in type II and type IV hyperlipoproteinaemia.[132] A double-blind controlled study showed that 300 mg/d are sufficient to change platelet MDA secretion.[130,220] These examples support the concept that antioxidant vitamins at sufficient concentration inhibit mechanisms favouring atherosclerosis. Several investigators actually have reported the beneficial effect of vitamin E therapy in patients with intermittent claudication.[276] A multicentre epidemiological study carried out in Finland, Scotland, Switzerland and Italy showed a clear inverse correlation between crude age-standardised mortality from ischaemic heart disease (higher in the northern countries) on the one hand and the median of plasma ascorbic acid ($r = 0{\cdot}74$), as well as the median of the cholesterol standardised vitamin E ($r = 0{\cdot}93$), on the other, but the correlations for β-carotene and selenium were rather weak.[6] Further epidemiological surveys disclose similar trends, showing the beneficial effects of diets or availability of foods which provide more vitamin C, or generally more antioxidants.[277]

Other Organ Systems

The nervous system is involved with aspects other than those mentioned above, in which the oxidant–antioxidant balance plays a major part. For example, spinal cord degeneration of varied intensity has been described by several authors in vitamin E malabsorption, with reversibility in the mildest cases after administration of the deficient vitamin, preferably by the parenteral route.[3] Vitamin E, selenium and vitamin C supplements have therefore been suggested in the early stages of Parkinson's disease,[278] Batten or Alzheimer syndromes.[279]

Gastro-enterology is another relevant field. Cholestatic patients, who have increased MDA erythrocyte levels and low serum vitamin E concentrations, require high doses of this vitamin, preferably by the intramuscular route.[280] Chronic pancreatitis depresses serum vitamin E and glutathione-peroxidase activity, whereas peroxide levels are increased. Ulcero-necrotic enterocolitis is mentioned in several papers dealing with free radicals and deficiency of the antioxidant defence.[13,281] Recurrence and malignant transformation of colon polyposis are narrowly linked to oxidant radicals and thus provide a promising field of research for carotenoids, vitamins E, C, selenium and glutathione.

Tumours

Cancer and tumours in general constitute a major threat to life and are severely stressful psychologically. In spite of real therapeutic progress, the large amount of research generated is fully justified. Most specialists agree as to the implication of free radicals in the degenerative mechanisms of carcinogensis, mainly disorders in DNA repair, which are at the heart of the problem.[164]

Food as well as life style may supply many carcinogenic substances, such as coffee, alcohol, pyrolysis products, nitrites, amines, fat, smoke, pesticides and most kinds of mutagens, which can generate active oxygen radicals or superoxides.[164] It is therefore not surprising that a growing number of epidemiological surveys aim at evaluating the protective effect of food antioxidants. Among these Peto *et al.* showed a clear-cut inverse relation between the plasma carotene level and the incidence of cancer[187] whereas another survey did not reveal any statistical difference.[282] A further epidemiological study carried out in Switzerland showed, after adjustment for age and smoking, an inverse relation between the mortality from bronchial or stomach cancer and serum β-carotene, as well as between gastro-intestinal cancers and vitamin C, or cholesterol standardised vitamin E.[283] The prospective data reported by Menkes *et al.* and including the antioxidants mentioned above and also selenium, are noteworthy.[188]

Several longitudinal studies evaluating the possible beneficial effect of long term carotene supplementation on the incidence of cancer, or the protection afforded by vitamin E or β-carotene in reducing the risk of lung cancer among heavy cigarette smokers, or among persons occupationally exposed to asbestos, are currently in progress.[284] Such studies were stimulated by the very promising results obtained in animal studies showing a prevention or a blocking effect by carotenoids,[172,258,285–287] vitamin E, cysteine and reduced glutathione.[288,289] While there are indications that tumours of various kinds increase the body requirement for ascorbic acid,[290] the extension of survival from an average of 50 to 210 days for terminal cancer patients supplemented with 10 g of ascorbic acid daily,[291] should however be interpreted more as a stimulus for further research than as a real therapeutic benefit for the patient. More practical human applications are being investigated intensively at the level of prevention in the field of premalignity or of post-surgical recurrence. For instance, reduction in the number and size of recurrent colon polyps was shown in patients taking 3 g/d of ascorbic acid in a randomised double-blind study.[292]

Xeroderma pigmentosum is a genetic proliferative dermatosis subject to malign transformation under the influence of sunlight. The positive results obtained in an open trial with carotenoids (β-carotene and canthaxanthin) on the incidence of malignment complications is a further confirmation of the potential contribution of these substances in the prevention or therapy of tumours.[256]

The effect of selenium is still of active research interest. In-vitro studies on the transformation by X-rays or by chemotherapy of C3H/10T-1/2 cells have shown that Se confers protection in part by inducing or activating cellular free radical scavenging systems, while vitamin E acts by peroxide breakdown,[288] the antioxidants thus acting complementarily. The antithrombotic role of antioxidants has also been suggested as a possible mechanism against tumour metastases.[293]

Only a few of the most striking experiments have been mentioned in this section but their results, as well as those of many others, justify dynamic research into free radical implications in malignity and premalignity, and the beneficial effect to be expected from the use of food antioxidants in human applications.

Miscellaneous

Many more histological or anatomical areas are affected by free radicals, and their necessary control by antioxidant systems. Dermatology, haematology, endocrinology and trophology have not been specifically mentioned but some of the diseases in which they are implicated have been cited in the text. Indeed all the membranes of all the cell organelles may be involved in oxygen free radical reactions, and research is justified wherever deviations from their normal metabolism generates disease, whatever the aetiology might be. The amount of evidence gathered from the in-vitro or in-vivo studies carried out and published so far, justifies continuing effort. Post-surgical tissue necrosis is no less implicated in free radical proliferation which has been shown to be decreased by vitamin C supplementation.[294]

This discussion was not intended to include nutritional deficiencies by insufficient intake (or absorption) of food antioxidants. Mention should be made, however, of Keshan disease which, unlike other nutritional deficiencies, is seldom described. Selenium is present at an average rate of 50–100 ng/ml in our daily food intake, thus representing 20–30 times the body requirement. In the Chinese province of Keshan, however, the opportunity has repeatedly been provided to observe cases of selenium deficiency. Children suffering from the disease present with multiple signs of peroxidation and their selenium blood concentration averages 15 ng/ml instead of 55. If early selenium supplementation is not administered, the disease will develop until it constitutes an irreversible myocardiopathy. Even at higher subnormal concentrations, Finnish authors have observed that values less than 45 ng/ml are accompanied by a threefold incidence of ischaemic cardiopathies. Not infrequently, cattle and other domestic animals suffer from the disease.[295]

RISK SITUATIONS

The preceding sections give an idea of the main situations involving distortions favouring oxidative reactions, and their complications. They are summarised below:

- —Malnutrition: both insufficiency (vegetables, fruits, cereals, bran, liver) and excess, of certain foods (smoked meat or fish, animal fats, fried meals).
- —Intoxication: cocaine addiction, tobacco, alcohol, long-term intake of pharmaceuticals, air pollution (traffic policemen, garage workers, asbestos miners).
- —Radiation (tropical regions mainly for whites and bald people, albinos, nuclear plant accidents).
- —Hyperoxygenation (premature infants, divers).
- —Inborn errors of metabolism and genetic disorders (haemoglobinopathies, porphyrias, G6PD deficiencies).
- —Metabolic deviations (hyperlipidaemia, diabetes).
- —Permanent or repeated stress.
- —Rheumatic diathesis.
- —Family history of cardiovascular disease.
- —Personal history of malignant disease or predisposition (operated cancer, colon polyposis).
- —Aged people.

DOSES AND SAFETY

The trials and experiments summarised in this chapter on the biological effects of food antioxidants often provided evidence of the need for high dose requirements, placing the utilisation of the substances studied at the level of drugs.

Thus, two types of situation may occur:

—An insufficient supply or availability of antioxidants is likely to result in a higher pathological risk, as shown by several epidemiological studies.[187,283] The intervention should then be to increase intakes until the blood concentration of antioxidants reaches normal values. This preventive approach will generally apply to doses only slightly higher than the recommended daily allowances (RDA).

—In contrast, situations of overt pathology or of continuous pro-oxidant stress require, as seen above, ranges which may reach a 100 times or more the RDA or the nutritional intake. The trials analysed thus refer to the administration of daily doses of one to several grams of ascorbic acid, several hundred milligrams of dl α-tocopherol or of carotenoids, or several hundred micrograms of selenium.

The requirements for high doses, and sometimes long-term administration, obviously raise the question of safety.[195]

As far as vitamin C, vitamin E and β-carotene are concerned, numerous animal and human studies have shown the considerable degree of safety of these antioxidants.[195] They are reviewed in several papers for vitamin C[296] and vitamin E,[297] or form a consensus of authors having used β-carotene at pharmacodynamic doses and who point out the absence of hypervitaminosis A.[149,150–152,190,191,210] Safety for non-provitamin A carotenoids has not been extensively tested except for canthaxanthin which, although having a low toxicity, was responsible for crystalline deposits in the macula of some people having ingested very high doses (>75 mg/d) for periods of one or more years (>25 g). The authors report either the absence of visual alterations or slight subjective decrease of adaptation to light variations, receding after drug discontinuance in spite of the remaining macular deposit.[198,200–202] This side effect is currently being further studied. The genetic toxicology of dietary flavonoids has been discussed by MacGregor.[298] The tolerance for selenium has been reviewed[299] and the Food and Nutrition Board of the US National Research Council has proposed the range of 50–200 μg/d as a safe and adequate dietary allowance.[143] Very little however is known about the safety of greater amounts.

CONCLUSIONS

There is now a consensus among scientists regarding the effects of uncontrolled oxygen radicals in the deterioration of health. Much however remains to be elucidated on the role of antioxidants and the limits of their potential. If they are purely classical antioxidants, might excesses of their action or of their concentration unduly interfere in necessary pro-oxidant metabolic phases? There is no evidence of this so far. If not, are some of them able to, without limit, control excessive generation of active oxygen radicals or peroxidation. Several experiments suggest this for vitamin E, which thus might rather be considered as a modulator of peroxidation.

The behavioral differences between various food antioxidants revealed by the studies discussed raise the question of their complementarity. It has been seen that ascorbic acid, which may exhibit antioxidant but also pro-oxidant activity (for instance, in presence of Fe), benefits from tocopherol intervention for its recycling. The synergistic action of a group of antioxidants has further been described in the 'tocopherol–selenium–glutathione–enzymes' system (see p. 207). In addition to the very rewarding work devoted to a better knowledge of the role of every antioxidant at the cellular level, the data accumulated therefore justify more research in the direction of synergistic effects as suggested between tocopherol and ascorbic acid[6] and, as a further extension between flavonoids and tocopherol or β-carotene. Supplementation of one type of antioxidant with another, may thus enhance beneficial effects, as is seen for instance with carotenoids and vitamin E.[190,191] Possible deleterious interactions, as reported for instance between selenium, vitamin C and vitamin A,[15,299] may also be better documented and understood.

Routine methods for the mass determination of oxidation and peroxidation levels further need to be developed and applied, allowing for routine individual checking and preventive interventions, both by clinical investigation of the aetiology and antioxidant supplementation.

33. Chatham, M. D., Sauder, L. R. & Kulle, T. J., Evaluation of the effect of vitamin C on ozone-induced bronchoconstriction in normal subjects. *Am. Rev. Resp. Dis.*, **129** (1984) A145.
34. Blondin, J., Baragi, V. K., Schwartz, E. R., Sadowski, J. & Taylor, A., Prevention of eye lens protein damage by dietary vitamin C. *Fed. Proc.*, **45** (1986) 478.
35. Tso, M. O. M., Woodford, B. J. & Lam, K. W., Distribution of ascorbate in normal primate retina and after photic injury; a biochemical, morphological correlated study. *Curr. Eye Res.*, **3** (1984) 181–91.
36. Bensch, K. G., Fleming, J. E. & Lohmann, W., The role of ascorbic acid in senile cataract. *Proc. Natl Acad. Sci. USA*, **82** (1985) 7193–6.
37. Wartanowicz, M., Panczenko-Kresowska, B., Ziemlanski, S., Kowalska, M. & Okolska, G. The effect of alpha-tocopherol and ascorbic acid on the serum lipid peroxide level in elderly people. *Ann. Nutr. Metab.*, **28** (1984) 186–91.
38. Niki, E., Saito, T. & Kamiya, Y., The role of vitamin C as an antioxidant. *Chem. Lett.*, **5** (1983) 631–2.
39. Walters, C. L., Influence of ascorbic acid on the formation of N-nitrosamines in foods, cosmetics and tobacco. In *Vitamin C (Ascorbic Acid)*, ed. J. N. Counsell & D. H. Hornig, Applied Science, London, 1982.
40. Mirvish, S. S., Wallcave, L., Eugen, M. & Shubik, P., Ascorbate–nitrite reactions: possible means of blocking the formation of N-nitroso-compounds. *Science*, **177** (1972) 65–8.
41. Abe, K., Kogure, K., Arai, H. & Nakano, M., Ascorbate induced lipid peroxidation results in loss of receptor binding in tryptophan but not in phosphate buffer; implications for the involvement of metal ions. *Biochem. Int.*, **11** (3) (1985) 341–8.
42. Campell, G. D., Steinberg, M. H. & Bower, J. D., Ascorbic acid induced hemolysis in G6PD deficiency. *Ann. Intern. Med.*, **82** (1975) 810.
43. Blum, H. F., *Photodynamic Action and Diseases Caused by Light.* Hafner, New York, 1964.
44. McCay, P. B., Vitamin E, interactions with free radicals and ascorbate. *Ann. Rev. Nutr.*, **5** (1985) 323–40.
45. Niki, E., Tsuchiya, R., Tanimura, R. & Kamiya, Y., Regeneration of vitamin E from the alphachromanoxyl radical by glutathione and vitamin C. *Chem. Lett.*, **6** (1982) 789–92.
46. Bascetta, E., Gunstone, F. D. & Walton, J. C., Electron spin resonance study of the roles of vitamin E and vitamin C in the inhibition of fatty acid oxidation in a model membrane. *Chem. Phys. Lipids*, **33** (1983) 207–10.
47. Barclay, L. R. C., Locke, S. J. & MacNeil, J. M., The autoxidation of unsaturated lipids in micelles; synergism of inhibitors vitamins C and E. *Can. J. Chem.*, **61** (1983) 1288–90.
48. Sternberg, L., Julicher, R. H. M., Bast, A. & Noordhoek, J., Effect of vitamin E on the balance between prooxidant and antioxidant activity of ascorbic acid in microsomes from rat heart, kidney and liver. *Toxicol. Lett. (Amst)*, **25** (2) (1985) 153–60.

49. Liotti, F. S., Bodo, M., Pezzetti, F., Guerrieri, P. & Menghini, A. R., Inhibition of the binding of 7,12, dimethylbenz(a)anthracene to DNA by ascorbic acid reduced glutathione and cysteine in chick embryo cells cultured in vitro. *Oncology* **43** (3) (1986) 183–6.
50. Olinescu, R., Milcoveanu, D., Nita, S., Pascu, E. & Urseanu, I., Inhibitory action of magnesium ascorbate on the formation of lipid peroxides by anthracyclin antibiotics. *Rev. Roum. Bioch.,* **23** (2) (1986) 127–33.
51. Kallner, A., Vitamin C man's requirement. In *Vitamin C (Ascorbic Acid)*, ed. J. N. Counsell & D. H. Hornig. Applied Science, London, 1982.
52. Deby, C., Tendances et recherches sur les médicaments antiradicaux libres. *Cah. Nutr. Diét. (Paris),* **22** (1) (1987) 77–81.
53. Maridonneau-Parini, I., Braquet, R. & Garay, R. P., Heterogeneous effect of flavonoids on potassium loss and lipid peroxidation induced by oxygen-free radicals in human red cells. *Pharmacol. Res. Commun.,* **18** (1) (1986) 61–72.
54. Ogasawara, H., Fujitani, T., Drzewiecki, G. & Middleton, E. Jr., The role of hydrogen peroxide in basophil histamine release and the effect of selected flavonoids: *J. Allergy, Clin. Immunol.,* **78** (2) (1986) 321–8.
55. Sousa, R. L. & Marbetta, M. A., Inhibition of cytochrome P-450 activity in rat liver microsomes by the naturally occurring flavonoid quercetin. *Arch. Biochem. Biophys.,* **240** (1) (1985) 321–8.
56. Buening, M. K., Fortner, J. G., Wood, A. W. & Conney, A. H., Activation and inhibition of benzo(a)pyrene and aflatoxin B-1 metabolism in human liver microsomes by naturally occurring flavonoids. *Cancer Res.,* **41** (1) (1981) 67–72.
57. Huang, M. T. *et al.,* Inhibition of the mutagenicity of Bay Region diol epoxides of polycyclic aromatic hydrocarbons by phenolic plant flavonoids. *Carcinogenesis,* **4** (12) (1983) 1631–8.
58. Yamazaki, H., Mori, Y., Toyoshi, K., Nagai, H., Koada, A. & Konishi, Y., A comparative study of the mutagenic activation of N-nitrosopropylamines by various animal species and man, evidence for a cytochrome P-450 dependent reaction. *Jpn J. Cancer Res.,* **77** (2) (1986) 107–17.
59. Koch, H. P. & Löffler, E., Influence of silymarin and some flavonoids on lipid peroxidation in human platelets. *Methods Find Exp. Clin. Pharmacol.,* **7** (1) (1985) 13–18.
60. Rosin, M. P., The influence of pH on the convertogenic activity of plant phenolics. *Mutat. Res.,* **135** (2) (1984) 109–14.
61. Crastes de Paulet, A., Un piégeur de radicaux libres: l'extrait de ginkgobiloba-preprint. *Journées sur les Radicaux Libres,* Bordeaux, 26–27.9.86, Société de Nutrition et Diététique de Langue Francaise, Paris.
62. Masquelier, J., Vin et radicaux libres. *Journées sur les Radicaux Libres,* Bordeaux, 26–27/9/1986. Preprint. Soc. Nutr. Diét., Paris.
63. Sorata, Y., Takahama, U. & Kimura, M., Prospective effect of quercetin and rutin on photosensitized lysis of human erythrocytes in the presence of hematoporphyrine. *Biochim. Biophys. Acta* **799** (3) (1984) 313–18.

by high performance liquid chromatography. *J. Liquid Chromatog.*, **7** (1984) 2611–30.
193. Tangney, C. C., Shekelle, R. D., Raynor, W., Gale, M. & Betz, E. P., Intra- and interindividual variations in measurement of beta-carotene, retinol and tocopherols in diet and plasma. *Am. J. Clin. Nutr.*, **45** (4) (1987) 764–9.
194. Ostrea, E. M., Balun, J. E. & Winkler, R., Serum levels of antioxidants in breastfed vs bottlefed infants: a possible role of breastmilk in protecting infants against oxygen toxicity. *Pediatr. Res.*, **15** (1981) 543.
195. Demopoulos, H. B. & Santomier, J. P., Safety of large-dose antioxidant regimens in man. *Med. Sci. Sports Exercise*, **17** (1985) 201–2.
196. Bagdon, R. E., Zbinden, G. & Studer, A., *Toxicol. Appl. Pharmacol.*, **2** (1960) 225–35.
197. Keating, J. P., Lell, M. E., Strauss, A. W., Zarkowsky, H. & Smith, G. E., Infantile methemoglobin caused by carrot juice. *N. Engl. J. Med.*, **288** (1973) 824–6.
198. Cortin, P., Corriveau, L. A., Rousseau, A. P., Tardif, Y., Malentant, M. & Boudreault, G. Maculopathie en paillettes d'or. *Can. J. Ophthalmol.*, **17** (1982) 103–6.
199. Boudreault, G., Cortin, P., Corriveau, L. A., Rousseau, A. P., Tardif, Y. & Malenfant, M. La rétinopathie à la canthaxanthine. 1. Etude Clinique de 51 consommateurs. *Can. J. Ophthalmol.*, **18** (1983) 325–8.
200. Meyer, J. J., Bermond, P., Pournaras, C. & Zoganas, L., Canthaxanthin, Langzeiteinnahmen und Sehfunktionen beim Menschen. *Deutsche Apothek. Ztng*, **125** (21) (1985) 1053–7.
201. Weber, U., Goerz, G. & Hennekes, R., Carotinoid-Retinopathie, 1. Morphologische und funktionelle Befunde. *Klin. Monatsbl. Augenheilkd.*, **186** (1985) 351–4.
202. Daicker, B., Schiedt, K., Adnet, J. J. & Bermond, P., Canthaxanthin retinopathy, an investigation by light and electron microscopy and physiochemical analysis. *Graefe's Arch. Clin. Exp. Ophthalmol.*, **225** (1987) 189–97.
203. Hittner, H. M. & Kretzer, F. L., Retrolental fibroplasia and vitamin E in the preterm infant; comparison of oral vs intramuscular and oral administration. *Pediatrics*, **73** (2) (1984) 238–49.
204. Dehan, M., Lindenbaum, A. & Niessen, F., Pédiatrie néonatale: les problèmes liés à l'hyperoxygénation. *Cah. Nutr. Diét.*, **22** (1) (1987) 15–20.
205. Speer, M. E., Blifeld, C., Rudolph, A. J., Chadda, P., Holbein, M. E. & Hittnes, H. M. Intraventricular haemorrhage and vitamin E in the very low birth weight infant; evidence for efficacy of early intramuscular vitamin E administration. *Pediatrics*, **76** (6) (1984) 1107–12.
206. Chiswick, M. L. *et al.*, Protective effect of vitamin E (DL-α-tocopherol) against intraventricular haemorrhages. *Br. Med. J.* (*Clin. Res.*), **287** (6385) (1983) 81–4.
207. Colburn, W. A. & Ehrenkranz, R. A., Pharmacokinetics of a single intramuscular injection of vitamin E to premature neonates. *Pediatr. Pharmacol.*, **3** (1) (1983) 7–14.

208. Reggiani, E. & Odaglia, G., Processi emocoagulativi nella malattia da decompressione e nel trattamento iperbarico. *Minerva Med.*, **72** (2) (1981) 1833–90.
209. Eales, L., The effects of canthaxanthin on the photocutaneous manifestations of porphyria. *S. Afr. Med. J.*, **54** (25) (1978) 1050–2.
210. Goertz, G. & Ippen, H., Carotinoid Behandlung von Lichtdermatosen. *Dtsch. Med. Wschr.*, **102** (29) (1977) 1051–5.
211. Rachmilewitz, E. I., Kornberg, A. & Acker, M., Vitamin E deficiency due to increased consumption in beta-thalassemia. *Ann. NY Acad. Sci.*, **393** (1982) 338–47.
212. Lachant, N. A. & Tanaka, K. R., Antioxidants in sickle cell disease; the *in vitro* effects of ascorbic acid. *Am. J. Med. Sci.*, **292** (1) (1986) 3–1a.
213. Natta, C. L., Zimmermann, C., Machlin, L. J. & Brin, M., Abnormal vitamin E and glutathione peroxidase levels in sickle cell anemia. *Proceedings of the 21st Annual Meeting of the Am. Coll. Nutr.*, Bethesda 1980. Allan R. Liss, New York 1981, p. 213.
214. Mavelli, I. *et al.*, Favism, a hemolytic disease associated with increased superoxide dismutase and decreased glutathione peroxidase activities in red blood cells. *Eur. J. Biochem.*, **139** (1) (1984) 13–18.
215. Sokol, R. J., Heubi, J. E., Iannaccone, S., Bove, K. E. & Balistreri, N., Mechanism causing vitamin E deficiency during chronic childhood cholestasis. *Gastroenterology*, **85** (5) (1983) 1172–82.
216. Hafez, M. *et al.*, Improved erythrocyte survival with combined vitamin E and selenium therapy in children with glucose-6-phosphate dehydrogenase deficiency and mild chronic hemolysis. *J. Pediatr.*, **108** (4) (1986) 558–61.
217. Schulman, J. D., Genetic disorders of glutathione and sulfur aminoacid metabolism. *Ann. Intern. Med.*, **93** (2) (1980) 330–46.
218. Powell, H. R., McCredie, D. A., Taylor, C. M., Burke, J. R. & Walker, R. G., Vitamin E treatment of haemolytic uraemic syndrome. *Arch. Dis. Child*, **59** (5) (1984) 401–4.
219. Ginter, E. & Bobek, P., The influence of vitamin C on lipid metabolism. In *Vitamin C* (*Ascorbic Acid*), ed. J. N. Counsell & D. H. Hornig. Applied Science, London, 1982.
220. Cordova, C. *et al.*, Influence of vitamin E on lipid peroxidation in man. *Haemostasis*, **12** (1–2) (1982) 43.
221. Fidanza, A., Andisio, M. & Mastroiacovo, P., Vitamin C and chlosterol. *Int. J. Vitam. Nutr. Res.*, (suppl.) **23** (1982) 153–71.
222. Creighton, M. O. & Trevithick, J. R., Cortical cataract formation prevented by vitamin E and antioxidants. Possible role of free radicals in cortical cataractogenesis. *Invest. Ophthalmol. Vis. Sci.*, (1979) (suppl.) 212, (ARVO Abstracts).
223. Beyer-Mears, A. & Farnsworth, P. N., Diminished diabetic cataractogenesis by quercetin. *Invest. Ophthalmol. Vis. Sci.*, (1979) (suppl.) 212–13, (ARVO Abstracts).
224. Bermond, P., Personal records of the author (200 diabetic patients).
225. Meerson, F. Z., Pavlova, V. I. & Kozobeinikova, E. N., Prevention of stress damage to the body by antioxidants and the beta-blocker inderal. *Vopr. Med. Khim.* (*USSR*), **26** (6) (1980) 827–32.

226. Shigeoka, A. O., Charette, R. P., Wyman, M. L. & Hill, H. R., Defective oxidant metabolic responses in neutrophils from stressed neonates. *J. Pediatr.* **98** (3) (1981) 392–8.
227. Meerson, F. Z. & Ustinova, E. E., Prevention of stress injury to the heart and its hypoxic contracture by using the natural antioxidant alpha-tocopherol. *Kardiologiia (USSR)*, **22–7** (1982) 89–94.
228. Gohil, K., Rothfuss, L., Lang, J. & Packer, L., Effects of exercise training on tissue, vitamin E and ubiquinone content. *J. Appl. Physiol.*, **63** (1987) 1638–41.
229. Hugues, R. E., Recommended daily amounts and biochemical roles; the vitamin C carnitine fatigue relationship. In *Vitamin C (Ascorbic Acid)*, ed. J. N. Counsell & D. H. Hornig. Applied Science, London, 1982.
230. Packer, L., Blood glutathione oxidation during human exercise. *J. Appl. Physiol.* (in press).
231. Manchester, D. K. & Jacoby, E. H., Resolution and reconstitution of human placental monooxygenase activity responsible to maternal cigarette smoking. *Dev. Pharmacol. Ther.*, **5** (3–4) (1982) 162–72.
232. Pacht, E., Kaskei, H. & Davis, W. B., Augmentation of lung vitamin E in young cigarette smokers. *Clin. Res.*, **33** (2) (1985) 470a.
233. Pacht, E., Kaseki, H. & Davis, W. B., Deficiency of vitamin E in the alveolar fluid of cigarette smokers; influence of alveolar macrophage cytotoxicity. *J. Clin. Invest.*, **77** (3) (1986) 789–96.
234. Schorah, C. J., Vitamin C status in population groups. In *Vitamin C (Ascorbic Acid)*, ed. J. N. Counsell & D. H. Hornig. Applied Science, London, 1982.
235. Brin, M., In *Nutrition and Drug Interrelations*, ed. J. N. Hathcock & J. Coon. Academic Press, New York, 1978, p. 138.
236. Bird, R. P., Draper, H. H. & Basrur, P. K., Effect of malonaldehyde–acetaldehyde on cultured mammalian cells. Production of micronuclear and chromosomal aberrations. *Mutal. Res.*, **101** (1982) 237–46.
237. Feron, V. J., Kruysse, A. & Woutersen, R. A., Respiratory tract tumours in hamsters exposed to acetaldehyde vapour alone or simultaneously to benzo(a)pyrene or diethylnitrosamine. *Eur. J. Cancer Clin. Oncol.*, **1b** (1982) 13–31.
238. Nordmann, R., Role des radicaux libres dans le métabolisme et la toxicité de l'alcool. *Cah. Nutr. Diét. (Paris)*, **22** (1) (1987) 63–5.
239. Levins, S. & Kidd, P. M., The free radical oxidant toxins of polluted air. In *Free Radicals and Antioxidant Adaptation*, Biocurrents Division, Allergy Research Group, San Leandro, CA, 1985, pp. 67–103.
240. Santamaria, L., Bianchi, A., Arnaboldi, I., Andreoni, L. & Bermond, P., Dietary carotenoids block photocarcinogenic enhancement by benzo(a)pyrene and inhibit its carcinogenesis in the dark. *Experientia*, **39** (1983) 1043–5.
241. Santamaria, L., Bianchi, A., Ravelto, C., Arnaboldi, A., Santagati, G. & Andreoni, L. Supplemental carotenoids prevent skin cancer by benzo(a)pyrene, breast cancer by PUVA, and gastric cancer by MNNG. Relevance in human chemoprevention. in *Vitamins and Cancer; Human Cancer Prevention by Vitamins and Micronutrients*, ed. F. L. Meyskens & K. N. Prasad. The Humana Press Inc., Cliffon, N. J., 1985.

242. Dougherty, J. J. & Hoekstra, W. G., Effects of vitamin E and selenium on copper-induced lipid peroxidation in vivo and on acute copper toxicity. *Proc. Soc. Exp. Biol. Med.*, **169** (1982) 201–8.
243. Basu, T. K., The influence of drugs with particular reference to aspirin or the bioavailability of vitamin C. In *Vitamin C (Ascorbic Acid)*, ed. J. N. Counsell & D. H. Hornig. Applied Science, London, 1982.
244. Ferradini, C., Radicaux libres, agression et moyens de défense. *Cah. Nutr. Diét. (Paris)*, **22** (1) (1987) 12–14.
245. Chaudière, J., La glutathion peroxydase (Se-GPX), un élément central du système de protection de nos cellules. *Journées sur les Radicaux Libres, Bordeaux*, **26–27** (9) (1986), preprint. Soc. Nutr. Diét. Paris.
246. Korallus, U., Harzdorl, C. & Lewalter, J., Experimental base for ascorbic acid therapy of poisoning by hexavalent chromium compounds. *Int. Arch. Occup. Environ. Health*, **53** (3) (1984) 247–56.
247. Pucci, G., Porpora, M. G., Redivo, A., Porta, R. P. & Villani, C., Possible prevention of adriamycin cardiotoxicity by alpha-tocopherol; a pilot study project. *Patol. Clin. Oster. Ginecol.*, **12** (5) (1985) 403–7.
248. Dasgupta, P. R., Dutta, S., Banerjee, P. & Majumbar, S., Vitamin E (alpha-tocopherol) in the management of menorrhagia associated with the use of intrauterine contraceptive devices. *Int. J. Fertil.*, **28** (1) (1983) 55–6.
249. Ohshima, H. & Batch, H., The influence of vitamin C on the in vivo formation of nitrosamines. In *Vitamin C (Ascorbic Acid)*, ed. J. N. Counsell & D. H. Hornig. Applied Science, London, 1982.
250. Watanabe, N., Shiki, Y., Morisaki, N., Saito, Y. & Yoshida, S., Cytotoxic effects of paraquat and inhibition of them by vitamin E. *Biochim. Biophys. Acta*, **833** (3) (1986) 420–5.
251. Flannigan, S. A., Tucker, S. B. & Key, M. M., Prophylaxis of synthetic pyrethroid exposure. *J. Soc. Occup. Med.*, **34** (1984) 24–6.
252. Chakroborty, D. *et al.*, Studies on L-ascorbic acid metabolism in rats under chronic toxicity due to organophosphorus insecticides; effects of supplementation of L-ascorbic acid in high doses. *J. Nutr.*, **108** (1978) 973–80.
253. Mayer, F. L., Mehrle, P. M. & Crutcher, P. L., Interaction of toxaphene and vitamin C in Channel catfish. *Trans. Am. Fisheries Soc.*, **107** (1978) 326–33.
254. Horio, F. & Yoshida, A., Effects of some xenobiotics on ascorbic acid metabolism in rats. *J. Nutr.*, **112** (1982) 416–25.
255. Duplan, J. F., Les radicaux libres: Leur rôle en radiobiologie. *Cah. Nutr. Diét. (Paris)* **22** (1) (1987) 21–2.
256. Gharbi, Reda, M., Xeroderma Pigmentosum. *Ther. Umschau (Switzerland)*, **39** (3) (1982) 193–201.
257. Macdonald, K., Holti, G. & Marks, J., Is there a place for β-carotene/canthaxanthin in phototherapy for psoriasis? *Dermatologica*, **169** (1984) 41–6.
258. Dormandy, T. L., An approach to free radicals. *Lancet*, **8357** (1983) 1010–14.
259. Emerit, J., Inflammation, cascade de l'acide arachidonique et radicaux libres. *Cah. Nutr. Diét. (Paris)*, **22** (1) (1987) 35–9.

260. Corey, E. J., Mehrotra, M. M. & Khan, A. V., Antiarthritic gold compounds effectively quench electronically excited singlet oxygen. *Science,* **236** (1987) 68–9.
261. Singh, V. N., Beta-carotene: Treatment of arthritis (a letter to the Editor of *Science*): *Science* (in press, 1987).
262. Grenier, J., Publication in preparation.
263. Szczepanska, I., Kwiatowska, J., Przybyszewski, W. M., Sitarska, E. & Malec, J., Amelioration of hydroxyurea-induced suppression of phagocytosis in human granulocytes by free radical scavengers. *Scand. J. Hoematol.,* **34** (1) (1985) 35–8.
264. Boogaerts, M. A., Van de Broeck, J., Deckmyn, H., Reolant, C., Vermylen, J. & Verwilghen, R., Protective effect of vitamin E on immune triggered granulocyte mediated endothelial injury. *Thromb. Haemost.,* **51** (1) (1984) 89–92.
265. Dabadie, H. & Paccalin, J., Aspects histopathologiques du vieillissement cerebral et pathologique. *Cah. Nutr. Diét. (Paris),* **22** (1) (1987) 51–3.
266. Sullivan, J. L., Superoxide dismutase, longevity and specific metabolic rate. *Gerontology,* **28** (1982) 242–4.
267. Heilbron, L. K., Nomura, A. & Stemmermann, G. N., Prospective study of serum uric acid and cancer. *J.Am. Med. Assoc.,* **116** (1982) 353–63.
268. Summerfield, F. W. & Tappel, A. L., Effects of dietary polyunsaturated fats and vitamin E on aging and peroxidative damage to DNA. *Arch. Biochem. Biophys.,* **233** (2) (1984) 408–16.
269. Thaw, H. H., Collins, V. P. & Brunk, U. T., Influence of oxygen tension, prooxidants and antioxidants on the formation of lipid peroxidation products (lipofuscin) in individual cultivated human gliol cells. *Mech. Ageing Dev.,* **24** (2) (1984) 211–23.
270. Menken, B. Z., Su, L. C., Ayaz, K. L. & Czallany, A. S., Organic solvent-soluble lipofuscin pigments and glutathione peroxidase in mouse brain and heart; effects of age and vitamin E. *J. Nutr.,* **116** (3) (1986) 350–5.
271. Halks-Miller, M., Kane, J. P., Beckstead, J. H. & Smuckler, E. A., Vitamin E enriched lipoproteins increase longevity of neurons *in-vitro. Brain Res.,* **254** (3) (1981) 439–47.
272. Meydani, M., Verdon, C. P. & Blumbers, J. B., Effect of vitamin E, selenium and age on lipid peroxidation events in rat cerebrum. *Nutr. Res.,* **5** (11) (1985) 1227–36.
273. Harland, W. A., Gilbert, J. D. & Brooks, C. J. W., Lipids of human atheroma. VIII Oxidized derivatives of cholesteryl linoleate. *Biochim. Biophys. Acta,* **316** (1973) 378–85.
274. Timiras, P. S., *Physiological Basis of Aging and Geriatrics.* Macmillan, New York (in press).
275. Musca, A., Cordova, C., Violi, F., Perrone, A., Alessandric, C. & Salvadori, F. Influence of vitamin E on platelet aggregation and lipid pattern. *Clin. Ter.,* **102** (3) (1982) 273–6.
276. Haeger, K., Long term study of alpha-tocopherol in intermittent claudication. *Ann. NY Acad. Sci.,* **393** (1982) 369.
277. Palgi, A., Association between dietary changes and mortality rates,

Israel 1949 to 1977; a trend-free regression model. *Am. J. Clin. Nutr.*, **340** (1981) 1569–83.

278. Cadet, J. L., The potential use of vitamin E and selenium in parkinsonism. *Med. Hypotheses,* **20** (1) (1986) 87–94.
279. Clausen, J., Demential syndromes and the lipid metabolism. *Acta Neurol. Scand.*, **70** (5) (1984) 345–55.
280. Cynamon, H. A., Isenberg, J. N. & Nguyen, C. H., Erythrocyte malonyldialdehyde release in vitro, a functional measure of vitamin E status. *Clin. Chim. Acta,* **151** (2) (1985) 169–76.
281. Droy-Lefaix, M. T., Drouet, Y., Géraud, G. & Schatz, B., Radicaux libres et tube digestif. *Cah. Nutr. Diét.* (*Paris*), **22** (1) (1987) 44–9.
282. Willett, W. C. *et al.*, Relation of serum vitamin A and E and carotenoids to the risk of cancer. *N. Engl. J. Med.*, **310** (1984) 430–4.
283. Gey, R. F. & Brubacher, G. B., Plasma levels of antioxidant vitamins in relation to ischemic heart disease and cancer. *Am. J. Clin. Nutr.*, **45** (1987) 1368–77.
284. Hennekens, C. H., Micronutrients and cancer prevention. *N. Engl. J. Med.*, **315** (20) (1986) 1288–9.
285. Seifter, E., Rettura, G., Padawer, J. & Levenson, S. M., Moloney murine sarcoma virus tumors in CBA/J mice, chemopreventive and chemotherapeutic actions of supplemental β-carotene. *JNCI,* **68** (5) (1982) 835–40.
286. Epstein, J. H., Effects of β-carotene on UV-induced tumor formation in the hairless mouse skin. *Photochem. Photobiol.*, **25** (1977) 211–13.
287. Mathews-Roth, M. M., Antitumor activity of β-carotene, canthaxanthin and phytoene. *Oncology,* **39** (1982) 33–7.
288. Borek, C., Ong, A., Mason, H., Donahue, L. & Biaglow, J. E., Selenium and vitamin E inhibit radiogenic and chemically induced transformations *in vitro* via different mechanisms. *Proc. Natl. Acad. Sci. USA,* **83** (5) (1986) 1490–4.
289. Perchellet, J. P., Owen, M. D., Posey, T. D., Orten, D. K. & Schneider, B. A., Inhibitory effects of glutathione level-raising agents and d-alpha-tocopherol on ornithine decarboxylase EC-4.1.1.17 induction and mouse skin tumor promotion by 12-0 tetradecanoyl-phenol-13-acetate. *Carcinogenesis,* **6** (4) (1985) 567–74.
290. Dickerson, J. W. T., Vitamin C and cancer. In *Vitamin C* (*Ascorbic Acid*), ed. J. N. Counsell & D. H. Hornig. Applied Science, London, 1982.
291. Cameron, E. & Pauling, L., Supplemental ascorbate in the supporting treatment of cancer: prolongation of survival times in terminal human cancer. *Proc. Natl Acad. Sci.*, **73** (1976) 3685.
292. Bussey, J. J. *et al.*, A randomized trial of ascorbic acid in polyposis. *Cancer* **50** (7) (1982) 1434–9.
293. McCarthy, M. F., An antithrombotic role for nutritional antioxidants, implications for tumor metastasis and other pathologies. *Med. Hypotheses,* **1974** (1986) 345–57.
294. Irvin, T. T., Vitamin C in surgical patients. In *Vitamin C* (*Ascorbic Acid*), ed. J. N. Counsell & D. H. Hornig, Applied Science, London, 1982.

295. Pucheu, A., Maladies cardiaques et carence en selenium. *Gazette Méd. (Paris)*, **93** (35) (1986) 1–2.
296. Hornig, D. H. & Moser, U., Safety of high vitamin C intakes. In *Vitamin C (Ascorbic Acid)* ed. J. N. Counsell & D. H. Hornig. Applied Science, London, 1982.
297. Chazan, J. B. & Szulc, M., Radicaux libres et vitamine E. *Cah. Nutr. Diét. (Paris)*, **22** (1) (1987) 66–76.
298. MacGregor, J. T., Genetic toxicology of dietary flavonoids. In *Genetic Toxicology of the Diet.* Alan R. Liss, New York, 1986, pp. 33–43.
299. Levander, O. A., Clinical consequences of low selenium intake and its relationship to vitamin E. *Ann. NY Acad. Sci.* **75** (1982) 70–80.
300. Yost, F. J. & Fridovich, I., Superoxide radicals and phagocytosis. *Arch. Biochem. Biophys.*, **161** (1974) 395–401.
301. Johnston, R. B., Koele, B., Webb, L., Kessler, D. & Rajogopalam, K. V., Inhibition of phagocytic bactericidal activity by superoxide dismutase: a possible role of superoxide anion in the killing of phagocytized bacteria. *J. Clin. Invest.*, **52** (1973) 44a.

Chapter 7

TOXICOLOGICAL ASPECTS OF ANTIOXIDANTS USED AS FOOD ADDITIVES

SUSAN M. BARLOW

Toxicology and Environmental Health Division, Department of Health, Hannibal House, Elephant and Castle, London SE1 6TE, UK

INTRODUCTION

The toxicology of antioxidants has become one of the more controversial areas in the continuing debate on the safety of food additives. In recent years, problems have arisen with the antioxidants butylated hydroxyanisole (BHA) and butylated hydroxytoluene (BHT) when new long-term studies showed that these compounds could produce tumours in animals. Chemicals which have been shown to cause cancer in long-term animal studies are normally not permitted as food additives, and regulatory authorities would have little hesitation in making such a decision in the case of any new chemical submitted to them for approval which appeared to be a genotoxic carcinogen. However, with BHA and BHT, not only did the new findings conflict with the results of earlier, negative, long-term studies, but also their significance for hazard assessment in man was surrounded by even more uncertainties than those which toxicologists are used to considering when extrapolating from animal experiments to man.

In the case of BHA, the increased incidence of tumours was seen in the forestomach of the rat, and subsequently in other rodents, and many have been quick to point out that the human species does not possess a forestomach. On the other hand, types of cell similar to those found in the rodent forestomach are found elsewhere in man, for example in the lining of the oesophagus. Furthermore it is a common

This contribution has been seen by colleagues in the Department of Health. Their comments and suggestions are gratefully acknowledged, but the views expressed are the author's own and not necessarily those of the Department.

empirical observation that individual carcinogens may produce tumours at differing sites in different species.

In the case of BHT, an increase in liver tumours was observed in a rat study continued beyond the normal 2-year duration of carcinogenicity bioassays because BHT seemingly improved the survival of the treated rats compared with the untreated controls. The tumours were observed only in rats dying or killed in the closing stages of this 144-week study, and it is doubtful whether any tumorigenic effect would have been observed if the study had been terminated, as is usual, at 102 weeks. The toxicologists were then left to ponder whether, had as many of the controls lived as long as the BHT treated animals, they too would have shown a similar incidence of liver tumours. Liver tumours occur spontaneously in rats and were seen, albeit at a low incidence, in the late surviving controls in the BHT study.

Other relevant information also has to be taken into account in considering whether BHA and BHT should continue to be acceptable for food use. These two antioxidants do not appear to be genotoxic in the studies carried out to date which have been well conducted, and this raises questions about the mechanism by which they cause tumours. In combination studies with initiators of tumours, BHA and BHT, in common with most other antioxidants, can act both as inhibitors and promoters of carcinogenicity, depending on the initiator used, the timing of antioxidant administration and the target site under consideration. Other antioxidants, viz. the ascorbates (vitamin C) and the gallates and their breakdown products, have mutagenic potential at high concentrations *in vitro,* whilst tertiary butylhydroquinone (TBHQ) has shown mutagenicity in some but not all in-vitro and in-vivo tests. Forestomach hyperplasia readily induced by BHA and enhanced by combination of BHA with other antioxidants, is also seen, to a lesser extent, after treatment with TBHQ or tocopherol (vitamin E) alone, and in the case of the latter has gone on to produce papilloma in the forestomach.

Some of these results have prompted questions as to whether at high concentrations these compounds may lose their antioxidant activity and acquire pro-oxidant activity, generating reactive oxygen species which have the capacity to damage DNA. As research continues it is becoming clear that antioxidants may share a number of toxic properties at high doses, and that knowledge of their mechanism(s) of action and the dose–response relationships for these effects are crucial to a scientific assessment of possible risks to humans consuming very low levels of antioxidants as food additives. It is against this background, and mindful of the possible deleterious effects on health of consumption of lipid oxidation products if antioxidants were not used,[1] that the majority of regulatory authorities have decided to continue to permit the use of antioxidants including BHA and BHT, whilst keeping the situation under review as new research results are generated.

The review of individual antioxidants which follows is not intended to be comprehensive, but attempts to highlight those aspects of the toxicology which are central to safety evaluation. Further information and reference to the studies mentioned above will be found in the review.

In any assessment of safety it is necessary to compare the doses at which toxic effects may occur with the estimated daily human intake. Accurate estimates of intake depend on up-to-date information about the use of antioxidants by the food industry and dietary habits, including the possibility that some individuals may have higher than average intakes because they are extreme consumers of particular items of food. Such information is difficult and expensive to acquire and extrapolations from data from other countries may not be valid since intakes are likely to differ between countries. Thus, in the absence of precise data, toxicologists making safety assessments often utilise 'worst case' assumptions about intakes. Such assumptions include, for example, that a particular antioxidant is used in all the foods that it may be used in, that it is used at the maximum permitted level of addition, and, a particularly conservative assumption for antioxidants, that there are no losses during processing. In this way, utilising a worst case assumption about intake gives an added margin of safety; it is likely to be at least one and possibly two orders of magnitude greater than the actual average intake. In the review which follows, any intake figures quoted are generally based on worst case assumptions.

In order that countries with differing patterns of use and intake may utilise safety evaluations produced by international bodies such as the Joint FAO/WHO Expert Committee on Food Additives (JECFA) and the European Commission (EC) Scientific Committee for Food (SCF), these bodies establish an acceptable daily intake (ADI) for an additive, rather than simply expressing a view as to whether the use of a particular additive is acceptable or not. The ADI is usually defined as the amount of chemical, expressed on a bodyweight basis, which it is considered can be consumed daily over a lifetime without causing harm. ADIs are devised by determining, from the range of toxicity tests carried out, which effect is the most sensitive and ascertaining the maximum dose at which that effect is no longer observed (the no-effect level). A further reduction or safety factor is then applied to the no-effect level to try to take into account any uncertainties in the data, possible differences in sensitivity between animals and humans and between individuals in the human population. In this manner, it is anticipated that there will be an adequate margin of safety for the consumer. The safety factor is arbitrary and variable, but in practice a factor of 100 is frequently used. A factor greater than 100 may be used if there is a gap in the data, or the nature or dose–response relationship of the toxic effect on which the ADI is based indicates particular caution should be exercised. A factor lower than 100 may be used, for example, if the ADI is based on human data. In the review which follows, JECFA and SCF ADIs are quoted when available.

ASCORBIC ACID AND ITS DERIVATIVES

L-Ascorbic acid (vitamin C), sodium and calcium L-ascorbate, and ascorbyl palmitate are all permitted antioxidants in EEC countries. Although a vitamin, ascorbic acid and its derivatives are not necessarily free of toxicity at high doses. However, their food additive uses contribute only a low proportion of the total daily intake of vitamin C from other sources (estimated to be 30–100 mg), and for this reason the SCF considers they are acceptable for food use with no ADI specified.[2] The JECFA has also agreed on 'ADI not specified' for ascorbic acid and its sodium, potassium and calcium salts[3] and an ADI of 0–1·25 mg/kg body weight (bw) for ascorbyl palmitate or ascorbyl stearate, or the sum of both if used together.[4] In considering the toxicity of ascorbates, studies on guinea pigs may be particularly relevant since, like humans, they are unable to synthesise vitamin C, whereas rats, mice and rabbits can. The structure of ascorbic acid is shown in Fig. 1.

Absorption, Metabolism and Excretion

There is no information on the metabolism of ascorbyl palmitate or stearate but it is assumed that they break down into ascorbic acid.

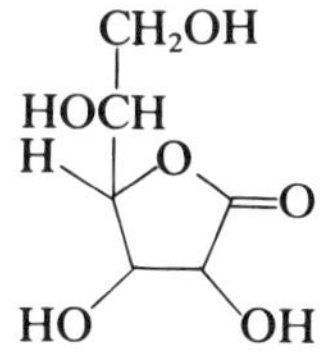

L-ascorbic acid

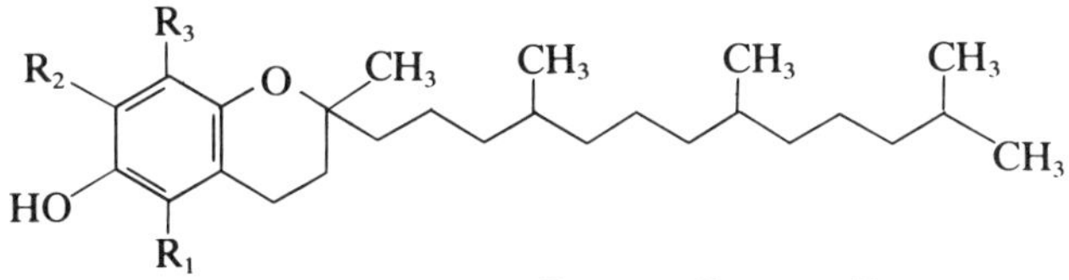

	R_1	R_2	R_3
α-tocopherol	CH_3	CH_3	CH_3
β-tocopherol	CH_3	H	CH_3
γ-tocopherol	H	CH_3	CH_3
δ-tocopherol	H	H	CH_3

FIG. 1. The structure of ascorbic acid and tocopherols.

Ascorbic acid and its salts are readily absorbed and metabolised. With repeated daily administration in humans, plasma levels show a rise until steady state conditions are reached, at which point the turnover rate is about 1 mg/kg bw/day and the half-life about 10 days.[3] In man oxalic acid is the major urinary metabolite. Other breakdown products which may be found in the urine include diketogulonic acid, lyxonic acid and xylonic acid.[5] Little or no metabolism to expired CO_2 occurs.[6]

It has been estimated that up to 50% of the normal urinary oxalic acid excretion of 25–50 mg/day may be derived from dietary vitamin C.[7] There has therefore been some speculation as to whether 'megatherapy', that is the taking of large doses of vitamin C of around 1–5 g daily, in the belief that it may prevent certain diseases, could substantially increase urinary oxalate excretion and precipitate the formation of renal stones.[8] Several studies have failed to show any significant increase in urinary oxalate excretion except at extremely high doses in excess of 4 g daily.[9–11] Other studies have claimed to show increased urinary oxalate excretion with intakes of ascorbic acid ranging from 400 mg to several g.[12–14] Subsequent studies suggest that even at high doses where increases in urinary oxalate of up to 70 mg/day can be demonstrated, the metabolic conversion of ascorbate to oxalate is limited, probably to less than 1% of the daily ingested dose, and that the metabolism of ascorbate may be saturated at intakes of around 200 mg/day.[15,16] Some individuals, however, may have an unusual capacity for conversion to oxalate; in a study in which 67 volunteers were screened, 3 (including a father and a son) excreted greatly increased amounts of oxalic acid, 600–700 mg daily, after ingesting 4 g of ascorbic acid daily for 7 days.[17] Thus, it remains possible, but unproven, that ingestion of very large doses of vitamin C might precipitate renal stone formation in susceptible individuals. Since hypercalcuric patients would be particularly susceptible to renal stone formation, the JECFA considered whether use of the calcium salt of ascorbic acid would increase the risk of crystalluria. However, they concluded that the intake of calcium from its food additive uses would be minor compared with total dietary intake of calcium.[3]

Short-Term Studies

Studies in which ascorbic acid was given to mice up to 1000 mg/kg bw/day and guinea pigs up to 2500 mg/kg bw/day for 6 or 7 days, including histological examination of various organs,[18] showed no adverse effects. Force-feeding guinea pigs, on an otherwise scorbutic diet, with 200 mg vitamin C daily for 112 days (equivalent to around 600 mg/kg bw/day), caused death in 4 out of 6 animals, attributed by the authors to liver toxicity, since there was macroscopic evidence of fat deposition and congestion in the liver.[19] The dose was said to be approximately 10–100 times greater than the varying estimates of guinea pigs' normal daily requirement for vitamin C.

Ascorbyl stearate at dietary doses up to 3000 mg/kg bw/day for 6 months was said to produce no adverse effects in rats.[4] Feeding 5% ascorbyl palmitate in the diet (equivalent to about 2500 mg/kg bw/day) for 9 months caused growth retardation, bladder stones and hyperplasia of the bladder epithelium, presumably due to urinary calcium oxalate precipitation. At 1000 mg/kg bw/day only slight growth retardation was observed. It is likely that the 'ascorbyl palmitate' used comprised 5–20% ascorbyl stearate and 95–80% ascorbyl palmitate.[4]

Reproduction

Early reports suggested that ascorbic acid caused abortion in guinea pigs at doses ranging from 25 to 1500 mg/kg bw,[20–22] and in rats at doses of 833 mg/kg bw given subcutaneously or 250 mg/kg bw orally.[23,24] However, more recent studies have shown no increase in abortion and no effects on litter size, viability or birth weight in guinea pigs given 400 mg/kg bw/day orally, or in rats, hamsters and mice given up to 1000 mg/kg bw/day orally throughout pregnancy.[25,26] The reason for these differences is not known. Published teratology studies however are lacking.

Mutagenicity

Ascorbic acid has been shown to be mutagenic at high concentrations *in vitro* to bacterial cells (*Salmonella typhimurium* TA 100 strain) and to mammalian Chinese hamster ovary (CHO) cells (HGPRT locus mutations), to cause chromosome aberrations in human fibroblasts, to increase sister-chromatid exchanges in Chinese hamster V79 cells, CHO cells and human fibroblasts, and to inhibit DNA synthesis in HeLa cells.[27–34] However, ascorbic acid did not show evidence of DNA damage or mutagenicity *in vivo* in tests employing doses of up to 10 000 mg/kg bw orally, including Chinese hamster bone marrow (sister-chromatid exchanges), rat dominant lethal test, guinea pig intrahepatic host-mediated test (using *Salmonella typhimurium* TA 100).[34,35,30] It also lacked mutagenic activity in the mouse lymphoma L5178Y/TK assay.[36] Further evidence from the in-vitro studies shows that mutagenic activity only occurs or is enhanced by addition of Cu^{2+} ions to the medium,[27,28,30,31] and is reduced or abolished by additions of catalase,[32,33] cysteine or glutathione[34] to the medium. These findings support the hypothesis that ascorbic acid is active *in vitro* due to its oxidation in the presence of oxygen and metal ions, such as Cu^{2+}, with the consequent generation of free radicals from hydrogen peroxide, which can damage DNA. A compound acting by this mechanism would not be expected to be mutagenic *in vivo* due to the efficient cellular mechanisms for mopping up free radicals. The negative in-vivo data are therefore also consistent with this hypothesis for ascorbate's mechanism of action.

Long-Term/Carcinogenicity Studies

Feeding of ascorbic acid to 26 rats/sex/group at 1000, 1500 or 2000 mg/kg bw/day for 2 years produced no adverse effects on haematology, blood enzyme activity, urinalysis, liver and kidney function tests, non-neoplastic lesions or tumour incidence.[3] In a more recent carcinogenesis bioassay, carried out under the auspices of the US National Toxicology Program, ascorbic acid was not carcinogenic when fed to rats or mice (50/sex/group) at 2·5% or 5·0% in the diet for 2 years.[37] These dietary levels were equivalent to mean intakes of 1255 and 2560 mg/kg bw/day for male rats, 1458 and 3051 mg/kg bw/day for female rats, 6515 and 12788 mg/kg bw/day for male mice, and 7186 and 14792 mg/kg bw/day for female mice.

In a 2-year study on a limited number of rats, ascorbyl palmitate, containing 5–20% ascorbyl stearate, fed at up to 0·5% of the diet (250 mg/kg bw/day) caused no adverse effects.[4]

The observations of promotion of urinary bladder cancer initiated by N-butyl-N-(4-hydroxybutyl)nitrosamine by feeding of high dietary levels (5%) of the sodium salts of ascorbic acid and isoascorbic acid, but not by the acids themselves, calcium ascorbate, ascorbyl dipalmitate nor ascorbyl stearate, suggest that the mechanism is related to the observed increase in urinary sodium concentration, rather than the antioxidant part of the molecule.[38–40]

Sodium ascorbate given at 5% in the diet to rats has also been shown to promote forestomach carcinogenesis initiated by N-methylnitrosourea (MNU) or N-methyl-N′-nitro-N-nitroso-guanidine (MNNG), to promote colon carcinogenesis initiated by 1,2-dimethylhydrazine, and to promote urinary bladder carcinogenesis initiated by MNU. It has no modifying effect on glandular stomach tumours initiated by MNNG, no effect on liver tumours initiated by N-ethyl-N-hydroxyethylnitrosamine or diethylnitrosamine, no effect on mammary tumours initiated by 7,12-dimethylbenz(a)anthracene, and no effect on thyroid tumours initiated by MNU.[41]

Whilst ascorbic acid does not protect against carcinogenesis from preformed nitrosamines, it does exert a protective effect when nitrate and nitrosatable amines are given, by inhibiting nitrosation reactions.[42]

Studies in Humans

Some studies have noted a diuretic effect in children and adults after daily doses of around 5 mg/kg bw/day, and glycosuria with doses of 30–100 mg/kg bw but these effects have not been confirmed in other large-scale, double blind studies in which daily doses up to 6000 mg have been taken.[3] No other adverse affects have been noted except for skin rashes in infants, and nausea, vomiting, diarrhoea, flushing of the face, headache, fatigue and disturbed sleep in adults taking doses around 6000 mg/day.[3]

Conclusions

The food additive uses of ascorbic acid and its derivatives are extremely unlikely to have any adverse effects. Mutagenic activity seen under certain conditions in vitro does not occur *in vivo* and long-term studies in animals have not shown any evidence of carcinogenicity. However, human studies indicate that ascorbic acid taken in very high doses of 1 g/day or more may cause adverse reactions and increase urinary oxalate excretion in some individuals.

TOCOPHEROLS

Tocopherols occur naturally in a wide variety of foodstuffs including cereals, nuts, seeds, fruits and vegetables. Oils and fats derived from these are particularly rich sources. They are present in only small amounts in animal tissues and dairy products. The structures of the tocopherols, α-, β-, γ-, and δ-tocopherol, vary only in the number of methyl groups bound to the hydroxylated benzene ring (see Fig. 1). In addition to their antioxidant activity, they also possess vitamin E activity, and intake from their use as food additives is very low compared with the intake of vitamin E from the diet as a whole. In EEC countries, extracts of naturally occurring mixed tocopherols, and α-, γ- and δ-tocopherol are permitted for use as food additives. The SCF has allocated tocopherols an 'ADI not specified' because of the low intakes from food additive use[2] and the JECFA in a recent re-evaluation of tocopherols allocated an ADI of 0·15–2 mg/kg bw.[43]

Absorption, Metabolism and Excretion

The majority of studies on absorption, metabolism and excretion have utilised the α-form of tocopherol. In rats, α-tocopherol is absorbed fairly rapidly from the gastrointestinal tract (32% in 6 h), whilst the other tocopherols, especially δ-tocopherol, are less readily absorbed.[44,45] In humans it has been shown that 55–97% of α-tocopherol is absorbed, the percentage absorbed being inversely proportional to the dose.[46] Tocopherols enter the gut mucosa, assisted by bile-induced formation of mixed micelles. They pass, largely unesterified, into the systemic circulation via the lymph.[47] Lipoproteins are the main carriers of α-tocopherol in human plasma.[48] Many tissues take up tocopherols, especially liver and adipose tissue, but the liver takes them up rapidly and depletes rapidly, whereas adipose tissue depletes only very slowly when dietary intake is deficient in vitamin E.[49] Most tocopherol is unchanged in the tissues, but there is limited metabolism in the liver and kidney. In humans, tocopherol is probably metabolised to tocopheronolactone.[50] The majority of an administered dose is excreted in the faeces and only up to 16% in urine.[51,52]

Short- and Long-Term Studies

Thyroid

A number of changes in thyroid gland parameters have been reported, but the effects are not consistent between or within species. Short-term oral administration of tocopherols at relatively low doses of 5–13 mg/kg bw/day to rabbits and guinea pigs was reported in experiments conducted 30–40 years ago to cause thyroid changes indicative of enhanced growth and activity, i.e. increased weight and volume of the gland, epithelial cell proliferation, hyperaemia and increased iodine uptake.[53,54] However, thyroid effects have only been seen in other species at much higher doses. Thyroid activity was reported to be depressed in rats and hamsters given tocopherols at 1000 mg/kg bw/day.[55,56] In a more recent study, rats given 125, 500 or 2000 mg/kg bw/day of α-tocopherol acetate orally by gavage for 13 weeks, showed elevations in serum thyroid stimulating hormone levels of 30–100%. However, the increases were not dose-related and no increases were found in organ weight, serum triiodothyronine or thyroxine, and no histopathological changes were seen in the gland.[57] Thyroid tumours were not observed in a 2-year rat study utilising oral doses of α-tocopheryl acetate up to 2000 mg/kg bw/day.[58] In humans, administration of doses of 400–500 mg of α-tocopheryl acetate, which are around 40 times the recommended daily intake of vitamin E and are borderline for producing other adverse effects (see below), have caused increased iodine uptake by the thyroid, slightly increased serum organic iodine levels, but, unexpectedly, a decreased metabolic rate.[54,59]

Liver

The liver does appear to be a target organ for tocopherols, but significant toxicity is only seen at very high dose levels. Biochemical changes have been reported following short-term administration of tocopherols to rats at doses around 500 mg/kg bw/day, including increased concentrations of ATP, ADP, coenzyme A, copper, calcium, phospholipids, total and esterified cholesterol.[60] Reports of effects on liver weight at these doses are conflicting; one study has reported increased liver weight and higher DNA concentrations, while other studies have reported no change or reduced liver weights following dietary administration of tocopherols.[60] More recent studies have reported increased relative liver weight in female rats only; in a 13-week study employing gavage dosing of α-tocopheryl acetate, it was seen at 500 or 1000 mg/kg bw/day, but not at 125 mg/kg,[57] and in a 2-year study, in which α-tocopheryl acetate was fed in the diet to give intakes of 500, 1000 or 2000 mg/kg bw/day, it was seen at 1000 mg/kg only.[58] Elevations in serum enzymes indicative of liver damage, i.e. alkaline phosphatase (AP), alanine aminotransferase (ALT), and aspartate aminotransferase (AST) were observed. In rats fed α-tocopheryl acetate in the diet for 8 weeks from weaning ALT and AST were elevated at intakes of around 3500 mg/kg bw/day, but not at intakes one-tenth of this or lower.[61] AP was elevated in female rats fed α-tocopheryl acetate in the diet at 500 or 1250 mg/kg bw/day for 16 months.[62] ALT was elevated transiently in males fed 500, 1000 or 2000 mg/kg bw/day and AP was elevated in both sexes fed 2000 mg/kg bw/day in a 2-year study.[58] In the latter study, 17% of treated males and 77% of treated females also showed evidence of histopathological change in agglomerations of vacuolated macrophages in the centroacini of the liver; the effect was not dose-related but was not seen in controls. Histopathological evidence of liver damage was not seen in the 13-week study of Abdo *et al.*,[57] despite liver enlargement in females at 500 and 2000 mg/kg bw; the authors attribute this absence of fatty change to the modifications of the diet made to protect against this change. In conclusion, none of these studies show severe hepatotoxicity, even at high oral doses.

Blood Clotting

An interesting feature of high-dose vitamin E toxicity is its interaction with vitamin K and consequent haemorrhagic effects. These have been reviewed in detail elsewhere.[63] Prolonged prothrombin time has been reported in chicks and rats given high oral doses of tocopherols in the presence of inadequate dietary vitamin K or in animals rendered vitamin K deficient by the administration of warfarin.[57,58,64–66] Overt haemorrhaging occurred at dose levels of 2000 mg/kg bw/day, but the effects on prothrombin time and haemorrhage could be prevented by vitamin K supplementation. A similar enhancing of anticoagulant activity by vitamin E has been observed in patients taking coumarin anticoagulants, or in post-myocardial infarction patients taking large doses (300 mg/day) of α-tocopherol alone for several months.[63] The mechanism of the interaction between vitamin E and vitamin K is not clear, though a number of suggestions have been put forward.[63,67] These include inhibition of platelet aggregation by vitamin E, structural antagonism to vitamin K by the vitamin E metabolite α-tocopherylquinone, antagonism by vitamin E of the vitamin K-dependent carboxylation reaction, and a mechanism related to vitamin E's antioxidant activity, since another antioxidant, butylated hydroxytoluene, is also haemorrhagic (see later).

Reproduction

There is a requirement for vitamin E in animal nutrition to achieve normal reproduction. At doses above the daily requirement some effects on the testis have been reported. Male mice given 10 mg of α-tocopheryl acetate/day for 20 days showed enlargement of testicular interstitial cells and marked enlargement of the smooth endoplasmic reticulum, suggesting enhanced steroid production.[68] Daily subcutaneous injections of 75 mg α-tocopheryl acetate in male hamsters for 8–30 days resulted in significantly lower testicular weights and a transient disruption of spermatogenesis compared with controls, but no alteration in testicular endocrine function.[69] Rats fed approximately 3500 mg/kg bw/day of α-tocopheryl acetate for 8 weeks from weaning, then mated, had reduced numbers of pups born alive, but no effects were seen at one-tenth this dose or below.[61] Female rats fed α-tocopheryl or α-tocopheryl acetate in the diet at total doses of 70–200 mg during gestation showed no increase in resorption rate, but the control incidence was high.[70] Standard teratology tests in rats, mice and hamsters at oral doses of 16, 74, 250, 345 and 1600 mg/kg bw/day of α-tocopherol acetate were without any adverse effects, but a similar study in rabbits was uninterpretable due to a higher maternal death rate in all groups.[71] Female rats given 22·5–2252 mg vitamin E/kg bw/day during pregnancy and lactation showed no effects on litter size, pup weight or pup survival, but eye abnormalities were seen in first and second generation offspring of dams which received 2252 mg/kg bw and in the second generation offspring of dams which received 900 or 450 mg/kg bw. There were no effects in the offspring of the 90 mg/kg bw group.[72] Thus, adverse effects on reproduction have only been seen at very high doses of tocopherols. In an early study in humans, oral administration of 30 mg of α-tocopherol daily to normal, adolescent girls for one or two menstrual cycles was said to increase urinary sex hormone excretion, whilst higher doses of 300 mg/day reduced sex hormone excretion and disturbed the cycle.[73] Increased urinary excretion of oestrogens and androgens has also been shown in women during vitamin E therapy.[74]

Mutagenicity

No conventional studies using tocopherols alone have been reported. However, several studies have shown that tocopherols may inhibit the genotoxic effects of other known mutagens.[67]

Long-Term/Carcinogenicity Studies

Tocopherols, in common with other antioxidants, show inhibitory activity against some known carcinogens, but not to the same extent as BHA or BHT. Vitamin E shows some protection against methylcholanthrene-induced tumours in the buccal pouch of hamsters.[75] and against dimethylhydrazine-induced colonic tumours in mice.[76] In dermal studies, tocopherols reduced tumour incidence in rats following initiation with 7,12-dimethylbenz(a)anthracene and promotion with croton oil,[77] and retarded growth of transplanted rat and mouse sarcomas.[78,79] Tocopherols also block N-nitroso compound formation in the stomach by competing for available nitrite,[80] an effect not shown by BHA or BHT.

In the only conventional long-term study with α-tocopheryl acetate alone, doses of 0, 500, 1000 or 2000 mg/kg bw/day were fed in the diet of rats for 104 weeks.[58] Apart from effects on the liver and blood clotting referred to earlier, there were no other adverse effects. There was a decrease in mammary fibro-adenomas with increasing dose. In a shorter-term study, in which α-tocopheryl acetate was administered daily by gavage in doses of 125, 500 or 2000 mg/kg bw, lung lesions diagnosed as adenomatous hyperplasia were observed at 13 weeks, present in all treated groups but increasing in incidence and severity in a dose-dependent manner.[57] No lung lesions have been seen in other studies in which administration of tocopherols was in the diet. A recent report has indicated that α-tocopherol fed at 1% in the diet to hamsters for 40 weeks induces forestomach lesions (hyperplasia and papilloma), but at a lower incidence than those caused by 1% BHA.[81] This is the first report of such an effect from tocopherols.

Hypervitaminosis E in Humans

Effects of high doses of vitamin E on the thyroid, blood clotting and urinary sex hormone excretion have been referred to earlier. The only adverse effects noted with excess vitamin E intake have been nausea, gastrointestinal disturbance, muscle weakness and fatigue. These occurred only at intakes of 500–2000 mg daily and were transient or disappeared on cessation of treatment.[82–86]

Two reports have indicated the possible susceptibility of premature infants to adverse effects when vitamin E is given therapeutically to prevent retrolental fibroplasia, haemolysis and haemorrhage, at doses which maintain supra-physiological serum levels of vitamin E. There was an increased incidence of neonatal sepsis and necrotising enterocolitis following oral or parenteral administration of vitamin E preparations.[87] An intravenous vitamin E product caused a syndrome of pulmonary deterioration, thrombocytopenia, liver failure and renal failure.[88] In these cases, however, it was not clear whether the effects were attributable to vitamin E or another component in the preparation.

Conclusions

The daily dietary intake of tocopherols from their food additive uses is estimated not to exceed 1 mg/person.[67] This is small in comparison with the intake of naturally occurring tocopherols in the diet (5–20 mg/person), and is considerably lower than the doses shown to cause adverse effects in animals or humans. The food additive uses of tocopherols are therefore extremely unlikely to pose any hazard to human health. However, some people are known to consume excessive amounts of vitamin E supplements, up to 1000 mg/day, and such doses are potentially toxic.

THE GALLATES

The gallate group of antioxidants comprises the propyl, octyl and dodecyl esters of gallic acid, and their structures are shown in Fig. 2. When the gallates were reviewed by the JECFA in 1980, a group ADI of 0–0·2 mg/kg bw was confirmed. This was based on a no-effect level of 50 mg/kg bw from rat reproduction studies, to which a higher than normal safety factor was applied.[89] The same ADI was established for the group by the SCF in 1978,[90] but has recently been revised upwards

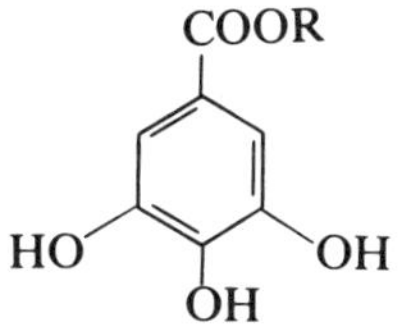

COOR

HO OH

OH

R

C_3H_7 Propyl
C_8H_{17} Octyl
$C_{12}H_{25}$ Dodecyl

Gallates

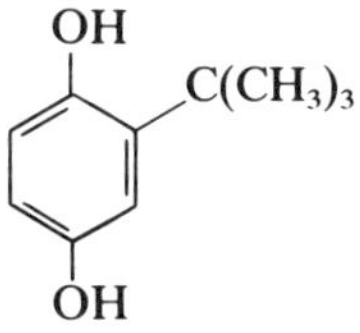

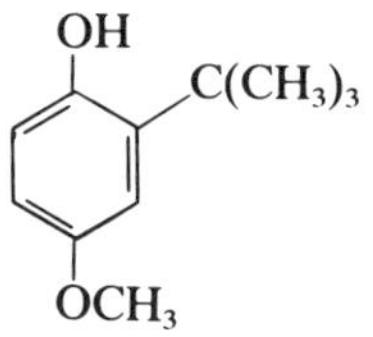

OH $C(CH_3)_3$ OH — TBHQ

OH $C(CH_3)_3$ OCH_3 — 3-BHA

$(CH_3)_3C$ OH $C(CH_3)_3$ CH_3 — BHT

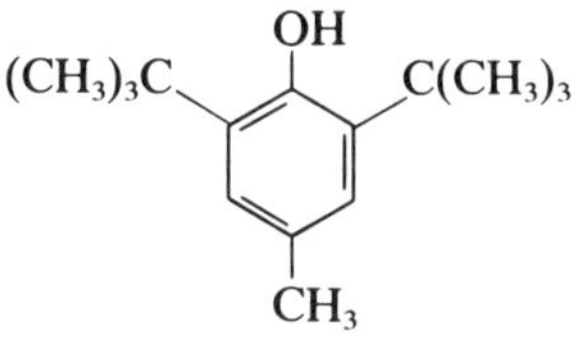

FIG. 2. The structure of the gallate group of antioxidants, TBHQ, BHA and BHT.

to 0–0·5 mg/kg bw following a re-evaluation of all available data.[2] However, the majority of the toxicity data relate to propyl gallate (PG) and what data there are on octyl gallate (OG) and dodecyl gallate (DG) suggest the toxicities of the gallates may differ, particularly with respect to critical effects on reproduction. In its most recent review, therefore, in 1986, the JECFA decided to consider each gallate separately. They were unable to establish ADIs for OG and DG because of insufficient information. An ADI of 0–2·5 mg/kg bw was allocated to PG, which does not seem to have the adverse effects on reproduction that OG and DG do.[91]

Absorption, Metabolism and Excretion

PG is well absorbed; after oral dosing in the rat over 70% of PG is absorbed from the gastrointestinal tract. The ester is hydrolysed to propyl alcohol and gallic acid and the latter methylated to yield 4-*O*-methyl gallic acid which is the main compound found in urine either free or conjugated with glucuronide. Gallic acid is also excreted in the urine in smaller amounts.[92] Similar urinary metabolites have been found in the rabbit together with pyrogallol.[93] There is little information about the metabolism of OG and DG apart from an early study indicating that absorption and hydrolysis of OG and DG is less than for PG and that unchanged compound comprises a higher proportion of the urinary components.[94]

Short-Term Toxicity Studies

Most of the short-term studies were carried out more than 20 years ago. Feeding of up to 2000 or 5000 mg/kg of diet of PG, OG or DG to rats, guinea pigs and dogs produced no adverse effects[94–98] except for a slight elevation in AST levels unaccompanied by any microscopical changes in the liver in rats and dogs fed the top dose of 5000 mg OG/kg of diet for 13 weeks.[98] Feeding of much higher doses to rats caused growth retardation but no pathological lesions (12 000 mg PG/kg of diet), or some deaths accompanied by kidney lesions (23 000 mg PG/kg of diet) or 100% mortality within a few days (25 000 or 50 000 mg DG/kg of diet).[95,99] A more recent unpublished study of Strik, referred to by van der Heijden,[98] confirmed that feeding of PG at 25 000 mg/kg of diet for 4 weeks produced hyperplasia in the tubuli of the outer kidney medulla, together with growth retardation, anaemia and increased liver enzyme activity. Strik also observed increased liver enzyme activities at 5000 mg/kg of diet, but not at 1000 mg/kg. Other studies confirm that at lower doses the gallates, in contrast to BHT, do not induce hepatic mixed function oxidases and may even inhibit some of these enzymes such as benzo(a)pyrene hydroxylase and ethoxycoumarin de-ethylase.[100,101]

Reproduction

Several multigeneration studies have been carried out involving PG and DG at doses up to 5000 mg/kg of diet and OG at up to 6000 mg/kg of diet.[94,98,102] At the top doses of OG and DG there were reductions in fertility, pup growth and pup survival. The latter were possibly due to poor maternal lactation and/or nursing since cross-fostering OG-treated pups to control dams markedly improved survival, whilst control pups reared by OG-treated dams showed poor survival. Reduced pup growth and survival were also seen in two of the OG studies at lower doses of 2500 and 3000 mg/kg of diet. An increase in gross kidney lesions was also found in the 2500 mg/kg group of the pups subjected to necropsy. Overall, these studies on the gallates indicate a no-effect level of 1000 mg/kg of diet, equivalent to around 50 mg/kg bw.

Teratogenicity studies are available only for PG. Dose levels of 4000, 10 000 and 25 000 mg/kg of diet were fed to rats from days 0–20 of pregnancy.[103] There were no adverse effects except for slight retardation of foetal development at the top dose level which was probably secondary to the marked suppression of maternal weight gain and food consumption also observed at that dose. One out of five and two out of five dams given 10 000 or 25 000 mg/kg respectively, which were allowed to litter out, cannibalised all their offspring. It will be recalled from the short-term toxicity studies that PG at such a high dose can cause death. The effects observed in this study are probably all attributable to maternal toxicity. In a rabbit study, doses of 2·5, 12, 54 or 250 mg/kg bw/day were given during organogenesis, with no significant observed effects.[104]

Mutagenicity

The mutagenicity of PG and its hydrolysis product gallic acid has been extensively investigated, but there are no studies available on OG or DG. In bacterial tests for point mutations, PG has been shown to be negative in the Ames test by six different laboratories using seven different strains of *Salmonella typhimurium* with and without metabolic activation,[105–111] and negative in the host-mediated assay.[105] In-vitro tests for chromosome damage were negative in human fibroblast (HE 2144) cells,[112] in human embryonic lung (W1–38) cells,[105] and onion root tip cells.[113] However, a dose-related increase in structural chromosome aberrations was reported in Chinese hamster fibroblast (CHL) cells without metabolic activation.[110] In-vivo tests for chromosome damage have also largely been negative; metaphase analysis in rat bone marrow[105] and a micronucleus test in mouse bone marrow[114] were negative, though a study reported in abstract only and containing no numerical data claimed an increase in micronuclei in the bone marrow of mice was found after feeding PG in the diet for 3 months up to levels of 5000 mg/kg of diet.[115] A rat dominant lethal study was negative.[105] Other tests indicative of DNA damage (*B. subtilis* 'rec' assay, mitotic recombination in yeast *in vitro* and in a host-mediated assay) were also negative.[109,105] In common with other phenolic antioxidants PG has been shown to inhibit or enhance the effect of known mutagens and carcinogens, depending on the conditions of the test.[106,107,113–117] Gallic acid has given consistently negative results in bacterial tests with and without metabolic activation.[118–123] However, it was clastogenic under certain circumstances in a test for chromosome damage *in vitro* in Chinese hamster ovary cells;[124] at a single concentration of 0·05 mg/ml without metabolic activation, 24% of metaphase cells had aberrations. This was reduced to 2% aberrant metaphases when rat liver S9 mix was added. In a further investigation by the same authors a concentrtaion of 0·01 mg/ml was not clastogenic unless transition metals (copper or manganese at 10^{-4} M) were also present. The only in-vivo study on gallic acid reported that gallic acid increased chromosome aberrations in mouse bone marrow metaphase cells following intraperitoneal injection of doses ranging from about 45 to 105 mg/kg bw.[125] However, there was no dose–response relationship and no negative or positive control values were reported. In tests indicative of DNA damage gallic acid was positive under certain conditions; in a study of mitotic gene conversion it was positive at neutral and alkaline pH, an effect which could be abolished by the addition of catalase,[126] and it induced double-strand breaks in Lambda-phage DNA in the presence of copper ions.[127] The enhanced activity of gallic acid at alkaline pH in yeast, the removal of this activity by catalase, and the enhancement by transition metals of chromsome breakage in Chinese hamster ovary cells and of double-strand breaks in DNA, suggest that the genotoxic activity of gallic acid seen in these in-vitro studies may involve autoxidation of gallic acid, leading to the generation of H_2O_2 and/or free radical species. Whether such free radicals or peroxide would be active in vivo in the presence of the normal cellular scavenging mechanisms for such agents is not known.

Long-Term/Carcinogenicity Studies

The majority of long-term studies have been carried out on PG. In an early study employing small groups of rats (10 animals/sex/group) doses of 0, 11·7, 117, 1170, 11 700 and 23 400 mg PG/kg of diet were fed for 2 years. The two highest doses resulted in reduced food intake, growth retardation and kidney damage, but there were no effects at the lower doses.[95] In another early study on PG, mice and rats given up to 10 000 mg/kg of diet for 2 years showed no adverse effects, but at the highest dose of 50 000 mg/kg patchy hyperplasia of the forestomach was observed.[128] However, it should be noted that in a recent study specifically investigating forestomach hyperplasia, no increase in cell proliferation in the rat was found when 20 000 mg/kg PG was fed in the diet for 9 days; a similar dose of BHA increased cell proliferation approximately five-fold.[129] In a later study, 25 mice/sex/group were given 0, 5000 or 10 000 mg PG/kg of diet for 21 months. No significant effects on food consumption, growth, survival, haematology, organ weights or histopathology were seen except for improved survival and reduced relative spleen weight in the top-dose males.[130]

More recent carcinogenicity bioassays of PG in rats and mice, conforming to present-day standards, utilised 50 animals/sex/group.[131] Doses of 0, 6000 or 12 000 mg PG/kg of diet were fed for 2 years. Dose-related reductions in body weight were seen in all treated groups. Some statistically significant increases in tumour incidence were seen in analyses for trend across the dose groups (i.e. malignant lymphoma in male mice; liver adenomas, but not adenomas plus carcinomas, in female mice; thyroid follicular-cell adenomas and carcinomas combined in male rats). However, the lack of consistency across the sexes and species, and similar incidences in historical control groups as in the high-dose groups for lymphomas and thyroid tumours, do not point to a conclusion that PG is carcinogenic. In addition, in the mouse study negative trends were seen for fibromas of the skin or subcutaneous tissue in males, and in the rat study tumours (mostly benign) of the preputial gland and pancreatic islet cells, and phaeochromocytomas were significantly increased in low-dose but not high-dose males, suggesting an anticarcinogenic effect. A protective effect of PG against dimethylbenz(a)anthracene-induced mammary tumours in rats has also been reported, the extent of protection varying with the type and amount of fat in the diet, PG being less effective in animals receiving diets high in polyunsaturated fats.[132–134] PG has also been shown to inhibit formation of N-nitroso compounds in the stomach[135] and gallic acid has been shown to inhibit the induction of mouse lung adenomas by morpholine and sodium nitrate, which combine to form N-nitrosomorpholine.[136]

The only long-term studies available on OG and DG were conducted over 30 years ago,[94] together with a study on PG. OG, DG and PG were each given at 0, 350, 2000 and 5000 mg/kg of diet for 2 years to rats. No adverse effects were observed, including tumour incidence.

Conclusions

The average intake of phenolic antioxidants (BHA, BHT, gallates) as a whole has been estimated as less than 1 mg/person/day.[2] Taking the gallates as a group, reproduction studies are the most sensitive indicator of toxicity and suggest a no-effect level of 1000 mg/kg of diet, equivalent to 50 mg/kg bw. In other studies, effects observed at higher doses up to 250 mg/kg bw were very minor. At doses above 500 mg/kg bw the gallates become acutely toxic and cause kidney damage. Propyl gallate and gallic acid are not mutagenic in the majority of in-vitro and in-vivo systems in which they have been tested, with the exception of some positive results obtained *in vitro* without metabolic activation. Several long-term studies do not indicate they are carcinogenic, nor do they appear to induce forestomach tumours. Against this toxicological background, the present use of the gallates as antioxidants in food is very unlikely to pose any hazard to human health. In relation to worker exposure via the skin, however, it is worth noting that the gallates have been shown to cause contact dermatitis in bakers and other workers handling them, and skin sensitisation in guinea pigs.[98]

TERTIARY-BUTYL HYDROQUINONE

Tertiary-butyl hydroquinone (TBHQ) is used as a food antioxidant in the USA and some other countries, but it is not permitted in the countries of the EEC, due to lack of adequate toxicological data.[137] The structure of TBHQ, shown in Fig. 2, resembles that of BHA and BHT, and in the 1970s the JECFA allocated a group ADI of 0–0·5 mg/kg bw for BHA, BHT and TBHQ.[138] However, BHA and BHT have since both been shown to be carcinogenic under some circumstances in animals (see later), and TBHQ has been shown to have equivocal mutagenic activity. Thus, the lack of adequate carcinogenicity data on TBHQ is an important gap in the data to be filled. As a result of these findings both the JECFA and the SCF now consider the toxicity of these three antioxidants separately. The SCF has not established an ADI for the reasons cited above. The JECFA has recently re-evaluated TBHQ[139] and concluded that there is some, albeit conflicting, evidence that TBHQ is mutagenic and that the only available long-term rat study is inadequate by present day standards. A temporary ADI of 0–0·2 mg/kg bw has therefore been allocated, based on a long-term dog study. Lifetime studies in two rodent species have been requested.

Absorption, Metabolism and Excretion

The fate of orally administered TBHQ is similar in rat, dog and man.[140] Absorption from the gastrointestinal tract and excretion in the urine are rapid. Rats eliminated single oral doses of 100–400 mg/kg bw mostly in the urine in 2–4 days as the 4-*O*-sulphate (57–80%), the 4-*O*-glucuronide (4%) and unchanged TBHQ (4–12%). The proportion of the dose eliminated in the urine and the rate of elimination were inversely proportional to the size of the dose. On prolonged feeding in the rat, the proportion of glucuronide in the urine increased. In the dog, single oral doses of 100 mg TBHQ/kg bw gave a similar pattern of urinary excretion to the rat, but with a higher glucuronide contribution (>20%), and little change with prolonged dosing. With prolonged administration there was very little, if any, accumulation of TBHQ in rat or dog tissues, including fat.

Humans given 2 mg/kg bw as a single dose in a high fat vehicle (30% corn oil) eliminated most of the dose in 2–3 days in the urine, mostly as the 4-*O*-sulphate (73–88%), the 4-*O*-glucuronide (15–22%) and less than 0·1% unchanged. However, in a low fat vehicle (10% corn oil), absorption was lower, and less than half the administered dose was eliminated in the urine.

Short-Term Studies

Effects on the Liver

There have been no conventional short-term studies with TBHQ. However, in view of the effect of the other related antioxidants, BHA and BHT, on the liver, special studies in this area have been undertaken with TBHQ. Unlike BHA and BHT, TBHQ does not produce significant liver enlargement, proliferation of endoplasmic reticulum or induction of mixed function oxidases. Some enzyme activities can be induced by TBHQ fed in the diet such as *p*-nitroanisole demethylase, aniline hydroxylase and bilirubin glucuronyl transferase, but the degree of induction was less than that seen after an equivalent dose of BHA.[140]

Reproduction

Several reproduction studies have been carried out in rats. Two are unpublished but have been summarised by van Esch[141] and one is published.[142] They comprise a three-generation, two-litters per generation study with a teratology phase, a one-generation study to follow up effects found in the three-generation study, and a separate teratology study with dosing of TBHQ during organogenesis only. In all studies TBHQ was fed in the diet up to a maximum of 5000 mg/kg of diet, equivalent to approximately 250 mg/kg bw/day. There was no evidence of teratogenic effects, but at the top dose in the three-generation study, pup mortality was increased and pup weight decreased. In the one-generation study, when the pups were 10 days of age, half the treated dams with their litters were switched from TBHQ diets to control diet, and half the controls were switched to high-dose TBHQ diet. As a result of this switch, pup weights recovered to control levels in formerly treated groups and pup weights in former control groups were reduced at 3–5 weeks of age. These data, together with the observations of reduced maternal food intake in the high-dose groups suggest that the diet may have been unpalatable when TBHQ content was high. The effects seen in the three- and one-generation studies may have been secondary to reduced food intakes, though this is unproven. The overall no-effect level, therefore, from reproduction studies was 1500 mg/kg of diet (approximately 75 mg/kg bw). No studies have been carried out on species other than the rat.

Mutagenicity

As for reproduction, the majority of mutagenicity studies, with the exception of that by Giri *et al.*,[143] are unpublished but have been summarised by van Esch.[141] They comprise a wide range of tests: the Ames test for gene mutations in bacteria was negative in five strains of *Salmonella typhimurium* with and without metabolic activation; of three tests for gene mutations in mammalian cells *in vitro*, one was positive with metabolic activation (L5178Y mouse lymphoma thymidine kinase assay), and two were negative (V79 Chinese hamster cells and Chinese hamster ovary HGPRT forward mutation assay); of four tests for chromosome damage in mammalian cells *in vivo*, one mouse micronucleus test was positive,[143] one negative, and a rat bone marrow cytogenetics assay was positive, a mouse bone marrow test negative; and finally, a rat dominant lethal assay for gene mutations in germ cells was negative. Thus, these tests do not rule out the possibility that TBHQ may be mutagenic *in vivo* and emphasise the need for well-conducted carcinogenicity studies.

Long-term/Carcinogenicity Studies

Long-term studies in the rat and dog are also unpublished but have been summarised by van Esch.[141] In the rat study doses of 0, 160, 500, 1600 and 5000 mg TBHQ/kg of diet were fed for 20 months. As in other studies, the top dose of 5000 mg/kg reduced food intake and weight gain. No other adverse effects were observed in the study, which included observations on haematology, clinical chemistry, urinalysis, organ weight and histopathology. However, interim kills and high mortality reduced the numbers of survivors at 20 months and many of the animals dying during the study were not examined. Thus, by present standards the study would be judged as inadequate to assess carcinogenic potential on the grounds of its rather short duration and poor survival rate.

In the dog study, doses of 0, 500, 1580 and 5000 mg/kg of diet were fed for 117 weeks, which again is of too short a duration in this species to assess carcinogenic potential. Limited histopathology revealed no abnormalities, but at the highest dose red blood cell counts, haemoglobin and haematocrit values were all reduced compared with controls, suggesting an effect on red cell destruction.

Since TBHQ is a phenolic antioxidant analogous to BHA it has been tested for its ability to induce cellular proliferation in the rat forestomach.[144] Feeding 2500 mg/kg of diet for 9 days had little, if any, effect. A higher dose of 10 000 mg/kg of diet caused a small but significant increase in cell proliferation of about 50% above the background rate; this compares with about a 350% increase for a similar level of BHA in the diet. In a similar short-term experiment in rats (duration not specified) feeding of 20 000 mg/kg of diet caused mild hyperplasia of the forestomach mucosa, with focally increased hyperplasia of basal cells, but this basal hyperplasia showed no tendency to differentiate. In the same study BHA caused a much more marked effect on the forestomach.[145] In the hamster, feeding of 5000 mg TBHQ/kg of diet for 20 weeks caused only a very slight, non-significant increase in mild and moderate hyperplasia of the forestomach compared with controls, whilst 10 000 mg BHA/kg of diet caused 100% severe hyperplasia and 60% papillomas of the forestomach.[146] Thus TBHQ may have a very limited capacity to induce forestomach hyperplasia at very high doses.

TBHQ has been shown to have weak promoting activity on urinary bladder carcinogenesis induced by N-butyl-N-(4-hydroxybutyl)-nitrosamine when fed at 2000 mg/kg of diet for 32 weeks to rats. Dietary TBHQ also elevated urinary pH, which is known to enhance urinary bladder carcinogenesis, but it is not known if this is the mechanism by which TBHQ is acting in this model.[147]

Also of relevance to carcinogenicity is the observation that TBHQ inhibits nitrosation reactions in the stomach. It can, therefore, protect against toxicity caused by simultaneous administration of nitrate and nitrosatable compounds, though it is unlikely to have any effect on preformed nitrosamines.[148]

Conclusions

TBHQ was accepted as suitable for food use in a number of countries at a time when there was only a very low level of concern about the toxicity of antioxidants in general, and on the basis of studies, some of which would not meet current standards in toxicity testing. More recent studies have indicated TBHQ may be mutagenic *in vivo,* though this requires further study. Unlike BHT it does not have any marked effects on the liver at high doses, and compared with BHA it has little or no effect on rodent forestomach hyperplasia. It is, therefore, unlikely to cause tumours in these organs. Nevertheless, there are no adequate long-term studies against which to judge the equivocal evidence of mutagenicity. Under these circumstances, it is understandable that, in a large number of countries, TBHQ is not permitted for food use.

BUTYLATED HYDROXYANISOLE

The structure of butylated hydroxyanisole (BHA) is shown in Fig. 2. Commercial BHA comprises two isomers, 3-tert-BHA (98%) and 2-tert-BHA (2%). Estimates of daily intake of BHA vary. Surveys have produced estimates of 4 mg/person in the Netherlands,[2] 5–12 mg/person on average and 11–28 mg/person at the 90th percentile in Canada[149] and 5–7 mg/person on average and 12–15 mg/person at the 90th percentile in the USA.[150] Thus in all cases, including extreme consumers, intakes will be less than 1 mg/kg bw/day. In the past, the JECFA has allocated a group ADI to the three phenolic antioxidants, BHA, BHT and TBHQ. However, they are now considered separately because of their differing toxicities. Initially, the JECFA allocated BHA a temporary ADI of 0–0·3 mg/kg bw, with a request for further studies to investigate oesophageal hyperplasia in pigs and monkeys fed BHA in the diet and for a multigeneration reproduction study.[151] In the light of further toxicological information, JECFA has now established an ADI of 0–0·5 mg/kg bw.[308] The SCF has allocated a temporary ADI, of 0–0·5 mg/kg bw. Average consumers will have BHA intakes at or below these ADIs. however, extreme consumers may occasionally have intakes slightly in excess of these ADIs.

Absorption, Metabolism and Excretion

BHA is well absorbed from the gastrointestinal tract of the rat, rabbit and dog. In rats and rabbits most of an oral dose is excreted in the urine within 24 h as the 4-*O*-glucuronide (50–80%) with smaller amounts present as the 4-*O*-sulphate.[152–154] In the dog, around 60% is excreted unchanged in the faeces and the remainder in the urine as the 4-*O*-sulphate with some tertiary butylhydroquinone (TBHQ), an unidentified phenol and a small proportion as the 4-*O*-glucuronide.[155] There is little evidence for tissue retention of BHA in rats, dogs, pigs or pullets.[154,156–159]

In humans, gastrointestinal absorption is also rapid,[160] but much lower doses in humans are required to produce a given plasma BHA level than in rats, an oral dose of around 0·5 mg/kg bw in humans being comparable to 200 mg/kg bw in rats.[159] This could have implications for risk assessment if the effects of BHA on the rat forestomach (see later) are mediated systemically and are thought to have implications for human health. Tissue storage of BHA and/or its metabolites in humans may also be greater than in the rat.[160]

When 0·5–0·7 mg BHA was given to male volunteers, 22–77% was recovered in the urine as the glucuronide, less than 1% as free BHA, and no demethylation or hydroxylation products were detected.[152] A much higher oral dose of 100 mg in male volunteers again showed little or no free BHA, with 46% as glucuronide and 26% as sulphate conjugates in the urine.[161] However, TBHQ excreted as glucuronide or sulphate conjugates was also identified in substantial quantity as a metabolite of BHA in this study[161] and in a later study.[159] Studies on rat liver microsomes *in vitro* have also shown that the 3-isomer was broken down to TBHQ,[162,164] in one case forming 54% of the metabolites. This has been confirmed *in vivo*; TBHQ conjugates accounted for 5–9% of urinary metabolites in rats given BHA orally, 3-BHA glucuronide being the major metabolite.[159,162] Smaller quantities of TBHQ and TBHQ sulphate were found in faeces.[164] Thus, TBHQ has been found as a metabolite in three species so far, including man. This finding may be of significance since TBHQ is probably mutagenic (see earlier section on TBHQ). The excretion of conjugates of BHA and TBHQ is similar in rats and humans, but other aspects of in-vivo metabolism and kinetics may differ markedly.

Short-Term Studies

The liver appears to be the target organ for BHA at high doses. In rats, increases in liver weight not accompanied by signs of hepatotoxicity have been observed in a number of studies.[165–172] The magnitude of the effect depends on dose and duration of treatment; the lowest observed effect levels were 500 ppm BHA in the diet (equivalent to about 25 mg/kg bw) or gavage doses of 100 mg/kg bw for 7 days. In dogs fed 250 mg/kg bw/day in the diet for 15 months, there was histopathological evidence of liver damage, but no liver effects were seen at lower doses of 5 or 50 mg/kg bw/day in the same study[158] or at doses of 0·3, 3, 30 or 100 mg/kg bw/day in the diet for 1 year in a separate dog study.[157] In monkeys, liver weight was increased after giving 500 mg/kg bw/day by gavage for 28 days and histologically accumulation of lipid droplets, cytomegaly, enlarged or fragmented nuclei and proliferation of smooth endoplasmic reticulum were seen.[173] A dose of 50 mg/kg bw/day in the monkey was without effect.

BHA increases the activity of a number of detoxifying enzymes in rat and mouse liver including epoxide hydrolase, glutathione-S-transferase, UDP-glucuronyl transferase, aryl hydrocarbon hydroxylase, biphenyl 4-hydroxylase, glucose-6-phosphate dehydrogenase, succinate dehydrogenase, guanylate cyclase, UDP glucose dehydrogenase and mixed function oxidases.[174–178] In the monkey there are differences from rodents; nitroanisole demethylase is increased and glucose-6-phosphatase is decreased by BHA,[173] whereas in rats activity of these enzymes is unchanged.[169,171]

Reproduction

No embryolethal, foetotoxic or teratogenic effects have been observed following administration of BHA by gavage at daily doses of 250–1000 mg/kg bw either prior to mating and throughout pregnancy or during organogenesis only in mice and rats, up to 120 mg/kg bw during organogenesis in hamsters and up to 400 mg/kg bw during organogenesis in rabbits.[179–182] Similarly, no reproductive or teratogenic effects were found in pigs fed BHA in the diet at levels up to 400 mg/kg bw/day prior to mating and up to day 110 of pregnancy.[183]

In postnatal studies, rats fed BHA in the diet at levels equivalent to 250, 125 and 62 mg/kg bw/day for 2 weeks before mating, throughout pregnancy and lactation, showed no effects except for increased offspring mortality up to 6 weeks of age at the top dose and a delay in development of the auditory startle reflex at the top two doses.[184] Mice exposed to BHA at 5000 ppm in the diet via their mothers during pregnancy and lactation, then directly up to 6 weeks of age showed increased exploration, decreased sleeping and decreased self-grooming and slower learning.[185] Rhesus monkeys were maintained on a diet providing an intake of 50 mg/kg bw/day each of BHA and BHT for 1 year, then bred and continued on the diet for a second year. Their offspring showed no clinical or behavioural abnormalities up to 2 years of age.[186]

Mutagenicity

BHA has not been shown to be mutagenic in any of the test systems used to date. It was negative in bacterial tests *in vitro* for point mutations with and without metabolic activation,[107,187–192] including in strains of *Salmonella typhimurium* specially developed to test oxidative mutagens such as hydrogen peroxide which are generated by rat liver microsomes in contact with BHA.[193] It was also negative in a host-mediated assay.[188] In mammalian cells, BHA did not cause gene mutation in rat liver epithelial cells,[192] Chinese hamster ovary cells with or without metabolic activation[194] or in Chinese hamster V79 cells with or without metabolic activation.[195] It was negative in the sex-linked recessive lethal mutation assay in *Drosophila*.[196,197] In tests for chromosome aberrations BHA was negative in Chinese hamster lung cells *in vitro*,[198] in Chinese hamster DON cells[199] and in human embryonic lung WI-38 cells.[187] Metabolic activation was not used in any of these in-vitro tests. However, in in-vivo tests BHA has been shown to be without clastogenic activity at doses up to 1500 mg/kg bw in cytogenetic analysis of rat bone marrow and in the rat dominant lethal assay.[188] BHA was also negative in in-vitro tests indicative of DNA damage or repair, that is, the *B subtilis* rec assay,[191] mitotic recombinations in yeast[188] and induction of sister-chromatid exchanges in Chinese hamster ovary cells,[198] Chinese hamster DON cells without metabolic activation and Chinese hamster V79 cells with metabolic activation.[195] It showed no evidence for DNA excision repair in an in-vitro rat hepatocyte unscheduled DNA synthesis assay.[191] A positive result has been reported in a yeast host-mediated assay.[188]

Carcinogenicity and Related Studies

Up until 1982, long-term studies on BHA using the oral route were negative, though it should be noted that none of these studies would be acceptable by current standards, being of too short a duration and/or deficient in the tissues chosen for histopathological examination.[166,200,201,307] Furthermore, there were no oral studies in the mouse.

In 1983, a study was published which showed that 2% BHA fed in the diet to Fischer 344 rats for 2 years induced papillomas of the forestomach in almost 100% of treated males and females, and squamous cell carcinomas of the forestomach in approximately 30% of treated males and females.[202] Most of the changes were situated close to the limiting ridge between forestomach and glandular stomach and there was little evidence of inflammatory infiltration, erosion or ulceration, indicating the effects were unlikely to be the result of non-specific irritation. Rats receiving a lower dose of 0·5% BHA had no carcinomas but a 20–25% incidence of forestomach hyperplasia was seen and one rat of each sex had a papilloma.

Subsequent studies have now replicated the above findings in the same or other strains of rat and in other rodents with a forestomach, and investigated the time-course, dose–response and reversibility of forestomach lesions induced by BHA. In the initial study by Ito *et al.*[202] BHA was given in a pelleted diet and incorporation of BHA into pellets may have modified the antioxidant. However, similar results were obtained when 2% BHA was given in a powdered diet to male F344 rats and 1% BHA caused papillomas of the forestomach but no carcinomas in 20% of the rats.[203] In other short-term experiments, forestomach hyperplasia has been seen whether BHA was given in dry, powdered diet, incorporated into powdered diet in corn oil, or incorporated into pelleted diets.[144]

In hamsters, 1% BHA in a powdered diet or 2% BHA in a pelleted diet fed for 24 weeks caused papillomas of the forestomach in all treated animals compared with none in controls, and submucosal growth, indicative of malignancy, was seen in the majority of the lesions.[204] Studies of longer duration in hamsters fed 1 or 2% BHA in the diet for 104 weeks showed forestomach papillomas in almost all treated animals and carcinomas in 7–10% of treated animals.[205] In mice doses of 0·5% or 1% BHA caused a low incidence of both forestomach papillomas (13·5% and 14·3%) and carcinomas (2·7% and 4·7%).[205] Mice given 1000 mg/kg bw/day by gavage for 4 weeks showed epithelial hyperplasia and hyperkeratosis of the forestomach remote from the limiting ridge.[145]

Studies of the timing of BHA-induced effects demonstrate that changes occur very quickly. When 1000 mg/kg bw/day was given by gavage to rats, oedema of the epithelium and increased mitotic activity of the forestomach were seen after the first dose and mild hyperplasia and hyperkeratosis with a marked increase in mitotic activity were seen after the second dose.[145] In contrast to dietary exposure, changes were observed remote from the limiting ridge following dosing by gavage. With dietary exposure to 2% BHA, after 1 week epithelial damage, hyperkeratosis and hyperplasia were seen.[145] In these studies epithelial change and inflammation occurred only to a small extent and hyperplasia was observed at separate sites. Thus the hyperplasia does not appear to be the result of hyper-regenerative activity following epithelial damage.

Reversibility studies indicate that the mild effects induced by feeding 2% BHA for 1 week followed by a 4-week recovery period are completely reversible whereas the more marked hyperplasia caused by feeding 2% BHA for 2 or 4 weeks does not completely regress during a 4-week recovery period. However, feeding 2% BHA for 6–15 months, followed by a 2 or 7 months recovery period, resulted in considerable regression of lesions after 2 months and almost complete regression after 7 months.[145] In a study in which mitotic activity in the forestomach was monitored by radiolabelled thymidine uptake, rats were fed BHA at levels of 0·1–2% in the diet for 13 weeks then returned to the control diet.[206] Mitotic activity returned to control levels within 1 week of returning to the control diet. The severe histopathological changes seen in the 2% dose group regressed more slowly, however, and minor changes were still apparent after 9 weeks of recovery. In a longer term study treatment with 2% BHA in the diet of rats for 3, 6, 12 or 15 months, followed by 12, 9, 3 or 0 months on control diet respectively, showed almost normal forestomach histology in the first two groups, but after 12 months BHA exposure and 3 months recovery, carcinoma or highly proliferating papillary growths were seen.[129]

A number of studies have been aimed at determining whether there is a threshold for BHA-induced effects on the forestomach. Using the index of radiolabelled thymidine uptake for detection of cell proliferation, studies in which, 0, 0·1, 0·25, 0·5 and 2% BHA were fed in the diet to F344 rats for 9 days or 13 weeks, showed 0·25% to be the no-effect level.[144,206] Using histological observations, a 90-day study in a different strain of rat, in which BHA in crystalline form was added to the diet, showed mild hyperplasia at the lowest dose used of 0·125%; in a second 90-day study in the same strain of rat, in which BHA was dissolved in oil then incorporated into the diet, 0·125% was a no-effect level.[145] In a 2-year study in which doses of 0, 0·125, 0·25, 0·5, 1 or 2% BHA were fed to F344 rats, papillomas or carcinomas were seen at 1 and 2% BHA and hyperplasia was dose-related with an incidence of 0, 2, 14, 32, 88 and 100% in increasing dose groups. The increase was significant at 0·25% BHA, suggesting 0·125% BHA is a minimal or no-effect level.[146] Thus, overall, 0·125% seems to be the minimal or no-effect level depending on how the BHA is given. These studies also indicate that short-term experiments using the thymidine labelling index are sufficient to show the threshold level applicable to longer-term exposure.

Since food-grade BHA is a mixture of two isomers, approximately 98% 3-BHA and 2% 2-BHA, the effect of the pure isomers has been investigated. Hamsters were fed 1% in the diet of food-grade BHA or one or the other isomer (purity > 99·9%) for 16 weeks, with interim kills at 1, 2, 3 and 4 weeks. By week 1 severe hyperplasia and by week 4 papillomas of the forestomach were seen in those fed food-grade BHA or 3-BHA but not in those fed 2-BHA or controls. At week 16 the highest number of papillomas were in the 3-BHA group, somewhat fewer in the food-grade BHA group and none in the 2-BHA group.[207] In the rat, however, when the BHA isomers were given by gavage at 1000 mg/kg bw/day for 10 days, 2-BHA was just as active as 3-BHA in induction of forestomach lesions.[145] Even if 2-BHA were generally less active than 3-BHA it would be difficult to utilise this since 3-BHA is the more active antioxidant.

In order to investigate the possible significance for man of the observed changes in rodent forestomach, other species, without a forestomach, have been studied. If the presence of high concentrations of BHA in the diet in contact with squamous epithelium is the problem, then the lower oesophagus and cardio-oesophageal junction might be at risk in species without a forestomach, though residence time in the oesophagus would be considerably shorter than in the forestomach. The rodent studies are notable for the lack of lesions in the oesophagus, which has been frequently studied along with the forestomach. Hyperplasia of the oesophagus has been reported in 1 of 19 rats given 2% BHA in the diet for 32 weeks and in a small number of mice fed 1% BHA in the diet for 80 weeks,[208] but otherwise the oesophagus has not been shown to be affected in rodents.

In beagle dogs of both sexes, fed 0, 1·0 or 1·3% BHA in the diet for 180 days, no proliferative or hyperplastic lesions were found in the stomach epithelium examined by light microscopy and no ultrastructural changes were seen in the lower oesophagus, cardio-oesophageal junction, cardia, body or pylorus of the stomach when examined by electron microscopy.[209] These results have been confirmed in another study in beagle dogs, in which males and females were fed 0, 0·25, 0·5 or 1·0% BHA in the diet for 6 months. Examinations of the oesophagus, stomach and duodenum by light microscopy revealed no hyperplasia or neoplastic changes and no significant increases in mean epithelial thickness of the oesophagus, fundic or pyloric areas of the stomach were seen, though the thickness of the gastric mucosa was slightly increased in the highest dose group. The mitotic index in the lower oesophagus was unchanged.[210]

Guinea pigs, given 1000 mg/kg bw/day by gavage for 4 weeks showed no changes in the stomach or lower oesophagus.[145] A study in the pig produced equivocal results. BHA was given to pregnant pigs in doses of 0, 50, 200 or 400 mg/kg bw/day from mating to day 110 of pregnancy. Proliferative and parakeratotic proliferative changes of the stomach stratified epithelium were observed in control and treated animals, changes known to be influenced by the type and grinding of the diet, and these may have masked any BHA effects. The oesophagus also showed proliferative and parakeratotic changes in the epithelium in a few pigs in the mid and high-dose groups, suggesting a possible effect of BHA.[211]

In a monkey study[212] BHA was given by gavage at 0, 125 or 500 mg/kg bw/day for 5 days/week for 20 days. The high dose group was then reduced to 250 mg/kg bw because of vomiting, and dosing continued in all groups for a total of 85 days. Examination of the whole length of the oesophagus and the stomach by light microscopy showed no evidence of hyperplasia. However, the mitotic index in the distal region of the oesophagus was increased 1·9 fold in the high dose group compared with controls. There was no effect on the mitotic index at 125 mg/kg bw. This increase in the mitotic index is equivalent to that seen in the forestomach after 0·5% BHA in the diet of rats. The gavage dose was delivered to the distal oesophagus, so this study indicates a very limited response to BHA in the monkey and it is not known if dietary administration would have any effect.

Research into the biochemical basis for the action of BHA on the forestomach has so far been inconclusive. BHA itself does not seem to be mutagenic (see earlier discussion) and no metabolites of BHA have been detected in the stomach contents of rats given a single intragastric injection of radiolabelled 3-BHA, apart from an unidentified minor polar metabolite.[213] About 95% of the radiolabel was recovered as the parent compound. In-vitro studies, which simulated stomach conditions similarly did not give rise to any detectable metabolites.[213] This suggests that it is BHA itself which is responsible for the initial hyperplasia. Whilst it cannot be ruled out that liver metabolites of BHA may be acting indirectly on the forestomach, microsomal metabolites seem unlikely to be the causal agents. For example, pretreatment with phenobarbitone does not enhance BHA-induced hyperplasia,[214] and the genotoxic metabolite TBHQ is much less potent than BHA in inducing forestomach hyperplasia.[145,146] However, any directly acting mechanism remains elusive. 3-BHA and 2-BHA do not bind covalently to DNA or RNA of rat forestomach or glandular stomach,[213] although there is considerably enhanced binding to protein in the forestomach compared to glandular stomach, liver or kidney.[213,215] Neither these studies nor those on mutagenicity exclude the possibility of certain types of direct damage to DNA, but this aspect has not yet been investigated. The possibility that BHA may act via binding to tissue thiols in forestomach epithelium is now being investigated following the observation that diethyl maleate, which depletes tissue glutathione, completely inhibits BHA-induced forestomach hyperplasia.[214] The possible mechanisms for this inhibition are several but diethyl maleate may be competing with BHA for binding to thiols.

Interactions with Known Carcinogens, Mutagens and Other Antioxidants

A considerable number of studies have investigated the possible interactions of BHA with known carcinogens and mutagens because the effects of BHA on enzyme systems and its antioxidant activity suggested that interactive effects were likely. BHA has been administered prior to, simultaneously with, or after treatment with other chemicals. Extensive reviews of these studies have appeared elsewhere.[146,216,217] When administered with carcinogens BHA has been shown to have enhancing, inhibiting or no effects on tumour incidence. When tested in combination with mutagenic chemicals, in most studies BHA reduced the DNA-damaging activity of chemicals which act indirectly as mutagens. However, BHA in combination with nitrite *in vitro*, under acidic conditions simulating those in the stomach, resulted in the formation of highly mutagenic products.[218]

This could have toxicological implications since many foods contain nitrite as a preservative or are rich in nitrate which may be converted by bacteria in the oral cavity to nitrite.

Of particular interest are the interactions of BHA with other chemicals in the rat forestomach, the target organ for toxicity.[146] Treatment of rats, for example, for 1 or 52 weeks with other antioxidants, such as ascorbate, propyl gallate, ethoxyquin and α-tocopherol, in combination with BHA in the diet, has been shown to enhance or reduce the forestomach hyperplasia induced by BHA alone, depending on whether the lesions are located in the prefundic region or the mid-region.[219,220] Thus, responses to antioxidants which are consumed in combination in the diet may be complex. BHA also acts synergistically in combination with other known inducers of forestomach papillomas and carcinomas, e.g. N-methyl-N′-nitro-N- nitrosoguanidine, N-methyl-N-nitrosourea and N,N′-dibutylnitrosamine.[146,221]

Conclusions

The only toxicological problem with BHA concerns its effects on the rodent forestomach and possible effects on related tissues in other species. Tests with other phenolic compounds have shown that structurally related compounds can also induce forestomach hyperplasia, but BHA seems to be one of the most active compounds. Studies with other species not possessing a forestomach including guinea pig, pig, dog and monkey have shown that at the maximum tolerated doses in these species BHA does not induce hyperplastic changes in the stomach, but the pig and monkey studies suggested that at high doses BHA may have a slight proliferative effect on the oesophagus. Thus the possibility that humans might also respond to BHA in this way cannot be ruled out. However, the rodent dose–response studies and the pig and monkey studies have all clearly shown that there is a threshold below which BHA does not seem to cause any increased mitotic activity or hyperplasia. In rodents the no-effect level is around 0·125% BHA in the diet, equivalent to around 62·5 mg/kg bw/day. In pigs 50 mg/kg bw and in monkeys 125 mg/kg bw were the no-effect levels. Bearing in mind that BHA has not been shown to be genotoxic, that clear thresholds seem to have been demonstrated for its effect on squamous epithelium and that these are two to three orders of magnitude above estimated 90th percentile intakes of BHA, it seems unlikely that the use of BHA as an antioxidant in foods presents any hazard to humans.

BUTYLATED HYDROXYTOLUENE

The structure of butylated hydroxytoluene (BHT) is shown in Fig. 2. Intakes of BHT have been estimated. Since BHA, BHT and the gallates may be used singly or in combination across a similar range of products, the intakes of BHT are unlikely to exceed those of BHA (i.e. <1 mg/kg bw/day) and almost certainly the intake of BHT itself will be less than the intake of BHA, due to much larger losses during processing and cooking of foods. In a recent survey in the Netherlands, no BHT was detectable in foods.[2] Ingestion of BHT decomposition products however, may occur.[222] The JECFA originally included BHT in a group ADI with BHA and TBHQ, but now considers them separately because of their differing toxicity profiles. At its most recent re-evaluation, a temporary ADI of 0–0·125 mg/kg bw was allocated,[223] based on a no-effect level of 25 mg/kg bw in a single-generation rat reproduction study. Further studies were requested to address the problem of hepatocarcinogenicity seen in one rat study after in-utero exposure (see later) and on the mechanism of BHT-induced haemorrhage in susceptible species. The SCF has established an ADI of 0–0·05 mg/kg bw.[2]

Absorption, Metabolism and Excretion

BHT is rapidly absorbed from the gastrointestinal tract. The highest tissue levels are found in fat, but accumulation in fat does not seem to occur with repeated administration in rodents, and the half-life in liver and fat is 7–10 days.[160,224] Samples of subcutaneous adipose tissue taken from UK adults consuming an estimated 1 mg of BHT daily contained 0·01–0·49 ppm, and samples from US adults consuming an estimated 2 mg of BHT daily contained 0·34–3·19 ppm.[225]

In contrast to BHA which is metabolised similarly in rat and man, the metabolism of BHT in man differs from that of the rat. The major route of degradation is oxidative metabolism mediated by the microsomal mono-oxygenase system. In rat, rabbit and monkey the methyl group attached to the benzene ring is oxidised, whereas in man oxidation of one of the tertiary butyl groups predominates. In mice, both types of oxidation readily occur. The metabolites can then be conjugated with glucuronic acid or glycine. In the rat, the major metabolites are 3,5-di-tert-butyl-4-hydroxy-benzoic acid (BHT-acid), both free and as the ester glucuronide, and S-(3,5-di-tert-butyl-4-hydroxybenzyl)N-acetyl cysteine, with lesser quantities of many other metabolites including BHT-alcohol, BHT-aldehyde, BHT-dimer, BHT-quinone and hydroquinone derivatives, being found in urine and faeces,[224,226,227] In faeces, free BHT-acid predominates in the rat. In the mouse, as well as oxidation of the *p*-methyl group, there is oxidation of the tert-butyl groups, resulting in products which are cyclised to some extent by reacting with the adjacent phenolic –OH group to give hemiacetals and lactones.[224] In the rabbit, BHT-acid is the major urinary metabolite, excreted both as the ester glucuronide and as the glycine conjugate, with BHT-alcohol, BHT-aldehyde and BHT-dimer also present.[228,229]

In the monkey, the major metabolite is the ester glucuronide of BHT-acid, with a rate of excretion similar to that of man.[230] Limited studies with 3 human volunteers have shown that 5-carboxy-7-(1-carboxy-1-methylethyl)-3,3-dimethyl-2-hydroxy-2,3-dihydrobenzofuran, as the glucuronide, is the major urinary metabolite. Free and conjugated BHT-acid was only a minor metabolite.[231,232]

Biliary excretion of BHT and metabolites has been reported in the rat,[160] rabbit and dog,[233] delaying the excretion of metabolites by enterohepatic recirculation. In the rat, 90% of a single oral dose was excreted in 4 days, up to 40% in the urine.[160,231] In the rabbit, 54% of an oral dose was excreted in the urine in 3–4 days.[234] In humans, one study showed approximately 50% of a single oral dose was excreted in the urine within 24 h and a further 25% excreted more slowly over 10 days,[309] but another study showed only 21% of a single oral dose was excreted in 72 h, suggesting that biliary excretion and enterohepatic recirculation also occurs in man.[232]

Short-Term Studies

Liver

Liver, lung and blood are the main targets for BHT toxicity. In rats, the effects of BHT on the liver are more marked than those of BHA. At around dietary levels of 1000 ppm (approximately 50 mg/kg bw/day) in the rat, body weight gain is reduced and absolute and relative liver weights increased.[167,235–237]

The liver hypertrophy is due at least in part to proliferation of smooth endoplasmic reticulum and these changes are reversible on cessation of dosing.[238,239] The increase in smooth endoplasmic reticulum and cytochrome P450 in the liver is accompanied by increases in associated enzymes including nitroanisole demethylase, thymidine kinase, ornithine decarboxylase, aniline hydroxylase, biphenyl-4-hydroxylase, aminopyrine demethylase, glutathione S-transferase, glutathione reductase, and epoxide hydrolase.[175,176,239–242] This enzyme stimulation is an important mechanism in the inhibitory action of BHT on the carcinogenic activity of other chemicals.

At higher doses of 500 mg/kg bw given via the diet or by gavage for 28 days in the rat, glucose-6-phosphatase levels are depressed indicating early liver damage.[169,243] and with gavage, but not with dietary administration, there is periportal hepatocyte necrosis, proliferation of bile ducts, fibrous and inflammatory cell reactions.[243] Even higher gavage doses of 1000 mg/kg bw or more cause centrilobular rather than periportal necrosis.[242,244] It has been suggested that the hepatic damage induced at these very high doses may be due to a reactive metabolite of BHT, 2,6-di-tert-butyl-4-methylene-2,5-cyclohexadienone (BHT-quinone methide). Quinone methide has been identified in the liver and bile of rats. It binds to glutathione, a possible detoxification pathway. When high doses of BHT are given, liver glutathione rapidly depletes, but the accompanying hepatic necrosis can be prevented by pretreatment with cobaltous chloride, which is an inducer of hepatic glutathione.[244–246] On the other hand, pretreatment with buthionine sulphoximine, an inhibitor of glutathione synthesis in the liver, increases levels of enzymes which are markers of liver damage in rats given doses of BHT which are normally non-toxic to the liver.[247] In the mouse, liver toxicity is not normally observed, even when high doses of BHT are given. However, if BHT is given in combination with buthionine sulphoximine, then liver damage is seen.[248]

In the monkey, in contrast to the rat, the effects of BHA on the liver are more pronounced than those of BHT.[173] In juvenile but not infant monkeys, BHT given by gavage at 500 mg/kg bw for 28 days, had no effect on liver weight, but histologically slight hepatocytomegaly, enlargement of hepatocyte nucleoli and moderate proliferation of endoplasmic reticulum were seen. Nitroanisole demethylase was increased and glucose-6-phosphatase activity decreased. A dose of 50 mg/kg bw was without effect. The lack of activity of the higher dose in infants may be due to absence of BHT-metabolising enzymes at that age.

Lung

Single intraperitoneal injections of 400–500 mg/kg bw BHT induce acute lung toxicity in the mouse, the effects including hyperplasia, hypertrophy and general disorganisation of cellular components. Lung weight may be increased. Biochemical responses of mouse lung to BHT include increases in DNA, RNA, protein and collagen synthesis, a transient sharp increase in thymidine kinase activity, increased activities of some key glycolytic enzymes, superoxide dismutase, glutathione peroxidase and reductase, glucose-6-phosphate dehydrogenase, and increased cyclic GMP levels.[249] The initial sequence of events involves infiltration of type I squamous epithelial cells, followed by multifocal necrosis and destruction of the blood barrier.[250–253] Testing of structural analogues to BHT and special metabolic studies on BHT itself suggest that lung damage is mediated by the same metabolite as may be active in liver, the quinone methide.[254]

Blood

A number of studies have shown that BHT can cause extensive internal and external haemorrhaging, in association with hypoprothrombinaemia, and these are dose-related. At high doses haemorrhaging is severe enough to cause death in some but not all strains of laboratory rats, mice and guinea pigs.[255–260] Rabbits, dogs and hamsters seem resistant. In susceptible animals the effect is seen with either intraperitoneal or dietary administration of BHT. In the most sensitive animal investigated so far, the young male Sprague–Dawley rat, 85 ppm BHT in the diet (7·5 mg/kg bw/day) caused a transient but non-significant reduction in the prothrombin index, 170 ppm (15 mg/kg bw/day) caused a transient, significant reduction, and 2500 ppm (200 mg/kg bw/day) caused a persistent hypoprothrombinaemia.[261] Investigations into the mechanism of BHT-induced haemorrhage have shown that it may be prevented by administration of vitamin K, as is also the case with vitamin E-induced haemorrhage.[256,262] Administration of BHT causes a reduction in vitamin K-dependent clotting factors.[263] It has been shown that BHT does not inhibit vitamin K synthesis in the gut, but that it may inhibit absorption of vitamin K from the gut or inhibit uptake of vitamin K by the liver.[264] In addition to reducing the activity of clotting factors, BHT treatment also alters platelet functions, vascular permeability, and the activities of the kallikrein–kinin system, and these changes may also play a minor role in the haemorrhagic effect.[265] Although the antioxidants vitamin E and ethoxyquin also cause haemorrhaging, a number of other antioxidants do not, suggesting it is not the antioxidant properties *per se* which cause haemorrhaging.[266] Several of the metabolites of BHT, BHT-alcohol, BHT-aldehyde, BHT-acid and the quinone methide have all been shown to increase clotting time to some extent,[259,266] but whether it is the parent compound or a metabolite which is mainly responsible for the effect is not known. Pretreatment with cobaltous chloride reduced the effects of BHT on clotting factors in the rat, suggesting that the quinone methide may well be involved.[267] Further confirmation comes from observations on oral administration of BHT-quinone methide itself to a strain of rat susceptible to the parent compound, BHT, and a strain of mouse which is not susceptible to BHT (perhaps because it does not normally form the quinone methide). Reduced plasma levels of vitamin K-dependent clotting factors occurred in both species.[268] This suggests that BHT-quinone methide is one of the agents involved. Identifying the causative agent may be important in view of the differences between rats and humans in the metabolism of BHT, though if the effect is largely at the point of gut absorption of vitamin K metabolic differences may be irrelevant.

Reproduction

An early study using small numbers of animals reported a 10% incidence of anophthalmia in litters of rats fed 1000 or 5000 ppm BHT in the diet for 5 months[166] but subsequent studies in the rat using daily doses of 500 or 750 mg/kg bw given by gavage before and/or throughout pregnancy have not confirmed this finding.[179] Single and multigeneration reproduction and teratogenicity studies, using doses of 300–3000 ppm in the diet of rats, 20–800 mg/kg bw/day or 1000 and 5000 ppm in the diet of mice, 3–280 mg/kg bw/day in hamsters, and 3–320 mg/kg bw/day in rabbits, have also not shown any increase in malformations in the offspring.[179,269–272] Significant effects observed in these studies were depression of growth rate of parents and offspring in a two-generation study on rats fed 3000 ppm BHT in the diet, but not at 300 or 1000 ppm, intrauterine deaths associated with maternal toxicity in rabbits given 3–320 mg/kg bw/day by gavage during embryogenesis, and prolonged time to birth of first litters, reduced pup numbers and pup weights 12 days after birth in mice fed 5000 ppm BHT in the diet, but not at 1000 ppm.

In studies focussing on postnatal development and behaviour, the offspring of rats fed 1250–5000 ppm BHT in the diet throughout pregnancy and lactation showed higher postnatal mortality at 2500 and 5000 ppm, reduced postnatal weight gain and delays in tests of motor development at 5000 ppm.[273] Offspring of rats given 5000 or 9000 ppm BHT in the diet showed reduced weight gain, developmental delays and increased cell death in the brain, but there was also evidence of maternal toxicity at these doses.[274]

Rhesus monkeys maintained on a diet containing BHA and BHT, providing an intake equivalent to 50 mg/kg bw/day for each antioxidant, were bred after 1 year and continued on the diet for a second year. Clinical and behavioural observations on the infants up to 2 years of age showed no abnormal findings.[186]

Overall, the reproduction studies indicate a no-effect level for BHT in the diet equivalent to 50 mg/kg bw/day.

Mutagenicity

Inconsistent effects have been reported in genotoxicity assays. In tests with lower organisms they were uniformly negative. BHT did not induce reverse mutations in several strains of *Salmonella typhimurium* including those specially developed to test oxidative mutagens, with or without metabolic activation.[107,190,193,275–282] It was also negative in the host-mediated assay in the mouse using *S. typhimurium* as indicator organism.[275] In yeasts, BHT did not induce gene conversion in *Saccharomyces cerevisiae in vitro* and did not increase recombination in a host-mediated assay in the mouse using *S. cerevisiae* as indicator organism.[275,276]

In-vitro tests in mammalian cells have produced variable results. BHT was weakly positive in a test for gene mutation at the HGPRT locus in hamster V79 cells,[283] but negative at the same locus in rat liver epithelial cells.[282] BHT was positive in the L5178Y tk^+/tk^- mouse lymphoma cell forward mutation assay with metabolic activation.[284] BHT did not induce DNA repair in rat hepatocytes or sister-chromatid exchanges in Chinese hamster ovary cells.[282] BHT did induce chromosome aberrations in cultured human lymphocytes[285] and in Chinese hamster ovary cells.[286]

In vivo, BHT was negative in tests for chromosome damage in the bone marrow and liver cells of rodents.[275,277,278,287] It was negative in a mouse specific locus assay.[288] Three dominant lethal studies in the mouse and one in the rat were negative,[275,277,289,290] but two other studies in rats reported positive effects with certain dosing regimens; one study found dominant lethal effects in weeks 3, 4 and 6 of breeding, after 5 consecutive daily oral doses of 30–500 mg/kg bw,[276] and the other reported a dominant lethal effect in the first week of breeding after 10 weeks of feeding 1333 or 4000 ppm BHT in the diet.[291] Subacute or dietary regimens employing lower doses than those mentioned above were negative. A weakly positive effect has been reported in a sperm abnormality test in mice given 250 mg/kg bw ip.[278] A heritable translocation assay in mice with BHT fed at 1000 ppm in the diet was negative.[290] Tests in *Drosophila* were negative for sex-linked recessive lethal muations, X-chromosome loss and translocations.[197,276,292] Thus the majority of in-vivo results suggest that BHT is not likely to be mutagenic at the very low levels of intake from the diet.

Long-Term/Carcinogenicity Studies

Administration of BHT at 7500 ppm in the diet to male BALB/c mice for 16 months considerably increased the incidence of lung tumours, slightly increased the incidence of stomach tumours (squamous cell carcinomas), and decreased the incidence of reticulum cell sarcomas in comparison with controls.[293] The incidence of hepatic cysts was also increased in the BHT-treated mice but liver tumours were not seen in either treated or control mice.

Another study,[294] using a different strain of mouse, confirmed these findings. CF1 mice of both sexes given BHT in the diet at 1000 ppm for the first 1 or 2 months, then at 1000, 2500 or 5000 ppm for 22–23 months, showed a dose-related increase in benign plus malignant tumours of the lung, and females showed an apparent increase in benign ovarian tumours, compared with controls.

In a National Cancer Institute study,[295] carried out according to the protocols of the US Carcinogenesis Testing Program, B6C3F1 mice of both sexes were exposed to 3000 or 6000 ppm BHT in the diet for 107–108 weeks. There was a dose-related reduction in body weight in both sexes and a significant dose-related trend towards better survival in treated males. In females, there was a significant increase in lung tumours (alveolar/bronchiolar adenomas and carcinomas) in the low-dose group (16/46), but not in the high-dose group (7/50) in comparison with controls (1/20). The incidences of other tumours did not differ significantly between groups. In the males, BHT administration was associated with a high incidence of non-neoplastic liver lesions (hepatocytomegaly, peliosis, hepatocellular degeneration and necrosis).

A similar study has been carried out[296] on B6C3F1 mice of both sexes, given 200, 1000 or notionally 5000 ppm (actually 4000 ppm) of

BHT in the diet for 96 weeks, followed by 8 weeks on a basal diet. There was a significant reduction in body weight in high-dose males and females. Survival was said to be unaffected by treatment. Tumours of the lung, liver and lymph nodes occurred at quite high incidences but there were no significant differences between treated and control mice of either sex for these or other tumours. Non-neoplastic findings related to BHT treatment were significantly increased incidences of lymphatic infiltration of the lung in females and of the urinary bladder in both sexes at the high dose only.

In more recent studies, BHT has been shown to cause liver tumours in male mice. BHT was fed to C3H mice at 0·05% or 0·5% in the diet for 10 months.[297] The incidence of liver tumours in C3H mice, which are prone to spontaneous liver tumours with age, was increased compared with controls fed a BHT-free diet or laboratory chow. However, the increase was only seen in males, not females, and was not dose-related (BHT-free control diet—5%; 0·05% BHT—58%; 0·5% BHT—28%). The incidence of lung tumours in C3H mice was increased in males at both dietary levels of BHT and in females at the top dose only. However the increases were not statistically significant and in the males were not dose-related.

In a study with B6C3F1 mice,[298] BHT was fed for 2 years at levels of 10 000 and 20 000 ppm in the diet, equivalent to average intakes of 1·6 or 3·5 and 1·7 or 4·1 g/kg bw/day in male and female mice respectively. The 2 year treatment period was followed by a 16-week recovery period. Survival in treated mice of both sexes was better than in controls, as has also been observed in rats treated with BHT.[299] B6C3F1 mice, especially males are spontaneously prone to liver tumours. Nevertheless, significant, dose-related increases in hepato-cellular adenomas and foci of alteration in the liver were seen in the males, but not the females. There were no treatment-related differences in any other tumour type. In previous studies on the B6C3F1 mouse,[295,296] liver lesions occurred but no significant, treatment-related increases in liver tumours were seen. However, the maximum doses used were about half of the lowest dose in this study.[298]

In an early study in rats[300] administration of BHT in the diet at levels of 2000, 5000, 8000 and 10 000 ppm for 2 years to Wistar rats of both sexes was shown to cause reduced body weight and some changes in relative organ weights in both sexes at the top dose only. No dose-related histopathological changes were seen (brain, heart, lungs, spleen, liver, kidneys and gastrointestinal tract only examined).

In the National Cancer Institute study,[295] F344 rats of both sexes were exposed to 3000 or 6000 ppm BHT in the diet for 105 weeks. There were dose-related reductions in body weights of both sexes but no significant effects on survival or tumour incidences at any site. There was a dose-related increase in focal alveolar histiocytosis in the lung in both sexes, but particularly marked in females.

In another study[301] Wistar rats of both sexes were exposed to 2500 or 10 000 ppm BHT in the diet for 104 weeks. Body weight was reduced for most of the study in both sexes at the high dose. Mortality was significantly increased amongst high-dose males after week 96. Increased relative liver weight was observed in all test animals. The overall incidence of tumours was slightly, but not significantly higher in treated groups of both sexes compared with controls. In the treated females, the incidence of hyperplastic nodules of the liver, pancreatic carcinomas and pituitary adenomas were higher than controls, but the increase only reached statistical significance in the case of pituitary adenomas in the lower dose group.

In 1983, preliminary findings of hepatocellular carcinomas in ageing, BHT-treated rats from a Danish two-generation study were published.[302] Further details have subsequently become available.[299] In this study, using Wistar rats, both sexes of the Fo generation were given BHT in a semisynthetic diet at levels of 25, 100 or 500 mg/kg bw/day from 7 weeks of age. The Fo rats were mated after 13 weeks of dosing and the Fo females continued to receive BHT at the same levels in the diet until the end of the lactation period. The F1 generation were given 25, 100 or 250 mg BHT/kg bw/day from weaning until 141–144 weeks of age. The top dose was lowered for the F1 generation because of kidney toxicity in the Fo females receiving 500 mg/kg. Survival in BHT-treated groups was better than in controls for both sexes from 104 weeks of age onwards (the time at which most carcinogenicity bioassays are usually terminated). Survival at 141–144 weeks of age in the top-dose group males and females was 44% and 39% respectively, compared with 16% and 17% in control males and females. Dose-related increases in the number of hepatocellular adenomas and carcinomas were statistically significant in male F1 rats in tests for heterogeneity or analysis for trend. The increase in hepatocellular adenomas and carcinomas in treated female F1 rats was only statistically significant for adenomas in the analysis for trend.

All hepatocellular tumours were detected when the rats were more than 2 years old and the majority were only found at the terminal post-mortem at 141–144 weeks. No increases were seen in nodular hyperplasia, adenoma or carcinoma of the liver in either sex given 25 mg/kg bw. The incidences of non-neoplastic lesions did not show any treatment-related effects, except as follows. In the males, the incidence of cysts and bile duct proliferation in the liver was increased, and in the females, the incidence of focal cellular enlargement in the liver was increased, in a dose-related manner.

Following the discovery of BHA-induced forestomach carcinogenesis in rodents, BHT, like other antioxidants, has been tested for its effects on the forestomach. In the hamster, BHT fed at 1% in the diet for 16 weeks caused no signs of hyperplasia in the forestomach, whereas an equivalent dose of BHA caused marked hyperplasia and papillomatous lesions by 16 weeks.[146] Similarly, in the rat, feeding 1% BHT in the diet for an unspecified period of time caused no forestomach hyperplasia whereas feeding 2% BHA for the same period of time caused pronounced forestomach lesions.[145] These results are in accordance with the lack of forestomach changes in the various long-term feeding studies on BHT, described earlier.

Interactions with Known Carcinogens and Mutagens

As with BHA, a large number of studies have been carried out to investigate the interactions of BHT with known carcinogens and mutagens. The results are too numerous to present here individually, but the interested reader is referred to reviews by Ito *et al.*,[146] by the International Agency for Research on Cancer,[303] and by Witschi.[304] BHT has been shown to enhance, inhibit or have no effect on the carcinogenic or DNA-damaging activities of other chemicals. The effects seen vary with species and strain of animal used, the carcinogen used, the target organs examined and the timing of the BHT dosing. If administered prior to, or concomitantly with other carcinogens, BHT generally reduced tumour incidence if it has any effect at all, but there are examples of enhancement by BHT. If administered after another carcinogen, BHT generally enhances tumour incidence if it has any effect, but again there are examples of the opposite type of effect, inhibition. With chemicals that cause mutagenic effects indirectly, i.e. it is not the parent compound which is active, BHT generally reduces their DNA-damaging capacity.

Of particular interest in relation to BHT are its interactions with lung, liver and forestomach carcinogens.[303] In strains of mice sensitive to induction of lung tumours by certain carcinogens BHT treatment at any time generally enhances tumour incidence, but in other, less sensitive strains it may have no effect. In rats, the incidence of liver tumours induced by other carcinogens may be enhanced, but more often it is reduced or unaffected by BHT administration. In rats and mice, tumours induced by forestomach carcinogens are usually unaffected or occasionally reduced in incidence by BHT treatment.

Conclusions

Putting aside the question of carcinogenicity, BHT is toxic at lower doses than any of the other antioxidants reviewed here, judging from animal data alone. This is due to its effects on blood clotting mechanisms, for which a minimum effect level of 7·5 mg/kg bw/day has been observed in rats, and is largely the basis for the low ADI set by the SCF of 0–0·05 mg/kg bw. However, estimated intakes of BHT itself are thought to be very low, within the SCF ADI. Intakes of BHT thermal decomposition products may be higher, but would not be likely to exceed 1 mg/kg bw/day as a worst case.

In view of the possibility that the quinone methide metabolite of BHT may be the cause of BHT's toxicity on a variety of systems, and the fact that BHT and BHA are sometimes used together as antioxidants, it is of interest that BHA has been shown to enhance the formation of BHT-quinone methide from BHT *in vitro* in the presence of microsomal peroxidase enzymes from a variety of mammalian tissues, including human lung.[305] However, to date, the metabolism of BHT in humans has been little studied and it is not known to what extent BHT-quinone methide may be formed in human tissues such as liver and lung *in vivo*.

There is no strong evidence that BHT is genotoxic *in vivo* and so it is unlikely that the tumours seen in mouse lung and rat and mouse liver are induced by genotoxic mechanisms. There is also no consensus among toxicologists as to whether BHT is carcinogenic or perhaps acting as a promoter, and an IARC working group has recently concluded that the Danish study showing increased liver tumours in rats could not be evaluated because of the differential survival among control and treated groups.[303] The group did consider that overall there was limited evidence for the carcinogenicity of BHT in experimental animals. Although mouse lung and rat liver are target sites for the acute and subchronic toxicity of BHT at high doses, there is as yet insufficient evidence as to whether prolonged exposure to lower doses may be producing milder cellular damage which in turn gives rise to tumours. The minimum dietary level of BHT which induces liver hyperplasia in the rat is around 50 mg/kg bw, which is above the minimal effect level for alterations in blood clotting and 1000 times above the current SCF ADI. Thus, it seems unlikely that the current dietary intakes of BHT are having any adverse effects. However, the reported use of BHT in high doses for herpes virus infections[306] and the use of BHT supplements may not be without adverse effects.

REFERENCES

1. Addis, P. B., *Food Chem. Toxicol.*, **24** (1986) 1021.
2. Haigh, R., *Food Chem. Toxicol.*, **24** (1986) 1031.
3. Joint FAO/WHO Expert Committee on Food Additives, *Tech. Rep. Ser. Wld Hlth Org.* **669** (1981) 52.
4. Joint FAO/WHO Expert Committee on Food Additives, *Tech. Rep. Ser. Wld Hlth Org.*, **539** (1974) 146.
5. Agricultural Research Council/Medical Research Council Committee Report, *Food Nutr. Res.*, **72** (1974), HMSO, London.
6. Baker, E. M., *Proc. Soc. Exp. Biol. Med.*, **109** (1962) 737.
7. Barness, L. A., *Ann. NY Acad. Sci.*, **258** (1975) 523.
8. McMichael, A. J., *Commun. Hlth Studies,* **2** (1978) 9.
9. Lamden, M. P. & Chrystowski, G. A., *Proc. Soc. Exp. Biol. Med.*, **85** (1954) 190.
10. Takenouchi, K., Aso, L., Kawase, K., Ichikawa, H. & Shioni, T., *J. Vitaminol,* **12** (1966) 49.
11. Takiguchi, H., Furuyama, S. & Shimazono, N., *J. Vitaminol.*, **12** (1966) 307.
12. Yamazoe, J., *Eiyo To Shokuryo,* **18** (1966) 342.
13. Tiselins, H. G. & Almgard, L. E., *Eur. Urol.*, **3** (1977) 41.
14. Kallner, A., *Acta Med. Scand.*, **20** (1977) 283.
15. Kallner, A., Hartmann, D. & Hornig, D., *Am. J. Clin. Nutr.*, **32** (1979) 530.
16. Schmidt, K. H., Hagmaier, V., Hornig, D. H., Vuilleumier, J. P. & Rutishauser, G., *Am. J. Clin. Nutr.*, **34** (1981) 305.
17. Briggs, M., *Lancet,* **i** (1976) 154.
18. Demole, V., *Biochem. J.*, **28** (1934) 770.
19. Ohno, T. & Myoga, K., *Nutr. Rep. Int.*, **24** (1981) 291.
20. Neuweiler, W., *Int. Z. Vitaminforsch,* **22** (1951) 392.
21. Mouriquand, G. & Edel, V., *C.R. Soc. Biol.*, **148** (1954) 1422.
22. Samborskaya, E. P., *Byull. Eksp. Biol. Med.*, **57** (1964) 105.
23. Samborskaya, E. P. & Ferdman, T. D., *Byull. Eksp. Biol. Med.*, **62** (1966) 96.
24. Fahim, M. S., Hilderbrand, D., Wilson, R., Harman, J. M. & Hall, D. G., Fifth International Congress in Pharmacology, 1972, San Francisco, Abstract p. 66.

25. Alleva, F. R., Alleva, J. J. & Balazs, T., *Toxicol. Appl. Pharmacol.*, **35** (1976) 393.
26. Frohberg, H., Gleich, J. & Kieser, H., *Arzneim. Forsch.*, **23** (1973) 1081.
27. Stich, H. F., Karim, J., Koropatnick, J. & Lo, L., *Nature*, **260** (1976) 722.
28. Stich, H. F., Wei, L. & Whiting, R. F., *Cancer Res.*, **39** (1979) 4145.
29. Omura, H., Shinohara, K., Maeda, H., Nonaka, M. & Murakami, H., *J. Nutr. Sci. Vitaminol.*, **24** (1978) 185.
30. Norkus, E. P., Kuenzig, W. & Conney, A. H., *Mutat. Res.*, **117** (1983) 183.
31. MacRae, W. D. & Stich, H. F., *Toxicology*, **13** (1979) 167.
32. Rosin, M. P., San, R. H. C. & Stich, H. F., *Cancer Lett.*, **8** (1980) 299.
33. Galloway, S. M. & Painter, R. B., *Mutat. Res.*, **60** (1979) 321.
34. Speit, G., Wolf, M. & Vogel, W., *Mutat. Res.*, **78** (1980) 273.
35. Chauhan, P. S., Aravindakshan, M. & Sundaram, K., *Mutat. Res.*, **53** (1978) 166.
36. Amacher, D. E. & Paillet, S. C., *Cancer Lett.*, **14** (1981) 151.
37. National Toxicology Program, Carcinogenesis bioassay of L-ascorbic acid (vitamin C) in F344/N rats and B6C3F1 mice (feed study). March 1983, NIH Publication No. 83-2503.
38. Fukushima, S., Imaida, K., Sakata, T., Okamura, T., Shibata, M. & Ito, N., *Cancer Res.*, **43** (1983) 4454.
39. Fukushima, S., Kurata, Y., Shibata, M., Ikawa, E. & Ito, N., *Cancer Lett.*, **23** (1984) 29.
40. Fukushima, S., Ogiso, T., Kurata, Y., Shibata, M. & Kakizoe, T., *Cancer Lett.*, **35** (1987) 17.
41. Ito, N., Hirose, M., Fukushima, S., Tsuda, H., Shirai, T. & Takematsu, M., *Food Chem. Toxicol.*, **24** (1986) 1071.
42. Cameron, E., Pauling, L. & Leibowitz, B., *Cancer Res.*, **39** (1979) 663.
43. Joint FAO/WHO Expert Committee on Food Additives, *Tech. Rep. Ser. Wld Hlth Org.*, **751** (1987) 18.
44. Weber, F., Gloor, U., Wursch, J. & Wiss, O., *Biochem. Biophys. Commun.*, **14** (1964) 189.
45. Pearson, C. K. & Barnes, M. M., *Int. Z. Vitaminforsch.*, **40** (1970) 19.
46. Schmandke, H., Sima, C. & Maune, R., *Int. Z. Vitaminforsch.*, **39** (1969) 296.
47. Johnson, P. & Pover, W. F. R., *Life Sci.*, **1** (1962) 115.
48. Takahashi, Y., Urono, K. & Kimura, S., *J. Nutr. Sci. Vitaminol.*, **23** (1977) 201.
49. Machlin, L. F. & Gabriel, E., *Ann. NY Acad. Sci.*, **393** (1982) 48.
50. Schmandke, H. & Schmidt, G., *Int. Z. Vitaminforsch.*, **38** (1968) 75.
51. Kelleher, J. & Losowsky, M. S., *Brit. J. Nutr.*, **24** (1970) 1033.
52. MacMahon, M. T. & Neale, G., *Clin. Sci.*, **38** (1970) 197.
53. Hüter, F., *Z. Naturforsch.*, **2B** (1947) 414.
54. Costa, A., Cetini, G., Monteferrario, P. & Volterrani, O., Vitamina E, Atti 3 Congr. Intern., Venice, 1955, p. 233.
55. Valenti, G. & Bottarelli, E., *Folia Endocrinol.*, **18** (1965) 318.

56. Czyba, J. C., Girod, C. & Durand, N., *C.R. Soc. Biol.*, **160** (1966) 2101.
57. Abdo, K. M., Rao, G., Montgomery, C. A., Dinowitz, M. & Kanalingam, K., *Food Chem. Toxicol.*, **24** (1986) 1043.
58. Wheldon, G. H., Bhatt, A., Kelber, P. & Hummler, H., *Int. J. Vit. Nutr. Res.*, **53** (1983) 287.
59. Greenblatt, I. J., *Circulation*, **16** (1957) 508.
60. Informatics Inc., GRAS (Generally Recognised As Safe) food ingredients—tocopherols. NTIS PB-221 237, 1973.
61. Dymsza, H. A. & Park, J., *Fedn Proc. Fedn Am. Socs Exp. Biol.*, **34** (1975) 321.
62. Yang, N. Y. & Desai, I. D., *J. Nutr.*, **107** (1977) 1410.
63. Anon., *Nutr. Rev.*, **41** (1983) 268.
64. March, B. E., Wong, E., Seier, L., Sim, J. & Biely, J., *J. Nutr.*, **103** (1973) 371.
65. Schrogie, J. J., *J. Am. Med. Assoc.*, **232** (1975) 19.
66. Corrigan, J. J. & Ulfers, L. L., *Am. J. Clin. Nutr.*, **34** (1981) 1701.
67. Tomassi, G. & Silano, V., *Food Chem. Toxicol.*, **24** (1986) 1051.
68. Ichihara, I., *Okajimas Folia Anat. Jap.*, **43** (1967) 203.
69. Czyba, J. C., *C.R. Soc. Biol.*, **160** (1966) 765.
70. Telford, I. R., Woodruff, C. S. & Linford, R. H., *Am. J. Anat.*, **110** (1962) 29.
71. Food and Drug Research Laboratories, Teratologic evaluation of FDA 71-58 (dl-alpha-tocopheryl acetate) in mice, rats, hamsters and rabbits'. NTIS PB-233 809, 1973.
72. Martin, M. M. & Hurley, L. S., *Am. J. Clin. Nutr.*, **30** (1977) 1629.
73. Winkler, H., *Zent. Gynakol.*, **67** (1943) 32.
74. Solomon, D., Strummer, D. & Nair, P. P., *Ann. NY Acad. Sci.*, **203** (1972) 103.
75. Shklar, G., *J. Natl Cancer Inst.*, **68** (1982) 791.
76. Cook, M. G. & McNamara, P., *Cancer Res.*, **40** (1980) 1329.
77. Shamberger, R. J. & Rudolph, G., *Experientia*, **22** (1966) 116.
78. Kagerud, A. & Paterson, H. I., *Acta Radiol Oncol.*, **20** (1981) 97.
79. Kurek, M. P. & Corwin, L. M., *Nutr. Cancer*, **4** (1982) 128.
80. Newberne, P. M. & Suphakarn, K., *Nutr. Cancer*, **5** (1983) 107.
81. Moore, M. A., Tsuda, H., Thamavit, W., Masui, T. & Ito, N., *J. Natl. Cancer Inst.*, **78** (1987) 289.
82. Welsh, A. L., *Arch. Dermatol.*, **65** (1952) 137.
83. Welsh, A. L., *Arch. Dermatol.*, **70** (1954) 181.
84. Hillman, R. W., *Am. J. Clin. Nutr.*, **5** (1957) 597.
85. Cohen, H. M., *Calif. Med.*, **119** (1973) 72.
86. Briggs, M., *New Engl. J. Med.*, **290** (1974) 579.
87. Johnson, L., Bowen, F. W., Abbasi, S., Herrmann, N., Weston, M., Sacks, L., Porat, R., Stahl, G., Peckham, G., Delivoria-Papadopoulos, M., Quinn, G. C. & Schaffer, D., *Pediatrics*, **75** (1985) 619.
88. Lorch, V., Murphy, D., Hoersten, L. R., Harris, E., Fitzgerald, J. & Sinha, S. N., *Pediatrics*, **75** (1985) 598.
89. Joint FAO/WHO Expert Committee on Food Additives, *Tech. Rep. Ser. Wld Hlth Org.*, **653** (1980) 16.

90. Scientific Committee for Food, Reports of the Scientific Committee for Food, 5th Series. Commission of the European Communities, Brussels, 1978.
91. Joint FAO/WHO Expert Committee on Food Additives, *Tech. Rep. Ser. Wld Hlth Org.*, **751** (1987) 15.
92. Booth, A. N., Masri, M. S., Robbins, D. J., Emerson, O. H., Jones, F. T. & De Eds, F., *J. Biol. Chem.*, **234** (1959) 3014.
93. Dacre, J. C., *J. N.Z. Inst. Chem.*, **24** (1960) 161.
94. Van Esch, G. J., *Voeding*, **16** (1955) 683.
95. Orten, J. M., Kuyper, A. C. & Smith, A. H., *Food Technol.*, **2** (1948) 308.
96. Tollenaar, F. D., *Proc. Pacific Sci. Congr.*, **5** (1957) 92 (published 1963).
97. Johnson, A. R. & Hegwill, F. R., *Aust. J. Exp. Biol. Med. Sci.*, **39** (1961) 353.
98. Van der Heijden, C. A., Janssen, P. J. C. M. & Strik, J. J. T. W. A., *Food Chem. Toxicol.*, **24** (1986) 1067. (Refers to unpublished studies: 13-week rat and dog studies by Hazleton Laboratories Inc., 1969; 4-week rat study by J. Strik, 1986; two-generation study in rats by Hazleton Laboratories Inc., 1970; three-generation rat study by Industrial Bio-Test Laboratories Inc., 1970).
99. Allen, C. S. & De Eds, F. D., *J. Am. Oil Chem. Soc.*, **28** (1951) 394.
100. King, M. M. & McCay, P. B., *Food Cosmet. Toxicol.*, **19** (1981) 13.
101. Depner, M., Kahl, G. F. & Kahl, R., *Food Chem. Toxicol.*, **20** (1982) 507.
102. Sluis, K. J. H., *Food Manuf.*, **26** (1951) 99.
103. Tanaka, D., Kawashima, K., Nakaura, S., Nagao, S. & Omori, Y., *Shokuhin Eisegaki Zasshi*, **20** (1979) 378.
104. Food and Drug Research Laboratories, Teratologic evaluation of FDA 71-39 (propyl gallate). NTIS PB-223-816, 1973.
105. Litton Bionetics Inc., Mutagenic evaluation of compound FDA 71-39, propyl gallate. NTIS PB-245-441, 1974.
106. Rosin, M. P. & Stich, H. F., *J. Environ. Pathol. Toxicol.*, **4** (1980) 159.
107. Shelef, L. A. & Chin, B., *Appl. Environ. Microbiol.*, **40** (1980) 1039.
108. Park, S. C., Kim, I. & Tchai, B. S., *Korean J. Biochem.*, **13** (1981) 180.
109. Morita, K., Ishigaki, M. & Abe, T., *J. Soc. Cosmet. Chem.*, **15** (1981) 243.
110. Ishidate, M., Sofuni, T., Yoshikawa, K., Hayashi, M., Nohmi, T., Sawada, M. & Matsuoka, A., *Food Chem. Toxicol.*, **22** (1984) 623.
111. Mortelmans, K., Haworth, S. & Lawlor, T., *Environ. Mutagen.*, **8** (Suppl 7) (1986) 1.
112. Sasaki, M., Sugimura, K., Yoshida, M. A. & Abe, S., *La Kromosomo*, **II**(20) (1980) 574.
113. Kaul, B. L., *Mutat. Res.*, **67** (1979) 239.
114. Raj, A. S. & Katz, M., *Mutat. Res.*, **136** (1984) 247.
115. Kamra, O. P. & Bhaskar, G., *Mutat Res.*, **53** (1977) 207.
116. Calle, L. M. & Sullivan, P. D., *Mutat. Res.*, **101** (1982) 99.
117. Ben-hur, E. & Green, M., *J. Radiat. Res.*, **22** (1981) 250.

118. Sugimura, T., Sato, S. & Nagao, M., Overlapping of carcinogens and mutagens. In *Fundamentals in Cancer Prevention*, ed. P. N. Magee, University of Tokyo Press, Tokyo, 1976.
119. Yoshikawa, K., Uchino, M. & Kurata, H., *Eisei Shikenjo Hokoku*, **94** (1976) 28.
120. Wang, C. Y. & Klemencic, J. M., *Proc. Am. Soc. Cancer Res.*, **20** (1979) 117.
121. Yamaguchi, T., *Agric. Biol. Chem.*, **45** (1981) 327.
122. Haworth, S., Lawlor, T., Mortelmans, K., Speck, W. & Zeiger, E., *Environ. Mutagen.*, Suppl. 1 (1983) 3.
123. Rashid, K. A., Baldwin, I. T. & Babish, J. G., *J. Environ. Sci. Hlth*, **B20** (1985) 153.
124. Stich, H. F., Rosin, M. P., Wu, C. H. & Powrie, W. D., *Cancer Lett.*, **14** (1981) 251.
125. Mitra, A. B. & Manna, G. K., *Indian J. Med. Res.*, **59** (1971) 1442.
126. Rosin, M. P., *Mutat. Res.*, **135** (1984) 109.
127. Yamada, K., Shirahata, S. & Murakami, H., *Agric. Biol. Chem.*, **49** (1985) 1423.
128. Lehman, A. J., Fitzhugh, O. G., Nelson, A. A. & Woodard, G., *Adv. Food Res.*, **3** (1951) 197.
129. Clayson, D. B., Iverson, F., Nera, E., Lok, E., Rogers, C., Rodrigues, C., Page, D. & Karpinski, K., *Food Chem. Toxicol.*, **24** (1986) 1171.
130. Dacre, J. C., *Food Cosmet. Toxicol.*, **12** (1974) 125.
131. National Toxicology Program, Carcinogenesis bioassay of propyl gallate (CAS No 121-79-9) in F344/N rats and B6C3F1 mice (feed study). NIH Publication No 83-1796, 1982.
132. Gammal, E. B., Carroll, K. K. & Plunkett, E. R., *Cancer Res.*, **27** (1967) 1737.
133. McCay, P. B., King, M. M. & Pitha, J. V., *Cancer Res.*, **41** (1981) 3745.
134. King, M. M. & McCay, P. B., *Cancer Res.*, **43** (1983) 2485S.
135. Astill, B. D. & Mulligan, L. T., *Food Cosmet. Toxicol.*, **15** (1977) 167.
136. Mirvish, S., Cardesa, A., Wallace, L. & Shubik, P., *J. Natl Cancer Inst.*, **55** (1975) 633.
137. Scientific Committee for Food, Report of the Scientific Committee for Food on Tertiary Butyl Hydroquinone. EEC Commission Document III/26/82, Brussels, 1982.
138. Joint FAO/WHO Expert Committee on Food Additives, *Tech. Rep. Ser. Wld Hlth Org.*, **617** (1978) 14.
139. Joint FAO/WHO Expert Committee on Food Additives, *Tech. Rep. Ser. Wld Hlth Org.*, **751** (1987) 17.
140. Astill, B. D., Terhaar, C. J., Krasavage, W. J., Wolf, G. L., Roudabush, R. L., Fassett, D. W. & Morgareidge, K., *J. Am. Oil Chem. Soc.*, **52** (1975) 53.
141. Van Esch, G. J., *Food Chem. Toxicol.*, **24** (1986) 1063.
142. Krasavage, W. J., *Teratology*, **16** (1977) 31.
143. Giri, A. K., San, S., Talukdor, G., Sharma, A. & Banerjee, T. S., *Food Chem. Toxicol.*, **22** (1984) 459.

144. Nera, E. A., Lok, E., Iverson, F., Ormsby, E., Karpinski, K. F. & Clayson, D. B., *Toxicology,* **32** (1984) 197.
145. Altmann, H.-J., Grunow, W., Mohr, U., Richter-Reichhelm, H. B. & Wester, P. W., *Food Chem. Toxicol.,* **24** (1986) 1183.
146. Ito, N., Hirose, M., Fukushima, S., Tsuda, H., Shirai, T. & Takematsu, M., *Food Chem. Toxicol.,* **24** (1986) 1071.
147. Tamano, S., Fukushima, S., Shirai, T., Hirose, M. & Ito, N., *Cancer Lett.,* **35** (1987) 39.
148. Astill, B. D. & Mulligan, L. T., *Food Cosmet. Toxicol.,* **15** (1977) 167.
149. Kirkpatrick, D. C. & Lauer, B. H., *Food Chem. Toxicol.,* **24** (1986) 1035.
150. National Academy of Sciences, 1977 survey of industry on the use of food additives. NAS, Washington, D.C., 1977.
151. Joint FAO/WHO Expert Committee on Food Additives, *Tech. Rep. Ser. Wld Hlth Org.,* **751** (1987) 12.
152. Minegishi, K. T., Watanabe, M. & Yamaha, T., *Chem. Pharm. Bull. Tokyo,* **29** (1981) 1377.
153. Dacre, J. C., Denz, F. A. & Kennedy, T. H., *Biochem. J.,* **64** (1956) 777.
154. Astill, B. D., Fassett, D. W. & Roudabush, R. L., *Biochem. J.,* **75** (1960) 543.
155. Astill, B. D., Mills, J., Fassett, D. W., Roudabush, R. L. & Terhaar, C. J., *J. Agric. Food Chem.,* **10** (1962) 315.
156. Francois, A. C. & Pihet, A., *Ann. Inst. Natl Rech. Agron.,* Ser. **D9** (1960) 195.
157. Hodge, H. C., Fassett, D. W., Maynard, E. A., Downs, W. L. & Coye, R. D., *Toxicol. Appl. Pharmacol.,* **6** (1964) 512.
158. Wilder, O. H. M., Ostby, P. C. & Gregory, B. R., *J. Agric. Food Chem.,* **8** (1960) 504.
159. Verhagen, H., Thijssen, H. H. W., ten Hoor, F., & Kleinjans, J. C. S., *Fd Chem. Toxicol.,* **27** (1989) 151.
160. Daniel, J. W. & Gage, J. C., *Food Cosmet, Toxicol.,* **3** (1965) 405.
161. El-Rashidy, R. & Niazi, S., *Biopharm. Drug Dispos.,* **4** (1983) 389.
162. Armstrong, K. E. & Wattenburg, L. W., *Cancer Res.,* **45** (1985) 1507.
163. Cummings, S. W., Ansari, G. A. S., Guengerich, F. P., Crouch, L. S. & Prough, R. A., *Cancer Res.,* **45** (1985) 5617.
164. Hirose, M., Hagiwara, A., Inoue, K., Ito, N., Kaneko, H., Saito, K., Matsunaga, H., Isobe, N., Yoshitake, A. & Miyamoto, J., *Toxicology.* **53** (1988) 33.
165. Karplyuk, I. A., *Vop. Pitan.,* **18** (1959) 24.
166. Brown, W. D., Johnson, A. R. & O'Halloran, M. W., *Aust. J. Exp. Biol. Med. Sci.,* **37** (1959) 533.
167. Johnson, A. R. & Hegwill, F. R., *Aust. J. Exp. Biol. Med. Sci.,* **39** (1961) 353.
168. Feuer, G., Gaunt, I. F., Goldberg, L. & Fairweather, F. A., *Food Cosmet. Toxicol.,* **3** (1965) 457.
169. Gaunt, I. F., Feuer, G., Fairweather, F. A. & Gilbert, D., *Food Cosmet. Toxicol.,* **3** (1965) 433.

170. Hiraga, K., Hayashida, S., Ichikawa, H., Yoneyama, M., Fijii, T., Ikeda, T. & Yano, N., *Tokyo Toritsu Eisei Kenkyusho Kenkyu Nempo,* **22** (1970) 231.
171. Gilbert, D., Martin, A. D., Gangolli, S. D., Abraham, R. & Goldberg, L., *Food Cosmet. Toxicol.,* **7** (1969) 603.
172. Ford, S. M., Hook, J. B., Arata, D. & Bond, J. T., *Fed. Proc.,* **36** (1977) 1116.
173. Allen, J. R. & Engblom, J. F., *Food Cosmet. Toxicol.,* **10** (1972) 769.
174. Cha, Y. & Bueding, E., *Biochem. Pharmacol.,* **28** (1979) 1917.
175. Cha, Y. & Heine, S., *Cancer Res.,* **42** (1982) 2609.
176. Creaven, P. J., Davies, W. H. & Williams, R. T., *J. Pharm. Pharmacol.,* **18** (1966) 485.
177. Creaven, P. J. & de Rubertis, F. R., *Cancer Res.,* **37** (1977) 4088.
178. Yang, C. S., Sydor, W., Martin, M. B. & Lewis, K. F., *Chem. Biol. Interact.,* **37** (1981) 337.
179. Clegg, D. J., *Food Cosmet. Toxicol.,* **3** (1965) 387.
180. Foods and Drugs Research Laboratories, Teratologic evaluation of FDA 71–24 (butylated hydroxyanisole) in mice, rats and hamsters. NTIS PB-221 783, 1972.
181. Food and Drugs Research Laboratories, Teratologic evaluation of compound FDA 71–24 (butylated hydroxyanisole) in rabbits. NTIS PB-267 200, 1974.
182. Hansen, E. & Meyer, O., *Toxicology,* **10** (1978) 195.
183. Hansen, E., Meyer, O. & Olsen, P., *Toxicology,* **23** (1982) 79.
184. Vorhees, C. V., Butcher, R. E., Brunner, R. L., Wootten, F. & Sobotka, T. J., *Neurobehav. Toxicol. Teratol.,* **3** (1981) 321.
185. Stokes, J. D. & Scudeter, C. L., *Dev. Psychobiol.,* **7** (1974) 343.
186. Allen, J. R., *Arch. Environ. Hlth,* **31** (1976) 47.
187. Litton Bionetics Inc., Mutagenic evaluation of compound FDA 71-24, butylated hydroxyanisole. NTIS PB-245 510, 1975.
188. Litton Bionetics Inc., Mutagenic evaluation of compound FDA 71-24, butylated hydroxyanisole. NTIS PB-245 460, 1974.
189. Ishidate, M., Sofuni, T., Yoshikawa, K., Hayashi, M., Nohmi, T., Sawada, M. & Matsuoka, A., *Food Chem. Toxicol.,* **22** (1984) 623.
190. Joner, P. E., *Acta Vet. Scand.,* **18** (1977) 187.
191. Morita, K., Ishigaki, M. & Abe, T., *J. Soc. Cosmet. Chem. Jap.* **15** (1981) 243.
192. Williams, G. M., *Food Chem. Toxicol.,* **24** (1986) 1163.
193. Hageman, G. J., Verhagen, H. & Kleinjans, J. C. S., *Mutat. Res.,* **208** (1988) 207.
194. Tan, E. L., Schenley, R. L. & Hsie, A. W., *Mutat. Res.,* **103** (1982) 359.
195. Rogers, C. G., Nayak, B. N. & Heroux-Metcalf, C., *Cancer Lett.,* **27** (1985) 61.
196. Prasad, O. & Kamra, O. P., *Int. J. Radiat. Biol.,* **25** (1974) 67.
197. Miyagi, M. P. & Goodheart, C. R., *Mutat. Res.,* **40** (1976) 37.
198. Ishidate, M. & Odashima, S., *Mutat. Res.,* **48** (1977) 337.
199. Abe, S. & Sasaki, M., *J. Natl Cancer Inst.,* **58** (1977) 1635.
200. Wilder, O. H. M. & Kraybill, H. R., *Fed. Proc.,* **8** (1949) 165.

201. Graham, W. D. & Grice, H. C., *J. Pharm. Pharmacol.*, **7** (1955) 1126.
202. Ito, N., Fukushima, S., Hagiwara, A., Shibata, M. & Ogiso, T., *J. Natl Cancer Inst.*, **70** (1983) 343.
203. Ito, N., Fukushima, S., Tamano, S., Hirose, M. & Hagiwara, A., *J. Natl Cancer Inst.*, **77** (1986) 1261.
204. Ito, N., Fukushima, S., Imaida, K., Sakata, T. & Masui, T., *Gann,* **74** (1983) 459.
205. Masui, T., Hirose, M., Imaida, K., Fukushima, S., Tamano, S. & Ito, N., *Jpn J. Cancer Res.*, **77** (1986) 1083.
206. Iverson, F., Lok, E., Nera, E. A., Karpinski, K. & Clayson, D. B., *Toxicology,* **35** (1985) 1.
207. Hirose, M., Masada, A., Kurata, Y., Ikawa, E., Mera, Y. & Ito, N., *J. Natl Cancer Inst.*, **76** (1986) 143.
208. Grice, H. C., *Food Chem. Toxicol.* **26** (1988) 717.
209. Ikeda, G. J., Stewart, J. E., Sapienza, P. P., Peggins, J. O., Michel, T. C., Olivito, V., Alam, H. Z. & O'Donnell, M. W., *Food Chem. Toxicol.*, **24** (1986) 1201.
210. Tobe, M., Furuya, T., Kawasaki, Y., Naito, K., Sekita, K., Matsumoto, K., Ochiai, T., Usui, A., Kokubo, T., Kanno, J. & Hayashi, Y., *Food Chem. Toxicol.*, **24** (1986) 1223.
211. Würtzen, G. & Olsen, P., *Food Chem. Toxicol.*, **24** (1986) 1229.
212. Iverson, F., Truelove, J., Nera, E., Wong, J., Lok, E. & Clayson, D. B., *Cancer Lett.*, **26** (1985) 43.
213. Hirose, M., Asamoto, M., Hagiwara, A., Ito, N., Kaneko, H., Saito, K., Takamatsu, Y., Yoshitake, A. & Miyamoto, J., *Toxicology.*, **45** (1987) 13.
214. Hirose, M., Inoue, T., Masuda, A., Tsuda, H. & Ito, N., *Carcinogenesis.*, **8** (1987) 1555.
215. de Stefany, C. M., Prabhu, U. D. G., Sparnins, V. L. & Wattenberg, L. W., *Food Chem. Toxicol.*, **24** (1986) 1149.
216. International Agency for Research on Cancer, Butylated hydroxyanisole. IARC Monographs, Vol. **40,** (1986), 123.
217. Wattenberg, L. W., *Food Chem. Toxicol.*, **24** (1986) 1099.
218. Mizuno, M., Ohara, A., Danno, G., Kanazawa, K. & Natake, M., *Mutat. Res.*, **176** (1987) 179.
219. Hirose, M., Hagiwara, A., Masui, T., Inoue, K. & Ito, N., *Cancer Lett.*, **30** (1986) 169.
220. Hirose, M., Masuda, H., Tsuda, H., Uwagawa, S. & Ito, N., *Carcinogenesis.*, **8** (1987) 1731.
221. Fukushima, S., Sakata, T., Tagawa, Y., Shibata, M.-A., Hirose, M. & Ito, N., *Cancer Res.* **47** (1987) 2113.
222. Warner, C. R., Brumley, W. C., Daniels, D. H., Joe, F. L. & Fazio, T., *Food Chem. Toxicol.*, **24** (1986) 1015.
223. Joint FAO/WHO Expert Committee on Food Additives, *Tech. Rep. Ser. Wld Hlth Org.*, **751** (1987) 14.
224. Matsuo, M., Mihara, K., Okuno, M., Ohkawa, H. & Miyamoto, J., *Food Chem. Toxicol.*, **22** (1984) 345.
225. Collings, A. J. & Sharratt, M., *Food Cosmet. Toxicol.*, **8** (1970) 409.

226. Yamamoto, K., Tajima, K. & Mizutani, T., *J. Pharm. Dyn.*, **2** (1980) 164.
227. Tajima, K., Yamamoto, K. & Mizutani, T., *Chem. Pharm. Bull.*, **29** (1981) 3738.
228. Akagi, M. & Aoki, I., *Chem. Pharm. Bull.*, **10** (1962) 101.
229. Aoki, I, *Chem. Pharm. Bull.*, **10,** (1962) 105.
230. Branen, A. L., *J. Am. Oil Chem. Soc.*, **52** (1975) 54.
231. Daniel, J. W., Gage, J. C. & Jones, D. I., *Biochem. J.*, **106** (1968) 783.
232. Wiebe, L. I., Mercier, J. R. & Ryan, A. J., *Drug Metab. Dispos.*, **6** (1978) 296.
233. El-Rashidy, R. & Niazi, S., *J. Pharm. Sci.*, **69** (1980) 1455.
234. Dacre, J. C., *J. N. Z. Inst. Chem.*, **24** (1960) 161.
235. Deichmann, W. B., Clemmer, J. J., Rakoczy, R. & Bianchine, J., *AMA Arch. Ind. Hlth,* **11** (1955) 93.
236. Day, A. J., Johnson, A. R., O'Halloran, M. W. & Schwartz, G. J., *Aust. J. Exp. Biol. Med. Sci.*, **37** (1959) 295.
237. Brown, W. D., Johnson, A. R. & O'Halloran, M. W., *Aust. J. Exp. Biol. Med. Sci.*, **37** (1959) 533.
238. Lane, B. P. & Lieber, C. S., *Lab. Invest.*, **16** (1967) 342.
239. Botham, C. M., Conning, D. M., Hayes, J., Litchfield, M. H. & McElligott, T. F., *Food Cosmet. Toxicol.*, **8** (1970) 1.
240. Kawano, S., Nakao, T. & Hiraga, K., *Jap. J. Pharmacol.*, **30** (1980) 861.
241. Kawano, S., Nakao, T. & Hiraga, K., *Jap. J. Pharmacol.*, **31** (1981) 459.
242. Saccone, G. T. P. & Pariza, M. W., *Cancer Lett.*, **5** (1978) 145.
243. Powell, C. J., Connelly, J. C., Jones, S. M., Grasso, P. & Bridges, J. W., *Food Chem. Toxicol.*, **24** (1986) 1131.
244. Nakagawa, K., Tayama, K. & Hiraga, K., *Toxicol. Lett.*, **31** (Suppl) (1986) P6–19.
245. Takahashi, O. & Hiraga, K., *Food Cosmet. Toxicol.*, **17** (1979) 451.
246. Tajima, K., Yamamoto, K. & Mizutani, T., *Chem. Pharm. Bull.*, **29** (1981) 3738.
247. Nakagawa, Y., *Toxicol. Lett.*, **37** (1987) 251.
248. Mizutani, T., Nomura, H., Nakanishi, K. & Fujita, S., *Toxicol. Appl. Pharmacol.*, **87** (1987) 166.
249. Kahl, R., *Toxicology,* **33** (1984) 185.
250. Marino, A. A. & Mitchell, J. T., *Proc. Soc. Exp. Biol. Med.*, **140** (1972) 122.
251. Saheb, W. & Witschi, H., *Toxicol. Appl. Pharmacol.*, **33** (1975) 309.
252. Omaye, S. T., Reddy, K. A. & Cross, C. E., *J. Toxicol. Environ. Hlth,* **3** (1977) 829.
253. Adamson, I. Y. R., Bowden, D. H., Cote, M. G. & Witschi, H., *Lab. Invest.*, **36** (1977) 26.
254. Mizutani, T., Ishida, I., Yamamoto, K. & Tajima, K., *Toxicol. Appl. Pharmacol.*, **62** (1982) 273.
255. Fujii, T., Nakagawa, Y., Fukumori, N., Sakamoto, Y. & Hiraga, K., *Jap. J. Pharmacol. (Suppl.)*, **25** (1975) 29.
256. Suzuki, H., Hiraga, K. & Nakao, T., *Tokyo Eiken Nempo,* **26** (1975) 49.
257. Takahashi, O. & Hiraga, K., *Tokyo Eiken Nempo,* **27** (1976) 86.

258. Takahashi, O., Takahashi, A., Hiraga, K. & Nakao, T., *J. Toxicol. Sci. Suppl.*, **2** (1977) 66.
259. Takahashi, O., Hayashida, S. & Hiraga, K., *Food Cosmet. Toxicol.*, **18** (1980) 299.
260. Takahashi, O. & Hiraga, K., *Toxicol. Appl. Pharmacol.*, **43** (1978) 399.
261. Takahashi, O. & Hiraga, K., *Food Cosmet. Toxicol.*, **16** (1978) 475.
262. Takahashi, O. & Hiraga, K., *J. Nutr.*, **109** (1979) 453.
263. Suzuki, H., Nakao, T. & Hiraga, K., *Toxicol. Appl. Pharmacol.*, **50** (1979) 261.
264. Suzuki, H., Nakao, T. & Hiraga, K., *Toxicol. Appl. Pharmacol.*, **67** (1983) 152.
265. Takahashi, O. & Hiraga, K., *Food Chem. Toxicol.*, **22** (1984) 97.
266. Takahashi, O. & Hiraga, K., *Toxicol. Appl. Pharmacol.*, **46** (1978) 871.
267. Takahashi, O., *Food Chem. Toxicol.*, **25** (1987) 219.
268. Takahashi, O., *Arch. Toxicol.*, **62** (1988) 325.
269. Johnson, A. R., *Food Cosmet. Toxicol.*, **3** (1965) 371.
270. Frawley, J. P., Kohn, F. E., Kay, J. H. & Calandra, J. C., *Food Cosmet. Toxicol.*, **3** (1965) 377.
271. Food and Drug Research Laboratories, Teratologic evaluation of FDA 71-25 (butylated hydroxytoluene—Ionol). NTIS PB-221-782, 1972.
272. Food and Drug Research Laboratories, Teratologic evaluation of compound FDA 71-25, butylated hydroxytoluene (Ionol) in rabbits. NTIS, PB-267-201, 1974.
273. Voorhees, C. V., Butcher, R. E., Brunner, R. L. & Sobotka, T. J., *Food Cosmet. Toxicol.*, **19** (1981) 153.
274. Meyer, O. & Hansen, E., *Toxicology*, **16** (1980) 247.
275. Stanford Research Institute, Study of the mutagenic effects of Ionol C. P. (butylated hydroxytoluene). NTIS PB-221-827, 1972.
276. Litton Bionetics Inc., Mutagenic evaluation of butylated hydroxytoluene. NTIS PB-245-487, 1975.
277. Hiraga, K., *Tokyo Metropol. Res. Lab. Public Hlth*, **32** (1977) 1.
278. Bruce, W. R. & Heddle, J. A., *Can. J. Genet. Cytol.*, **21** (1979) 319.
279. Bonin, A. W. & Baker, R. S. U., *Food Technol. Aust.*, **32** (1980) 608.
280. Kinae, N., Hashizume, T. & Makita, T., *Water Res.*, **15** (1981) 17.
281. Reddy, B. S., Hanson, D., Matthews, L. & Sharma C., *Food Chem. Toxicol.*, **21** (1983) 129.
282. Williams, G. M., Shimada, T., McQueen, C., Tong, C. & Brat, S. V., *The Toxicologist*, **4** (1984) 104.
283. Paschin, Y. V. & Bahitova, L. M., *Mutat. Res.*, **137** (1984) 57.
284. McGregor, D. B., Brown, A., Cattanach, P., Edwards, I., McBride, D. & Caspary, W. J., *Environ. Molec. Mutagen.*, **11** (1988) 91.
285. Sciorra, L. J., Kaufmann, B. N. & Maier, R., *Food Cosmet. Toxicol.*, **12** (1974) 33.
286. Patterson, R. M., Keith, L. A. & Stewart, J., *Toxicol. in vitro.*, **1** (1987) 55.
287. Harman, D., Curtis, H. J. & Tilley, J., *J. Gerontol.*, **25** (1970) 17.
288. Cumming, R. B., Walton, M. F., Kelly, E. M. & Russel, W. L., *Mutat. Res.*, **115** (1983) 293.

289. Epstein, S. S. & Shafner, H., *Nature,* **219** (1968) 385.
290. Scheu, C. W., Cain, K. T., Rushbrook, C. J., Jorgenson, T. A. & Generoso, W. M., *Environ. Mutagen.,* **8** (1986) 357.
291. Stanford Research Institute, Study of the mutagenic effects of butylated hydroxytoluene (71-25) by the dominant lethal test. NTIS PB-278-026, 1977.
292. Kamra, O. P., *Int. J. Radiat. Biol.,* **23** (1973) 295.
293. Clapp, N. K., Tyndall, R. L., Cumming, R. B. & Otten, J. A., *Food Cosmet. Toxicol.,* **12** (1974) 367.
294. Brooks, T. M., Hunt, P. F., Thorpe, E. & Walker, A. T., Unpublished results listed in *Fed. Register,* **42** (1974) 27603.
295. National Institutes of Health, Bioassay of butylated hydroxytoluene (BHT) for possible carcinogenicity. NCI-CG-TR-150, 1979.
296. Shirai, T., Hagiwara, A., Kurata, Y., Shibata, M., Fukushima, S. & Ito, N., *Food Chem. Toxicol.,* **20** (1982) 861.
297. Lindenschmidt, R. C., Tryka, A. F., Goad, M. E. & Witschi, H. P., *Toxicology,* **38** (1986) 151.
298. Inai, K., Kobuke, T., Nambu, S., Takemoto, T., Kou, E., Nishina, H., Fujihara, M., Yonehara, S., Suehiro, S., Tsuya, T., Horiuchi, K. and Tokuoka, S., *Jpn J. Cancer Res.* (*Gann*). **79** (1988) 49.
299. Olsen, P., Meyer, O., Billie, N. & Würtzen, G., *Food Chem. Toxicol.,* **24** (1986) 1.
300. Deichmann, W. B., Clemmer, J. J., Rakoczy, R. & Bianchine, J., *AMA Arch. Ind. Hlth,* **11** (1955) 93.
301. Hirose, M., Shibata, M., Hagiwara, A., Imaida, K. & Ito, N., *Food Cosmet. Toxicol.,* **19** (1981) 147.
302. Olsen, P., Billie, N. & Meyer, O., *Acta Pharmacol. Toxicol.,* **53** (1983) 433.
303. International Agency for Research on Cancer, Butylated hydroxytoluene. IARC Monograph, Vol. 40, 1986, p. 161.
304. Witschi, H. P., *Food Chem. Toxicol.,* **24** (1986) 1127.
305. Thompson, D. C. & Trush, M. A., *Food Chem. Toxicol.,* **24** (1986) 1189.
306. Shlian, D. M. & Goldstone, J., *New Engl. J. Med.* **314** (1986) 648.
307. Graham, W. D., Teed, H. & Grice, H. C., *J. Pharm. Pharmacol.,* **6** (1954) 534.
308. Joint FAO/WHO Expert Committee on Food Additives, *Tech. Rep. Ser. Wld Hlth. Org.* **776** (1989) 14.
309. Daniel, J. W., Gage, J. C., Jones, D. I. & Stevens, M. A., *Food Cosmet. Toxicol.,* **5** (1967) 475.

INDEX

Abietic acid, 162
Acceptable daily intake (ADI), 255–6, 261, 266, 267
Acetaldehyde, 221
Acetobacter suboxidans, 123
Active Oxygen Method (AOM), 50
Addictive compounds, 221–2
Aesculetin, 175
Aflatoxin B_1, 201
Ageing, 225
Air pollution, 222
Alanine aminotransferase, 263
Alcohol effects, 221–2
Alkaline phosphatase, 263
Alkyl radicals, 4
Alkylperoxy radicals, 4
Alzheimer syndrome, 227
Amines, synergistic effect of, 89–91
Amino acids, 185–7
  antioxidant activity, 128
  synergistic effects, 91–2
Aniline hydroxylase, 202
Antioxidants
  action in vitro, 1–18
  biological systems, 196–217
  detection and determination methods for, 23–50
  EEC approach on safety, 23
  essential requirements of, 101
  estimation of activity, 52
  evaluation methods, 50–2
  general rules for use of, 100–1
  losses of, 23
Antioxidants—*contd.*
  most commonly used, 21
  multiple functions, with, 14–15
  permitted for food use, 101–35
  primary, 4–8
    examples of, 20
  radical trapping, 7–8
  review of types, 20
  role in food systems, 19–64
  secondary (preventive), 8–15, 20
  see also Natural antioxidants; Synthetic antioxidants
Apocarotenal, 159, 160
Arteriosclerosis, 144
Arthritis, inflammation, 224
Ascorbates, 155
  synergistic effects, 136–8, 138
Ascorbic acid, 11, 113–27, 141, 151, 158, 164, 197–200, 228, 232, 256–61
  absorption, metabolism and excretion, 256–8
  analytical methods, 46–7
  biosynthesis of, 120–1
  chemical properties, 115
  chemical synthesis, 121
  contents of foods, 120
  decomposition pathways, 118
  erythorbic acid, versus, 113–16
  extraction from natural sources, 121
  industrial methods for, 124–5
  long-term carcinogenicity studies, 259–60

Ascorbic acid—*contd.*
mechanisms of antioxidant contribution of, 199
mutagenicity, 259
occurrence in nature, 117–20
physical properties, 115, 116
production of, 121
pro-oxidant effect, 140–1
properties of, 113
reactions of, 116–17
reproduction studies, in, 258–9
short-term studies, 258–9
sterochemical configuration, 114
studies in humans, 260
synergistic effect of, 86–7
see also Vitamin C
Ascorbic acid-dehydroascorbic acid redox system, 119
Ascorbyl palmitate, 11, 12, 136, 143, 145, 158
solubility of, 127
synergistic effects, 138
synthesis and properties of, 125–7
Aspartate aminotransferase, 263
Atherosclerosis, 226
ATP, 206
Autoxidation
fatty acids, 142
initiation of, 1–3
mechanism of, 2
propagation and termination of, 3–4
Δ^5-avenasterol, 13

B. subtilis, 269, 279
Batten syndrome, 227
Beer, 159–60
Benzo(a)pyrene, 201, 222
Benzo(a)pyrene hydroxylase, 202
Benzoic acid, 5
Beverages, 158–60
Biological effects, 193–251
Biological oxidative environment, 194–6
1,2-bis(γ-tocopherol-5′-yl)ethane (α-TED), 69
Brussels sprouts, 174
Butylated hydroxyanisole (BHA), 253, 254, 255, 276–85, 295
absorption, metabolism and excretion, 277
BHT mixture, and, 76–7, 84
carcinogenicity and related studies, 279–84
combined antioxidant, as, 92–4
degradation of, 68–9
interactions with known carcinogens, mutagens and other antioxidants, 284–5
mutagenicity, 279
oxidation products, 79–81
reproduction studies, in, 278–9
short-term studies, 278–9
Butylated hydroxytoluene (BHT), 187, 253–5, 265, 275, 286–97
absorption, metabolism and excretion, 286–7
BHA mixture, and, 76–7, 84
blood, in, 289–90
combined antioxidant, as, 92–4
degradation of, 66–8
interactions with known carcinogens and mutagens, 295–6
liver, in, 287–8
long-term/carcinogenicity studies, 292–5
lung, in, 288–9
mutagenicity, 291–2
oxidation products, 79
PG mixture, and, 77
reproduction studies, in, 290–1
short-term studies, 287–92
Butylhydroquinone (TBHQ), 254
2-*t*-butyl-4-methoxyphenol (BHA), 5

Caffeic acids, 182
Cancer, 228–9, 253, 260, 265, 270–1, 279–84, 292–5
Canthaxanthin, 220, 231
Capillary column gas–liquid chromatography (CCGLC), tocopherols, 31

Cardiovascular diseases, 226–7
Carnosic acid, 133
  antioxidant activity of, 134
  synergistic effects, 138
Carotene, 217
B-Carotene, 8, 12, 149, 159, 160, 211–13, 228, 231, 232
Carotene discoloration method, 51
Carotenoids, 148–9, 211–18, 223, 228, 229, 232
Catalase, 12
Cellular lesions, 195
Chalcones, 175, 183, 203
Chelating agents
  colorimetric methods, 49
  examples of, 20–1
Chelators, analytical methods, 47–50
Chewing gum, 162–4
Chickens, 154
Chlorophyll, 7
Cholesterol, 144–6
  oxidation products of, 145
Chromatographic methods
  synthetic antioxidants, 40–5
  tocopherols, 29–37
Cigarette smoking, 221, 228
Cinnamic acids, 175, 181, 182
Citric acid, 10, 11, 158
  gas liquid chromatography (GLC), 49
  synergistic effect of, 86
Citrostadienol, 13
Cod liver oil, 147
Collagen, 224
Colophony, 162, 164
Colorimetric methods
  chelating agents, 49
  synthetic antioxidants, 38–40
  tocopherols, 28
Confectionery products, 162–4
Conjugated double bonds (CDB), 215
Cooking oil, 143
Crude vegetable oils, 128
Cysteine, 228

Deep frying oils, 143
2,6-di-*t*-butyl-4-methylphenol (BHT), 6
2,6-dichlorophenol-indophenol
  ascorbic acid, 46
  oxygen scavengers, 46
Diethylnitrosamine, 260
Differential pulse polarography
  ascorbic acid, 46
  oxygen scavengers, 46
Dihydrocaffeic acid, 183
Dihydrochalcones, 175
Dihydroquercetin, 179
1,2-dihydroxybenzene derivatives, 6
Dimethyl sulphoxide (DMSO), 226
7,12-dimethylbenz(a)anthracene, 260
1,2-dimethylhydrazine, 260
Dipalmitoyl phosphatidyl ethanolamine, 88
DNA damage, 269–70, 284
Dose requirements, 231
DPPH, 86, 92–3
Drosophila, 279, 292

EDTA, 10, 141
Electrochemical methods, tocopherols, 28–9
Environmental contaminants, 222
Enzymatic browning
  fruits, 155–7
  vegetables, 155–7
Enzymic antioxidants, 12
  examples of, 20
Erythorbic acid, 11, 116
  ascorbic acid, versus, 113–6
  sterochemical configuration, 114
Ester gum, 163
Ethoxyquin
  determination of, 49–50
  high performance liquid chromatography (HPLC), 50
Ethyl protocatechuate (EP)
  antioxidant activity, 82
  degradation of, 73–5
N-ethyl-N-hydroxyethylnitrosamine, 260

Ethylenediaminetetraacetic acid (EDTA), 10, 141
European Economic Community (EEC), 255–6
Extraction procedures, synthetic antioxidants, 38

Fats, 142–4
Fatty acids, autoxidation, 142
Fish and fish products, 154–5
Flavanones, 175
  antioxidant activity of, 178
Flavones, 175, 183
  antioxidant activity of, 177
Flavonoids and flavonoid compounds, 129–30, 173–95, 200–4, 231
  antioxidant activity, 174, 180
  mutagenic effects of, 130
  sources of, 179
  structure–activity relationships, 176–9
  structures of, 175
  synergistic effect, 188
Flavonols, 174, 202
  antioxidant mechanism, 176
Flavonones, 203
Flavour
  oxidation effects, 160
  threshold of oxidation products, 99
Fluorometric methods
  ascorbic acid, 46
  oxygen scavengers, 46
Food toxins, 223
Free radicals, 195, 196, 229
Fruits, 155–8
  enzymatic browning, 155–7
Fucosterol, 13

Gallates, 266–72
  absorption, metabolism and excretion, 267–8
  long term/carcinogenicity studies, 270–1
  mutagenicity, 269–70
  reproduction studies, in, 268–9
  short-term toxicity studies, 268–70
Gallic acid, 183
Gas–liquid chromatography (GLC)
  citric acid, 49
  synthetic antioxidants, 41
  tocopherols, 29–31
Gastro-enterology, 227
Ginkgo biloba, 202
Glucose-6-phosphate dehydrogenase (G6PD) deficiency, 199, 220
Glutathione, 200, 203, 216, 218, 228
Glutathione-peroxidase activity, 227
Gold compounds, 224
Guaiac gum, 129
Guaiaconic acids, 129

Haemoglobinopathies, 220
Hammett equation, 5
Herbicides, 223
Herbs, 187–8
  antioxidants, as, 131–2
  examples of, 21
Hesperitin, 202
High density lipoprotein (HDL), 220, 226
High performance liquid chromatography (HPLC)
  ascorbic acid, 46
  ethoxyquin, 50
  oxygen scavengers, 46
  synthetic antioxidants, 45
High-performance liquid chromatography (HPLC), tocopherols, 31–7
HRP/H_2O_2/halide system, 225
Hydrogen peroxide, 12
Hydrogen radicals, 65
Hydroperoxides, 2–4, 8, 10, 12, 107, 193, 226
Hydroquinone, 6–7, 140
Hyperlipoproteinaemia, 220
Hyperoxygenation, 219
Hypersiderosis, 220
Hypervitaminosis A, 231
Hypervitaminosis E, 266

Immunity, 225
Induction Period (IP), 50–1

Infection, 225
Insecticides, 223
Intrauterine contraceptive devices, 223
Iron preparations, 222
Isoascorbic acid, 11
Isoflavones, 175, 184–5

Joint FAO/WHO Expert Committee on Food Additives (JECFA), 255–6, 258, 261, 266, 267

Kallikrein-kinin system, 289
Kämpferol, 130
Keshan disease, 229–30
Krill, antioxidant principle of, 128

Lard, 172
  antioxidant activity in, 83–5, 88
  antioxidant tests in, 136
Larrea divaricata, 129
Lecithins, analytical methods, 47–50
D-limonene, 160
Linoleic acid, 107
Lipid oxidation, 20
Lipid peroxidation, 207, 221, 225
Lipid peroxides, 226
Lipid radicals, 5, 6
Lipid-soluble substances, 99
Lipids, 99
Lipoproteins, 261
Low density lipoprotein (LDL), 220, 226

Maillard reaction products, 14–15, 135
Malonaldehyde (MDA), 196
Malonyldialdehyde (MDA), 202, 206–10, 216, 222, 226, 227
Margarines, 10
Meat and meat products, 149–54
  colour change in, 151
  cured meat, 152–3
Meat and meat products—*contd.*
  fresh meat, 150–1
  rancidity problem, 153
Melanoidins, 91–2
Metabolic disorders, 219–21
Metal ions, 9
2-methoxyphenols, 7
Methyl linoleate, antioxidant activity in, 87
Methyl oleate, antioxidant activities in, 83–5
Methyl silicone, 13–14
N-methyl-N′-nitro-N-nitroso-guanidine (MNNG), 260
Methyllinoleate, vitamin A-palmitate retention, 147
N-methylnitrosourea (MNU), 260
Methyl-tocopherylquinone, 107
Milk powder, 155
Mycoplasma pneumoniae, 209
Myocardial infarction, 144
Myoglobin molecule, 149

Natural antioxidants, 21, 99–170
  commercially exploited, 101–2
  commercially, not exploited, 171–91
  methods of functioning, 173
  pro-oxidant effect, 138–41
  protection of foods with, 141–65
  reasons for non-exploitation, 188–9
  sources of, 172
  substances of interest, 135–8
  synergistic effects, 135–8
Nephelometric (turbidimeter) methods, phosphatides, 48
Nervous system, 227
Nitrosamines, 154
N-nitrosamines, 223
Nordihydroguaiaretic acid (NDGA), 129
Nuts, protection of, 157

Oat products, 128
Off-flavours, 10, 153–4
Oils, 142–4

Orange drinks, 159, 160
Orange oil, 160
Oxidation in disease, 218
Oxidation reactions, 193
Oxygen radicals, 225
Oxygen scavengers
  analytical methods, 11–12, 46–7
  examples of, 20

Pancreatitis, 227
Paper chromatography (PC), synthetic antioxidants, 40
Parkinson's disease, 227
Peppermint oil, 161
Peptides, 185–7
  antioxidant activity, 128
Peroxidation procedures, 194, 195
Peroxyl radicals, 8
Pesticides, 223
Phenol, 5, 6
Phenolic acids, 181–3
Phenolic antioxidants, 7, 65–98
  antioxidant activity of oxidation products of, 79–85
  degradation of, 65–76
  mixed, 76–7
  pro-oxidant effect, 138–41
  synergism with non-phenolic compounds, 85–92
Phenoxyl radical, 6, 7
Phosphates, 10
Phosphatides
  analytical methods, 47–50
  nephelometric (turbidimeter) methods, 48
Phosphatidyl choline (PC), 14, 88
Phosphatidyl ethanolamine (PE), 14, 88
Phosphatidyl inositol (PI), 14
Phosphatidylcholine, high performance liquid chromatography (HPLC), 49
Phospholipids, 14
  analytical methods, 47–50
  synergistic effect of, 87–9
  thin-layer chromatography, 48
Phosphoric acid derivatives, 10
Phosphorus, qualitative and quantitative determination of, 48
Pigs, 154
Plate diffusion method, 52
Polydimethyl siloxane, 13–14
Polyphosphates, 10
Pork products, 154
Pork sausages, 151
Potato powders, deterioration of, 157
Preserves, oxidative deterioration, 157
Pro-oxidant effect, 138–41
Propyl gallate, (PG)
  antioxidant activity of oxidation product, 83
  BHA mixture, and, 77
  combined antioxidant, as, 92–4
  irradiation of, 76
Prostaglandins, 219
Proteins, antioxidant activity, 128
Provitamin D_3, 148

Quercetin, 173, 179, 202
Quinone, 6–7

Radiation, 223–4
Rancidity, 172
Rancimat test, 50
Recommended daily allowances (RDA), 231
Reducing agents, 11–12
Reductones, 14
Resonance-stabilised carbon-centred radical, 8
Respiratory reactions occurring in biological cell, 193
Retinoids, 213
Reversed phase chromatographic procedures, synthetic antioxidants, 45
Rhamnetin, 130
Rheumatism, 224
Rhodopseudomonas spheroides, 214
Rosemary, 132, 160
  antioxidant activity of, 134

Rosemary—*contd.*
processes for recovery of, 133
synergistic effects, 138
Rosmaric acid, 133
Rutin, 130, 202

Saccharomyces cerevisiae, 291
Safety issues, 231
Sage, 132
Salami, stabilization of, 153
Salmonella typhimurium, 259, 269, 274, 279, 291
Sarcina lutea, 214, 215
Sarcina phytoene, 215
Sarcina phytofluene, 215
Sausages, 151, 153
Schaal oven test, 50
Scientific Committee for Food (SCF), 255–6, 266
Selenium, 218, 226, 227, 229–32
Sequestering agents, 9–10
examples of, 20–1
Sesamol
antioxidant activity, 81
oxidation of, 73
Sesamol dimer, antioxidant activity, 81
Singlet oxygen quenchers, 12
Sodium citrate, 155
Sodium erythorbate, 11
Soft drinks, 158–9
Soybean oil, antioxidant activity in, 83, 88
Soybean products, 127–8
Soybeans, 181, 182, 184–5
Spectrophotometric methods
ascorbic acid, 46
oxygen scavengers, 46
Spices, 132, 187–8
antioxidant activity of, 132–4
antioxidants, as, 131–2
commercial products, 134
examples of, 21
freedom from odour and taste, 135
manufacture of, 132
protection factor in various foods, 131
Spinal cord degeneration, 227
Sterol antioxidants, 13–14
Stress effects, 221
Superoxide dismutase, 12, 207
Superoxide radicals, 9, 12
Synergistic effects, 85–94
see also under specific items
Synergists, analytical methods for, 47–50
Synthetic antioxidants
analytical methods for, 37–45
chromatographic methods, 40–5
colorimetric methods, 38–40
extraction procedures, 38
gas liquid chromatography (GLC), 41
high performance liquid chromatography (HPLC), 45
paper chromatography (PC), 40
reversed phase chromatographic procedures, 45
thin-layer chromatography (TLC), 40
UV visible spectrophotometric methods, 38–40

Tertiary-butyl hydroquinone (TBHQ), 13, 272–6, 277, 284
absorption, metabolism and excretion, 272–3
antioxidant activities of oxidation products, 82
liver, in, 273
long-term/carcinogenicity studies, 274–6
mutagenicity, 274
oxidation of, 75
reproduction studies, in, 273–4
short-term studies, 273–4
Thalassaemia, 220
Thin-layer chromatography, (TLC)
phospholipids, 48
synthetic antioxidants, 40
tocopherols, 29–30
Thiocyanate method, 51–2

Thromboxane A, 219
Titrimetric methods
ascorbic acid, 46
oxygen scavengers, 46
Tocols, 26–37, 102
structure of, 102
Tocopherol-selenium-glutathione enzyme system, 207–11, 232
5-(δ-tocopherol-5-yl)-δ-tocopherol (δ-TBD), 70–1
Tocopherols, 20, 102–12, 158, 199, 200, 204–6, 210–11, 221, 223, 232, 261–6
absorption, metabolism and excretion, 261–2
analytical methods for, 26–37
antioxidant activity of, 106–9
antioxidant-pro-oxidant properties, 138–40
approximate content in vegetable oils, 105
biosynthesis of, 109
blood clotting, 263–4
capillary column gas–liquid chromatography (CCGLC), 31
chemical synthesis, 111–12
chromatographic methods, 29–37
colorimetric methods, 28
commercial production, 111
commercial products, 109
electrochemical methods, 28–9
extraction from natural sources, 111
gas–liquid chromatography (GLC), 29–31
high-performance liquid chromatography (HPLC), 31–7
liver, in, 262–3
long-term/carcinogenicity studies, 265
mutagenicity, 265
natural occurrence in foods, 104
oxidation of, 69–73
properties of, 106
reproduction studies, in, 264
short-term studies, 262–5
Tocopherols—*contd.*
structures of, 102
synergistic effects, 136–8
thin-layer chromatography (TLC), 29–30
thyroid gland, in, 262
voltammetric/polarographic techniques, 28–9
Tocopheroquinone, 5
5-(γ-tocopheroxy)-γ-tocopherol (γ-TED), 70–2
Tocopheroxyl radical, 12
Tocopheryl acetate, 108
Tocopheryl radicals, 136
Tocopheryl semiquinone radical, 106–7
Tocoquinone, 12
Tocotrienols, 26–37, 102
approximate content in vegetable oils, 105
natural occurrence in foods, 104
structure of, 102
Toxic substances, 221–3
Toxicological aspects, 253–307
Trimethylamine oxide (TMAO), 89
Trioctylamine, antioxidant activity, 91
Trisomy-21, 225
Trolox-C, 91
Tumours, 228–9, 253, 265, 270–1, 279–84, 292–5
Turkeys, 154

UV spectrophotometry
ascorbic acid, 46
oxygen scavengers, 46
synthetic antioxidants, 38–40

Vegetables, 155–8
enzymatic browning, 155–7
Vitamin A, 147, 155, 211, 213, 232
cod liver oil, in, 147
oxidation of, 146
Vitamin A-palmitate, 108
Vitamin B_1, 146

Vitamin C
  analytical methods, 46–7, 113, 136, 146, 164, 198–200, 218, 220, 221, 223–7, 229, 231, 232, 254
  levels found in human organs and breast milk, 120
  see also Ascorbic acid
Vitamin D, 146, 147, 155
Vitamin D_3, stabilization of, 148
Vitamin E, 102, 136, 154, 199, 200, 204–11, 215, 218–21, 223–8, 231, 232, 263–6
  analytical methods for, 26–37
  biological activity of, 104
Vitamin K, 4, 263, 289, 290
Vitamin P, 130
Vitamins, oxidation of, 146
Voltammetric/polarographic techniques, tocopherols, 28–9

Warfarin, 263
White spot, 151

Xanthine oxidase, 12
Xeroderma pigmentosum, 229